高等学校规划教材

Chemical Engineering Thermodynamics

化工热力学

王英龙 主编　钟立梅　董殿权　李龙 副主编

化学工业出版社

·北京·

内容简介

《Chemical Engineering Thermodynamics》（化工热力学）共 8 章，第 1 章介绍了化工热力学的用途、研究内容、研究特点和基本定律；第 2 章交代了纯物质的相态变化、纯物质的 p-V-T 关系、气体的状态方程和对比态原理及其应用；第 3 章详细讨论了热力学性质间的关系和热力学性质的计算；第 4 章介绍了剩余性质的定义，阐述了多组分混合物的热力学、混合物的实际热力学行为，不同二元混合物的混合摩尔体积、偏摩尔吉布斯能、偏摩尔体积和焓的实验测定，从实验数据计算无限稀释部分摩尔焓、混合物中组分的吉布斯能和逸度的估计以及偏摩尔吉布斯能和逸度；第 5 章全面介绍了相平衡判据的数学表达式、化学势和逸度及其在相平衡建模中的应用，讲述了测定液体和固体逸度、分布系数、相对挥发性以及热力学一致性检验。第 5 章主要涉及了相平衡的相关定律和方程；第 6 章解释了热机的不可逆性的比率、系统的㶲变化、㶲在压缩过程中发生变化、㶲递减原理与㶲破坏以及㶲衡算及㶲效率；第 7 章介绍了用简单模型分析制冷循环以及卡诺循环和它在工程中的价值，并对蒸汽和联合动力循环、卡诺蒸汽循环、制冷循环和热泵系统进行了相应的解释；第 8 章讨论了化学反应平衡基础、化学反应的平衡准则、平衡常数和工艺参数等条件对化学平衡组成的影响。

《Chemical Engineering Thermodynamics》注重理论原理与实际应用的结合，不仅能够为读者提供丰富的热力学基础知识，还能为经验丰富的化工工程师提供所需的专业知识。本书附有大量的例题，并且都系统地给出了解答步骤。读者能够通过本书迅速获取化工热力学的知识内容，适合自学，同时也是学习和掌握专业英语的高效途径。

《Chemical Engineering Thermodynamics》（化工热力学）可作为化工及相关专业的本科生和研究生学习化工热力学的教材，也可供化工专业的过程开发、合成、优化等领域的科研人员参考。

图书在版编目（CIP）数据

化工热力学 = Chemical Engineering Thermodynamics：英文 / 王英龙主编. —北京：化学工业出版社，2022.10
高等学校规划教材
ISBN 978-7-122-41471-7

Ⅰ. ①化… Ⅱ. ①王… Ⅲ. ①化工热力学-高等学校-教材-英文 Ⅳ. ①TQ013.1

中国版本图书馆 CIP 数据核字（2022）第 085892 号

责任编辑：刘俊之　汪　靓　　　　文字编辑：汪　靓
责任校对：赵懿桐　　　　　　　　装帧设计：韩　飞

出版发行：化学工业出版社 (北京市东城区青年湖南街 13 号　邮政编码 100011)
印　　装：三河市双峰印刷装订有限公司
787mm×1092mm　1/16　印张 20　字数 496 千字　2022 年 9 月北京第 1 版第 1 次印刷

购书咨询：010-64518888　　　　　　　　　售后服务：010-64518899
网　　址：http://www.cip.com.cn
凡购买本书，如有缺损质量问题，本社销售中心负责调换。

定　　价：69.00 元　　　　　　　　　　　　　　　版权所有　违者必究

《Chemical Engineering Thermodynamics》编写成员

主　编： 王英龙

副主编： 钟立梅　董殿权　李　龙

编　写： 王英龙　钟立梅　董殿权　李　龙

　　　　　　夏　力　陶少辉　王伟文

Preface

Chemical engineering thermodynamics is the theoretical basis of chemical process research, development and design, and is also a required course for chemical majors in colleges and universities. The concept of chemical engineering thermodynamics is rigorous and theoretical, which often makes many students confused by the derivation of complex and lengthy mathematical formulas and abstract concepts. In order to meet the needs of the reform in teaching of chemical engineering thermodynamics. Based on many years of teaching experience, the editor continues to reform and innovate teaching contents and methods. By introducing practical cases to enhance the charm of the course, introducing relevant international cutting-edge research content to broaden students' horizons, and strengthening the connection between disciplines in combination with experimental phenomena and other core courses of chemical engineering, and the teaching effect of chemical thermodynamics is improved. Focusing on the transformation and effective utilization of material and energy, the narrative strives to go from shallow to deep. Combined with well-designed examples and exercises, it emphasizes the connection between the basic principles of thermodynamics and practical applications. It organically infiltrates the elements of ideological and political education, and cultivates students' thermodynamic knowledge system and basic literacy. It incorporates actual cases of thermodynamic knowledge in chemical production and uses chemical process simulation software to stimulate students' interest in learning and innovative spirit.

The book consists of eight chapters, the first chapter mainly introduces the chemical engineering thermodynamics, research content and some basic concepts of chemical engineering thermodynamics. The second chapter introduces the p-V-T relation and thermodynamic equation of state of pure fluid, introduces a practical case involving thermodynamics knowledge, explains it by using the principle of chemical thermodynamics, stimulates students' interest in chemical thermodynamics, and enables students to understand the importance of chemical thermodynamics through practical cases. The third chapter and fourth chapter describe the thermodynamic properties and calculation of fluids, and introduce how to express physical quantities that cannot be directly measured by Maxwell relations and residual properties. Emphasis is placed on the understanding of basic concepts of multicomponent thermodynamics such as partial molar properties, mixing variables, ideal solutions and excess properties, which guide students to deal with complex mathematical formulas, and make the thermodynamics knowledge of rigorous theory and abstract concept easy to understand. In the fifth chapter, the properties and criteria of phase equilibrium, the mutual calculation of T, p, x and y, and the phase equilibrium and expression of mixtures are introduced. By introducing concepts such as ideal work, lost work and exergy, the sixth chapter explains that energy has both "quantity" and "quality", so as to understand the maximum utilization limit of energy in the chemical process, cultivate students' awareness of effective use of energy and the concept of correct and rational use of energy. The seventh chapter is the introduction of the thermodynamic cycle. Through the study of this chapter, students' consciousness of energy conservation and emission reduction can be cultivated. Chapter eight discusses chemical equilibrium, the calculation of equilibrium constant, the influence of temperature and pressure on equilibrium conversion and the equilibrium of chemical reactions in complex systems. By introducing the knowledge of chemical reaction equilibrium into the example, the general chemical equilibrium law is obtained through rigorous formula derivation.

Due to the limitation of the editor's level, the deficiencies in the book are unavoidable, and we sincerely hope readers can give criticism and correction for further modification.

Editor
2022.4.

Chapter 1

This chapter briefly introduces the category and function of chemical thermodynamics, focusing on the storage and transfer of energy. Energy is stored through substance in three forms: internal energy, potential energy and kinetic energy, and transmitted in two forms: work and heat. Through a brief description of substance and energy, the conservation of energy is derived, and the basic laws of thermodynamics are explained, namely the first law of thermodynamics, the second law of thermodynamics, the third law of thermodynamics and the zero law of thermodynamics. The application of chemical thermodynamics in social life and the research methods and characteristics of chemical thermodynamics are introduced. Discuss the basic concepts such as system, state, balance, process and cycle, and further enhance the understanding of the strength and breadth of the system.

Chapter 2

This chapter mainly introduces the concept of pure substance, discusses the physical phase transition process, explains various property diagrams, and p-V-T phase diagrams of pure substance. Phase diagram is an intuitive expression of p-V-T relationship, i.e. qualitative description, and the equation of state is an analytical expression of p-V-T relationship, i.e. quantitative description. Many other thermodynamic properties need to be calculated by using the p-V-T data of fluid and the basic thermodynamic relationship. This chapter explains the famous equation of state, such as cubic equation of state (vdW, RK, SRK, PR), virial equation of state, multi parameter equation of state, etc. Among them, SRK and PR equations of state have been widely used in industry because of their simplicity and accuracy.

Chapter 3

The thermodynamic properties of fluid are divided into directly measurable properties such as p, V and T and indirectly measurable properties such as S, U, A and G. These properties are the indispensable basis for chemical process calculation, analysis and chemical plant design. The main purpose of this chapter is to solve the problem of material transformation and the change of thermodynamic properties such as H, S, U, G caused by the change of substance state. This chapter mainly introduces how to express the physical quantities that indirectly measurable physical quantities into directly measurable physical quantities through Maxwell relation and residual properties, and introduces in detail the thermodynamic differential equations, thermodynamic calculations and thermodynamic data commonly used in the chemical industry.

Chapter 4

This chapter mainly introduces the thermodynamic properties of real solutions. The properties of the solution must be related to the properties of the components constituting the solution. Therefore, this chapter focuses on the understanding of the basic concepts of multicomponent thermodynamics, such as partial molar properties, mixed variables, ideal solution and excess properties. These basic concepts describe the relationship between the thermodynamic properties of solution and components, and further study the calculation of thermodynamic properties such as fugacity coefficient and activity coefficient of real solution. The Gibbs-Duhem equation is introduced to describe the dependence of partial molar properties of components in solution, it can be used to test the correctness of the thermodynamic property data of the mixture measured in the experiment, or to calculate the partial moles of one component from the partial moles of another component.

Chapter 5

This chapter mainly introduces *Raoult's law* and the modified *Raoult's law*, which lays the foundation for the modeling of vapor liquid equilibrium of mixtures. The properties and criterion of phase equilibrium, the expression of phase equilibrium of mixtures, the mutual calculation of T, p, x, and y and methods of phase equilibrium calculation are introduced. Through the excess molar Gibbs free energy model, the activity coefficient is used to correlate, predict and quantify the deviation from the ideal solution behavior. Wilson equation, Van Laar Equations and Regular Solution Theory, Relationship between activity coefficient and temperature and pressure, Van der Waals one-fluid mixing rules ware introduced. Based on the Gibbs-Duhan equation, methods for deriving thermodynamic consistency of experimental data ware derived.

Chapter 6

This chapter introduces the concepts of ideal work, lost work and exergy. By describing the three transfer forms of heat, work and quality of exergy, combined with the thermal power conversion process of heat engine and the kinetic energy and electric energy conversion of wind power generation process. This chapter analyzes the energy conversion, transfer and effective utilization in the process of chemical production. It also discusses the utilization and grade change of energy in the transformation process, the degree of effective energy utilization and the causes of energy loss, reveals the size and causes of energy loss. And provides basis for effectively reducing production energy consumption, economically and reasonably utilizing energy and scientific energy conservation, which is of great significance to green sustainable development.

Chapter 7

This chapter mainly introduces the application of mass, energy and entropy balance in thermodynamic cycle system, as well as the working principle and content of steam compression cycle and refrigeration cycle. Steam power cycle and refrigeration cycle utilize the state changes of working medium in thermal equipment through thermal processes such as heat absorption, expansion, heat release and compression, so as to realize the mutual transformation between thermal energy and mechanical energy. Considering many factors affecting the process of chemical thermal cycle, the methods to improve energy efficiency are discussed. At the same time, this chapter discusses the design of steam power cycle and refrigeration cycle in the form of cases. Through the combination of knowledge points and examples, emphasis on the key points and difficulties, it cultivate students' ability to solve practical problems based on thermodynamic knowledge.

Chapter 8

This chapter mainly has a clear understanding of whether chemical reactions can proceed in the direction of products, the limits to which they may proceed, and the conditions that may affect the limits of the reaction, under conditions of certain temperature, pressure, and composition. And it introduces the chemical equilibrium, the calculation of equilibrium constant, the influence of temperature and pressure on the equilibrium conversion, the chemical reaction equilibrium of complex systems. The deviation of ideal gas and ideal solution model, and Gibbs free energy is combined. The knowledge of chemical reaction equilibrium is introduced through examples. And the general law of chemical equilibrium is deduced through rigorous formulas. The difficulty of the examples is progressive from shallow to deep and guides readers' thinking.

Contents

Chapter 1 Introduction — 1
1.1 The Category of Chemical Engineering Thermodynamics — 1
1.2 The Role of Thermodynamics in Chemical Engineering — 2
1.3 Fundamental Law of Thermodynamics — 3
1.4 Application of Chemical Engineering Thermodynamics — 5
1.5 The State and System — 7

Chapter 2 The Physical Properties of Pure Substances — 10
2.1 Pure Substance — 10
2.2 Phases of Pure Substance — 10
2.3 Phase-change Processes of Pure Substances — 11
2.4 Property Diagrams for Phase-Change Processes — 13
    2.4.1 The $T\text{-}V$ Diagram — 14
    2.4.2 The $p\text{-}V$ Diagram — 15
    2.4.3 The $p\text{-}T$ Diagram — 17
    2.4.4 The $p\text{-}V\text{-}T$ Surface — 17
2.5 Equation of State — 22
    2.5.1 The Ideal-Gas Equation of State — 22
    2.5.2 Nonideality of Gases — 23
2.6 Other Equations of State — 23
    2.6.1 The van der Waals Equation of State — 24
    2.6.2 Redlich-Kwong (RK) Equation of State — 25
    2.6.3 The Soave-Redlich-Kwong (SRK) Equation of State — 25
    2.6.4 Peng-Robinson (PR) Equation of State — 26
    2.6.5 Virial Equation of State — 26
    2.6.6 Multiparameter Equation of State — 27
2.7 Principle of Corresponding States and Generalized Association — 33
    2.7.1 Principle of Corresponding States — 34
    2.7.2 Principle of Corresponding States with Two Parameters — 34
    2.7.3 Principle of Corresponding States with Three Parameters — 35

2.7.4 Generalized Compressibility Factor Graph Method	35
2.7.5 Generalized Virial Coefficient Method	36
2.8 Application of Aspen Plus in Calculation of Thermodynamic Equation of State	39
EXERCISES	43
REFERENCES	44

Chapter 3 Thermodynamic Properties of Pure Fluids

	45
3.1 Mathematical Relationship between Functions	45
3.1.1 Partial Differentials	45
3.1.2 Partial Differential Relations	47
3.1.3 Fundamental Thermodynamic Relation	48
3.2 The Maxwell Relations	49
3.3 The Clapeyron Equation	51
3.4 General Relations for dU, dH, dA, and dG	52
3.5 Joule-Thomson Coefficient	58
3.6 The ΔH, ΔU, and ΔS of Real Gas	60
3.7 Application of Aspen in Thermodynamic Properties	62
CONCLUSION	64
EXERCISES	66
REFERENCES	68

Chapter 4 The Thermodynamics of Multicomponent Mixtures

	69
4.1 Excess Property	70
4.2 Properties Change on Mixing	71
4.3 Partial Molar Gibbs Free Energy	78
4.4 Gibbs-Duhem Equation	79
4.5 The Experimental Measurement of Partial Molar Volume and Enthalpy	82
4.6 Gibbs Free Energy and Fugacity of a Component in a Mixture	89
4.6.1 Ideal Gas Mixture	89
4.6.2 Ideal Mixture and Excess Mixture Properties	91
4.6.3 Partial Molar Gibbs Free Energy and Fugacity	95
4.7 Application of Aspen Plus to Thermodynamic Properties of multicomponent Mixtures	100
CONCLUSION	103
EXERCISES	103
REFERENCES	105

Chapter 5 Phase Equilibrium 106

5.1 Phase Equilibrium for a Single-Component System 106
    5.1.1 Mathematical Models of Phase Equilibrium 106
    5.1.2 Fugacity and Its Use in Modeling Phase Equilibrium 117
5.2 Vapor-Liquid Equilibrium 121
    5.2.1 Motivational Example 121
    5.2.2 *Raoult's Law* and the Presentation of Data 123
    5.2.3 Mixture Critical Points 131
    5.2.4 Lever Rule and the Flash Problem 132
5.3 Theory and Model of Vapor Liquid Equilibrium of Mixtures: Modified *Raoult's law* Method 134
    5.3.1 Examples of Incentives 134
    5.3.2 Phase Equilibrium of Mixture 135
    5.3.3 Fugacity of Mixture 138
    5.3.4 Gamma-Phi Modeling 142
    5.3.5 *Raoult's law* Revisited 143
    5.3.6 *Henry's law* 144
5.4 Wilson and Van Laar Equation 155
    5.4.1 Wilson Equation 155
    5.4.2 Relationship between Activity Coefficient and Temperature and Pressure 157
    5.4.3 Van Laar Equation and Regular Solution Theory 160
    5.4.4 Van Der Waals One-Fluid Mixing Rules 161
5.5 Supplementary Simulation Examples 166
    5.5.1 Vapor-Liquid Equilibrium Calculations Using Activity Coefficient Models 166
    5.5.2 Vapor-Liquid Equilibrium Calculations Using an Equation of State 179
    5.5.3 Prediction of Liquid-Liquid and Vapor-Liquid-Liquid Equilibrium 192
EXERCISES 196
REFERENCES 198

Chapter 6 Energy Analysis of Chemical Process 200

6.1 The Definition of Entropy Exergy 200
6.2 Exergy (Work Potential) Associated with Kinetic and Potential Energy 201
6.3 Reversible Work and Irreversibility 203
6.4 Second-law Efficiency 204
6.5 Exergy Change of a System 206
    6.5.1 Exergy of a Fixed Mass: Nonflow (or Closed System) Exergy 206
    6.5.2 Exergy of a Flow Stream: Flow (or Stream) Exergy 208
6.6 Exergy Transfer by heat, work, and mass 212

    6.6.1 Exergy Transfer by Heat, Q 212
    6.6.2 Exergy Transfer from Work, X_{work} 213
    6.6.3 Exergy Transfer by Mass, m 214
6.7 The Decrease of Exergy Principle and Exergy Destruction   214
6.8 Exergy Balance: Closed Systems   219
6.9 Exergy Balance: Control Volumes   227
    6.9.1 Exergy Balance for Steady-Flow Systems   228
    6.9.2 Second-Law Efficiency of Steady-Flow Devices   230
6.10 Chemical Process Energy Analysis and Aspen Plus   233
EXERCISES   233
REFERENCES   235

Chapter 7 Thermodynamic Processes and Cycles   237

7.1 Chemical Process Design   237
7.2 Real Heat Engines   239
    7.2.1 Comparing the Carnot Cycle with the Rankine Cycle   240
    7.2.2 Design Variations in the Rankine Heat Engine   241
7.3 The Vapor-Compression Cycle   245
7.4 Power Cycle and Refrigeration Cycle   246
    7.4.1 Thermodynamic Cycles   246
    7.4.2 Property Diagrams   248
    7.4.3 The Carnot Cycle and Its Value in Engineering   248
    7.4.4 Air-standard Assumptions   250
    7.4.5 Rankine Cycle: The Ideal Cycle for Vapor Power Cycles   251
    7.4.6 Energy Analysis of the Ideal Rankine Cycle   252
    7.4.7 The Ideal Re-heat Rankine Cycle   254
    7.4.8 The Ideal Regenerative Rankine Cycle   255
7.5 Second-Law Analysis of Vapor Power Cycles   255
    7.5.1 Combined Gas-Vapor Power Cycles   257
    7.5.2 Refrigeration Cycles   258
    7.5.3 Refrigerators and Heat Pumps   260
    7.5.4 The Reversed Carnot Cycle   261
7.6 Application of Thermodynamic Processes and Cycles in Aspen Plus   262
EXERCISES   267
REFERENCES   270

Chapter 8 Chemical Reaction Equilibrium   271

8.1 Motivational Example: Propylene from Propane   272

8.2 Chemical Reaction Stoichiometry	278
8.2.1 Extent of Reaction and Time-Independent Mole Balances	279
8.2.2 Extent of Reaction and Time-Dependent Material Balances	281
8.3 The Equilibrium Criterion Applied to a Chemical Reaction	282
8.3.1 The Equilibrium Constant	282
8.3.2 Accounting for the Effects of Pressure	285
8.3.3 Accounting for Changes in Temperature	286
8.3.4 Reference States and Nomenclature	292
8.4 Multiple Reaction Equilibrium	293
8.5 Summary	297
8.6 Chemical Reaction Equilibrium Simulation	298
EXERCISES	301
REFERENCES	303

Chapter 1

Introduction

1.1 The Category of Chemical Engineering Thermodynamics

"Thermodynamics is a funny subject. The first time you go through it, you don't understand it at all. The second time you go through it, you think you understand it, except for one or two small points. The third time you go through it, you know you don't understand it, but by that time you are so used to it, it doesn't bother you anymore."

Chemical engineering thermodynamics is a branch subject formed by the application of the basic laws of thermodynamics to chemical engineering, mainly including thermochemical, phase equilibrium and chemical equilibrium theory. When the basic laws of thermodynamics are applied to engineering fields, such as power plants, gas turbines and freezers, engineering thermodynamics is formed. The main research contents are the basic thermodynamic properties of engineering and the working process of various plants, so as to explore the way to improve energy conversion efficiency. Students who make a preliminary study of the subject will find that there is a difficult challenge from the beginning. It needs a certain degree of memory and thinking organization ability to deal with the continuous emergence of a wide range of new concepts, new words and new symbols. After crossing the first obstacle, there is a greater challenge waiting for them, i.e., to develop reasoning ability in the context of thermodynamics, so as to be able to apply the principles of thermodynamics to solve practical problems.

Based on chemical thermodynamics and engineering thermodynamics, chemical engineering thermodynamics is formed with the development of chemical industry. It combines the strong points of chemical thermodynamics and engineering thermodynamics, but it is much more complex than them. On one hand, with the emergence of chemical engineering unit operations such as distillation, absorption, extraction, crystallization, adsorption and various types of reaction devices, the relationship among temperature, pressure, phase composition and various thermodynamic properties of the multi-component system becomes the essential data in the research, development and design. The acquisition of data requires not only thermodynamic principles, but also thermodynamic theoretical models from low pressure to high pressure, including critical region, from no-polarity to polarity to hydrogen bond formation, from small molecules, ions to macromolecules and biological macromolecules (Tian et al., 2012). We also need to solve the corresponding complex calculation problems, which are far beyond the content of conventional chemical thermodynamics. On the other hand, the energy consumption in chemical engineering production accounts for a high proportion of the production cost, and the industrial process involved is very complex. Therefore, it is more necessary to study the effective utilization of energy, including low-grade energy. And the establishment of the thermodynamic analysis method suitable for the chemical engineering process is also necessary.

1.2 The Role of Thermodynamics in Chemical Engineering

Chemical engineering thermodynamics is the theoretical basis of chemical process research, development and design. A chemical process mainly includes chemical reaction process, product separation and purification process, and chemical thermodynamics plays a very important role in solving the two problems of reaction and separation in chemical process. As shown in Fig. 1-1, chemical reactions almost never produce pure products. The mixture left in the reactor usually contains not only the desired product, but also by-products and unused raw materials. Therefore, some chemical or physical process is required to separate the material left in the reactor. But what does the "Chemical Reaction" and "Separation Process" in Fig. 1-1 represent? The exact process refers to the answers to the following questions.

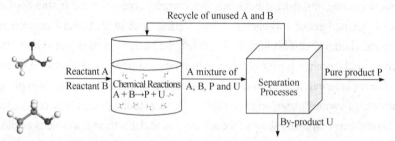

Fig.1-1 Reaction and separation in the preparation of chemical products

- How many different chemical reactions are required?
- Can all reactions be carried out in one reactor or does each reaction require separate equipment? How high is the purity of the product you need to achieve? At what temperature and pressure does each device operate?
- How to separate them?
- Can all the separation be done with a single device? Or do you need a few obvious separation steps?

Answering these questions is fundamental to the practice of chemical engineering, and process designers find that each process presents unique challenges and opportunities.

- How much raw material and energy does it take to produce 10 million pounds of this product per year?
- What method can be used to separate the product from by-products and unused raw materials?
- How much energy is needed to heat stream of process to the required temperature of 300 ℃?
- How can reactor conditions be optimized to achieve maximum yield of desired product?

Chemical engineering thermodynamics is a subject that studies the thermal properties and laws of substances from a macroscopic perspective. Energy is a measure of the distribution of substances over time and space. It is used to describe the ability of a physical system to do work. In thermodynamics, we are concerned with how energy is stored and transferred. Substances store energy in three forms: internal, potential and kinetic energy. Energy can be transferred in two other forms: work and heat (Wang et al., 2019).

The 10th International Conference of Metrology in 1954 defined the thermodynamic temperature unit. It selected the triple point of water as the basic point and defined its temperature as 273.16 K.

In 1967, the 13th International Metrology Conference adopted the name of Kelvin (symbol K) instead of "degree kelvin" (symbol K). Its form is defined as the thermodynamic temperature unit Kelvin, that is, 1/273.16 of the thermodynamic temperature of the three-phase point of water. Temperature intervals or differences are denoted by the unit Kelvin and its symbol K.

T_0 = 273.15 K is the thermodynamic temperature of the freezing point of water, which differs by 0.01 K (Kelvin) from the thermodynamic temperature of the triple point of water.

Heat is a kind of material called "caloric". Heat calorie is a kind of massless gas. After the material absorbs heat, the temperature will rise, and the heat from the high temperature flow to the low temperature flow, can also pass through the pores of the solid or liquid.

In 1798, the British scientist Rumford proposed "Discussion on the Source of Heat Generated by Friction".

In 1843, Joule made experiments and theorized that heat was only a form of energy, and gave the equivalent value of thermal work.

The proportional constant is universal and independent of the system. James Joule measured it in 1845 and 1847 and described it as a thermal mechanical equivalent.

In 1841, he published *On Force in the Inorganic World*, which stated the law of conservation of energy and measured the equivalent value of heat and work. In 1853, Joule and Thomson finally completed the precise formulation of the laws of conservation and transformation of energy.

1.3 Fundamental Law of Thermodynamics

The First Law of Thermodynamics

Rudolf Julius Emanuel Clausius proposed in 1850 that in thermodynamic processes involving closed systems, the increase of internal energy was equal to the difference between the heat accumulated by the system and the work done. George put forward the definition of heat in 1907. When energy does not flow from one system or part of a system to another through mechanical work, the energy transmitted in this way is called heat.

The first law of thermodynamics is the law of conservation of energy. The transformation and conservation of energy is not only the objective law of nature, but also one of the core contents of natural science. It widely permeates various disciplines and is closely related to various industries and daily life. It reflects the essence of substance movement and interaction at a deeper level.

Chemical thermodynamics is the basis of energy analysis of chemical engineering processes. Chemical production often requires strict control of temperature and pressure. How to abide by the law of conservation of energy, use the law of energy transfer and transformation to ensure the appropriate process conditions is the key to the success of chemical engineering production (Ma et al., 2011).

The total energy of an isolated system is constant, energy can be changed from one form to another, and can neither be created nor destroyed.

$$\Delta U = Q - W \tag{1-1}$$

The change in internal energy of a closed system ΔU is equal to the heat added to the system minus the work done by the system to the environment. Equivalently, perpetual motion machines of the first kind (machines that produce work with no energy input) are impossible.

Second Law of Thermodynamics

Rudolf Julius Emanuel Clausius put forward in 1854 that heat could not be automatically transferred from cold source to heat source. And Lord Kelvin put forward in 1854 that it was impossible to absorb heat from a single source and turn it into useful work without causing other changes.

In a natural thermodynamic process, the sum of the entropies of the interacting thermodynamic systems increases. Equivalently, perpetual motion machines of the second kind (machines that spontaneously convert thermal energy into mechanical work) are impossible. The entropy of an isolated system can only increase or stay constant as it reaches its limit. Its mathematical expression is,

$$\Delta S_t \geqslant 0 \tag{1-2}$$

The entropy of an isolated system never decreases, and that's the principle of entropy increase. Several mathematical expressions of the second law of thermodynamics are as follows,

$$\int \frac{\delta Q}{T} \leqslant 0 \quad \text{(Cycle process)} \tag{1-3}$$

$$\Delta S_{sys} \geqslant \int \frac{\delta Q}{T} \quad \text{(Closed system)} \tag{1-4}$$

$$\Delta S_t = \Delta S_{sys} + \Delta S_{sur} \geqslant 0 \quad \text{(Isolated system)} \tag{1-5}$$

The second law of thermodynamics reveals that different forms of energy have qualitative differences in their transfer and conversion capacities, and different forms of energy cannot be unconditionally transformed into each other. Therefore, in the process of energy transfer and conversion, there are certain characteristics of direction, condition and limit. Various expressions of the second law of thermodynamics essentially say that "spontaneous processes are irreversible". In some cases, the feasibility of the process can be judged intuitively. However, for in-depth research, quantitative description is more necessary. This requires the second law of thermodynamics, i.e., the principle of entropy and entropy increase.

Third Law of Thermodynamics

The entropy of a system approaches a constant value as the temperature approaches absolute zero. It is impossible to cool an object to zero of absolute temperature. No system can reduce its temperature to absolute zero through limited steps.

Zeroth Law of Thermodynamics

Ralph Howard Fowler first proposed the zeroth law of thermodynamics in 1931, that heat could not be transferred from a cold source to heat source unless some relevant changes occur at the same time. In 1939, it was proposed that if the two components were in thermodynamic equilibrium with the third component, they would be in thermodynamic equilibrium with each other.

1.4 Application of Chemical Engineering Thermodynamics

Thermodynamics can be applied in engineering thermodynamics, chemical thermodynamics, chemical engineering thermodynamics and other fields. We only discuss the application in the field of chemical engineering thermodynamics.

(1) Describe the law of energy conversion and determine the maximum efficiency of energy conversion to target energy.

For example, a polyvinyl alcohol plant in the United States consumes a lot of energy, especially the energy consumption of the separation section accounts for 65% of the whole plant. Using the results of thermodynamic phase equilibrium, the content of acetaldehyde in the raw material was reduced from 0.7% to 0.4%, and the operation cost was saved by 50%.

(2) Determine the possibility and direction of chemical reaction, the reaction equilibrium conditions and the equilibrium state of the system.

For example, the thermodynamic calculation of the effect of temperature and pressure on the transformation between graphite and diamond not only points out the necessary conditions for the manufacture of synthetic diamond, but also puts forward the hypothesis of geological conditions for the formation of diamond in nature.

(3) The possibility and direction of phase transformation, the condition of phase equilibrium and the state of the system are determined.

(4) Describe the law of state change and state properties(Fig.1-2).

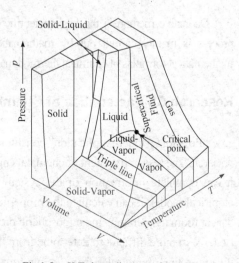

Fig.1-2 p-V-T phase diagram of pure matter

$$p = \frac{RT}{V-b} - \frac{a}{V^2} \tag{1-6}$$

There are two kinds of problems in the preparation of chemical products.

(1) Reaction problems:

Feasibility analysis: what kind of products can be obtained by the interaction of A and B?

Chemical equilibrium: the optimum conditions (such as pressure, temperature, etc.) and yield for preparing product C.

(2) Separation problem: How to obtain pure product C through phase equilibrium and effective utilization of energy?

These problems are the theoretical basis of chemical process research, development and design.

Phase Equilibrium, Effective Utilization of Energy

The research methods of classical thermodynamics include classical thermodynamics and molecular thermodynamics (statistical thermodynamics). These research methods have the following characteristics:

(1) The state of the system is described by macroscopic physical quantity, and the interaction between systems is investigated from a macroscopic point of view.

(2) Based on the direct observation and experiment of a large number of macroscopic phenomena, the universal basic laws of thermodynamics are summarized.

(3) The relationship between macroscopic properties is based on the three basic laws of thermodynamics and obtained by using mathematical methods.

(4) It only focuses on the equilibrium problem (equilibrium state), but does not involve the microstructure and consider how to reach the equilibrium.

Research Methods of Chemical Thermodynamics

Deductive method is the basic scientific method of chemical thermodynamics. The deductive process is mainly carried out by mathematical methods, which leads to the complexity of the mathematical formulas of chemical thermodynamics and the rigorous and abstract theoretical concepts.

Research Characteristics of Chemical Engineering Thermodynamics

"Principle-Model-Application" constitutes the "three elements" of chemical thermodynamics research. The research content of chemical engineering thermodynamics has the characteristics that it can complete information of the system calculated from local experimental data and semi empirical model; it can calculate the properties under harsh conditions from the physical property data at room temperature and atmospheric pressure; it can calculate the pure material property data (v, H, S, g) more difficult to determine from the easily obtained physical property data (p, V, T, x); it can use the mixing rules from the pure material information and the method can take the ideal state as the standard state and correct it to deal with the real state (Chang et al., 2005).

1.5 The State and System

Thermodynamic state refers to a group of states that describe a thermodynamic system. As long as there are enough known state parameters in a thermodynamic system, other state parameters have been determined, and the number of state parameters required depends on the complexity of the system. State is the macroscopic physical state of a system at a certain moment. The macroscopic physical quantity describing the state of the system is called thermodynamic variable, also known as state function. Thermodynamic variables can be divided into intensity and breadth. Intensive properties are those that are independent of the mass of a system, such as temperature, pressure, and density. Extensive properties are those whose values depend on the size or extent of the system(Fig.1-3).

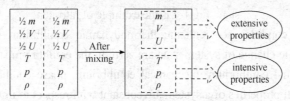

Fig.1-3 Criterion to differentiate intensive and extensive properties

A system is a macro-object composed of a large number of micro particles in a given range. A closed system consists of a certain amount of matter. No substance can cross its boundary, i.e., there is no substance exchange with the external environment, but there is an energy exchange. Open system is a system with both energy exchange and substance exchange with the environment. Simple compressible system is a thermodynamic system composed of compressible fluid. There are no external force fields such as electricity, magnetism, gravity, motion and surface tension, no chemical reaction, and only volume change work is exchanged between the system and the external environment. The state of a simple compressible system is completely determined by two independent intensive properties. Table 1-1 is some examples.

Table 1-1 Examples of common intensive and extensive properties

Intensive Property: Symbol and SI Units		Extensive Property: Symbol and SI Units	
Specific Volume = $\hat{V}$	m³/kg	Volume = V	m³
Density = ρ	kg/m³	Mass = M	kg
Specific Enthalpy = $\hat{H}$	J/kg	Enthalpy = H	J
Specific Internal Energy = $\hat{U}$	J/kg	Internal Energy = U	J
Specific Entropy = $\hat{S}$	J/(kg · K)	Entropy = S	J/K
Temperature = T	K	Kinetic Energy = K.E.	J
Pressure = p	Pa	Potential energy = P.E.	J

The substance system that exchanges mass and energy with the system is called the environment. The real or supposition surface that separates the system from its surroundings is called a boundary. The boundary of the system can be fixed or movable.

Thermodynamic Equilibrium

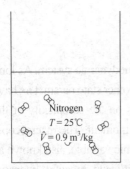

Fig.1-4 The state of nitrogen is fixed by two independent, intensive properties

Equilibrium state is that a system is in thermodynamic equilibrium and its macroscopic properties do not change with time, under the condition of no external influence. The necessary conditions for achieving thermodynamic equilibrium (namely, thermal equilibrium, force equilibrium, phase equilibrium and chemical equilibrium) are that all potential differences, such as temperature difference, pressure difference, chemical potential difference, causing changes in the system state are zero. So, an equilibrium system is a system that exists without any change in state. Fig.1-4 is the example of nitrogen.

The necessary condition for realizing thermodynamic equilibrium is that all potential differences causing the change of system state are zero. In the equilibrium state, the system shall generally meet thermal equilibrium, mechanical equilibrium, phase equilibrium and chemical equilibrium. When all properties of a system are constant with respect to time, the system is in a stable state. When there is no driving force to change the state attributes of a system, it is in equilibrium. It should be pointed out that when the system reaches an equilibrium state, the molecules that make up the system are in constant motion, but the average effect of the molecular motion does not change with time, so the macroscopic state is unchanged. The equilibrium state is essentially a kind of dynamic equilibrium (Chen et al., 2011).

Process and Cycle

Any change that a system undergoes from one equilibrium state to another is called a process, and the series of states through which a system passes during a process is called the path of the process(Fig.1-5).

Process refers to the sum of all states that a system experiences when it changes from one equilibrium state to another. Processes can be classified into different categories. According to the change law of a state function in the process, there are isobaric process, isothermal process, isovolumetric process, isentropic process and so on. According to reversibility, there are reversible process and irreversible process.

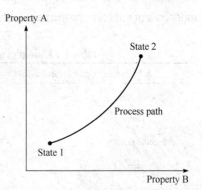

Fig.1-5 A process between State 1 and State 2 and the Process path

Reversible process is a very important concept in thermodynamics. It is defined as: after a process is completed, if the process is reversed, the system and environment involved in the process can completely return to their original state without any change, this process is called reversible process. It is an ideal process that can only be tended to but cannot be realized. The reversible process is carried out under the condition that the difference between force and resistance is infinitely small. Irreversible process: a unidirectional process must leave some traces after it occurs, which cannot be completely eliminated by any method. It is called irreversible process in thermodynamics. It can be said that all practical processes are irreversible.

After a series of state changes, the system finally returns to its original state, and the whole change is called cycle. Circulation can be divided into positive circulation and reverse circulation. The thermal cycle that converts heat energy into mechanical energy is called positive cycle. The heat engine used in engineering uses positive cycle. Energy consumption forces heat to be removed from a low-temperature object and transferred to a high-temperature object. If the cycle effect is to consume mechanical energy to force heat from low temperature to high temperature, the cycle with this effect is called reverse cycle. Refrigeration and heat pump work in reverse cycle.

Chapter 2

The Physical Properties of Pure Substances

Learning Objectives

- The significance of points, lines and surface in p-V and p-T phase diagrams, especially the critical points; The historical evolution, status and equation form of each equation of state; Eccentricity factor, three parameters corresponding state principle; Z is solved by vdW, RK, SRK, PR, virial and generalized equation of state.
- Calculation of liquid p-V-T; Selection and use of equation of state.
- The ideas of "principle of corresponding state" and "generalization" play an important role in chemical thermodynamics.

2.1 Pure Substance

Pure substance refers to a kind of substance only composed of the same chemical substance (or molecule). For example, nitrogen, helium water, and carbon dioxide are all pure substances. However, a pure substance is not necessarily a single chemical element or compound. A mixture of various homogeneous chemical elements or compounds can also be called pure substances such as air, it is a mixture of several gases, but it is usually considered to be pure substance. The mixture of two or more phases of pure substance is still pure substance, if the chemical composition of each phase is the same. For example, because the two phases have the same chemical composition, a mixture of ice and liquid water is also a pure substance. But due to different components in the air condense at different temperatures and specific pressures, the mixture of liquid air and gaseous air cannot be called a pure substance.

2.2 Phases of Pure Substance

Substances may exist in different phases under different conditions(Fig.2-1). In solid phase, the attractive forces of molecules on each other are large, the distance between them is small, and the molecules are kept in a fixed position. In liquid phase, the intermolecular force is weaker than that in solid, but still stronger than that in gas. As for gas phase, the molecules are so far away from each other that there is no molecular order. Gas molecules can move freely in space and constantly collide with each other and the container wall where they are located.

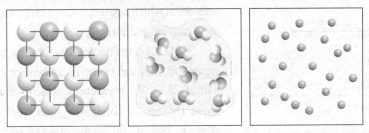

Fig.2-1 The arrangement of atoms in different phases

2.3 Phase-change Processes of Pure Substances

All pure substances exhibit the same general behavior. In many practical cases, two phases of a pure substance coexist in equilibrium. Water exists as a mixture of liquid and vapor in the boiler and the condenser of a steam power plant. The refrigerant turns from liquid to vapor in the freezer of a refrigerator. Water is used as a common substance to demonstrate the basic principles involved (Sun et al., 2020).

Compressed liquids (Subcooled liquid) Liquids that can be compressed are called Compressed liquids.

Saturated liquids A saturated liquid is a liquid that is in equilibrium with a gas at a given pressure and temperature.

When water is heated to 100 ℃, water is still liquid. But if you keep heating, some liquid water will begin to evaporate. That is, the phase transition from liquid to steam is about to occur. The liquid to evaporate is called saturated liquids.

Saturated vapor Saturated vapor is the vapor with a temperature of 100 ℃ under an atmospheric pressure. The temperature cannot rise any more. It is the vapor in the saturated state.

During the boiling process, the only change we will observe is the significant increase in volume and the stable decrease in the amount of liquid, which is the result of more liquid turning into steam.

Superheated vapor If the saturated vapor continues to be heated, the temperature will rise and The saturated vapor become superheated vapor.

As shown in Fig. 2-2, State 1 is compressed liquids (subcooled liquid), State 2 is saturated liquid. In the middle of the vaporization line (State 3), the cylinder contains equal amounts of liquid and vapor. When heating is continued, the evaporation process continues until the last drop of liquid evaporates (State 4). The state at this time is saturated vapor. Substances between State 2 and State 4 are called saturated liquid-vapor mixtures because the liquid-vapor phases coexist in equilibrium in these states. In State 5 (superheated vapor), the temperature of vapor is higher than the boiling point. If some heat is removed, the temperature may decrease, but condensation will not occur as long as the temperature remains above the boiling point.

Saturation Temperature and Saturation Pressure

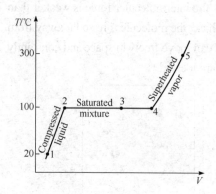

Fig.2-2 T-V constant pressure water heating process diagram

At a given pressure, the temperature at which a pure substance changes phase is called the saturation temperature T_{sat}. Likewise, at a given temperature, the pressure at which a pure substance changes phase is called the saturation pressure p_{sat}.

Table 2-1 gives partial data for saturated pressure of water and saturated vapor. It can be seen from this table that at 25 ℃, the pressure at which water changes its phase (boiling or condensing) must be 3.17 kPa. For water to boil at 250 ℃, the pressure of water must be maintained at 3976 kPa. Similarly, water can be frozen if the pressure is lowered below 0.61 kPa. Saturation tables that list the saturation pressure against the temperature (or the saturation temperature against the pressure) are available for practically all substances.

Table 2-1 Saturation (or vapor) pressure of water at various temperatures

Temperature T/℃	Saturation Pressure p_{sat}/kPa	Temperature T/℃	Saturation pressure p_{sat}/kPa
−10	0.260	30	4.25
−5	0.403	40	7.38
0	0.611	50	12.35
5	0.872	100	101.3(1 atm)
10	1.23	150	475.8
15	1.71	200	1554
20	2.34	250	3973
25	3.17	300	8581

A phase change process, such as melting a solid or vaporizing a liquid, requires a lot of energy. The energy absorbed or released during the phase transition is called latent heat. Specifically, the energy absorbed by a substance during melting, called the latent heat of fusion, is equal to the energy released during freezing. Similarly, the energy absorbed in the process of vaporization, called the latent heat of vaporization, is equal to the energy released during condensation. The amount of latent heat depends on the temperature or pressure at which the phase transition occurs. For example, at a pressure of 1 atm, the latent heat of water fusion is 333.7 kJ/kg and that of evaporation is 2256.5 kJ/kg.

In the phase transition process, pressure and temperature are obviously correlated, that is, $p_{sat}=f(T_{sat})$. Fig. 2-3 shows the relationship between T_{sat} and p_{sat}. It

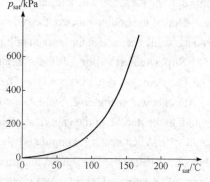

Fig.2-3 The liquid-vapor saturation curve of a pure substance (numerical values are for water)

can be seen from the figure that p_{sat} increases with the increase of T_{sat}, and the curve is called liquid-gas saturation curve.

Atmospheric pressure and the boiling point of water decrease with elevation. For example, at an altitude of 2,000 meters, the standard atmospheric pressure is 79.50 kPa. At this atmospheric pressure, the corresponding water boils at 93.3℃, while it is well known that when at the sea level, the boiling point is 100℃. Table 2-2 shows the data of boiling point of water changing with altitude at standard atmospheric pressure (101.325 kPa). The boiling point drops by about 3℃ for every 1000 meters of elevation.

Table 2-2 Variation of the standard atmospheric pressure and the boiling (saturation)

Elevation/m	Atmospheric pressure/kPa	Boiling point/℃
0	101.33	100.0
1000	89.55	96.5
2000	79.50	93.3
5000	54.05	83.3
10000	26.50	66.3
20000	5.53	34.7

The boiling point of nitrogen at atmospheric pressure is −196℃. Therefore, nitrogen is usually used in cryogenic scientific research or medical treatment. The experiment in Fig. 2-4 is completed by placing the test chamber in a liquid nitrogen bath open to the atmosphere. Any heat transfer from the environment to the test section is absorbed by nitrogen, and the nitrogen is isothermal evaporated to maintain the laboratory temperature at −196℃. The entire test section must be highly insulated to minimize heat transfer and thus reduce liquid nitrogen consumption (Tian et al., 2020).

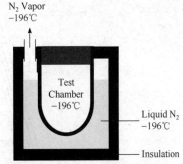

Fig.2-4 Liquid nitrogen is exposed to the atmosphere at a temperature of −196℃, so it remains −196℃ in the laboratory

2.4 Property Diagrams for Phase-Change Processes

Before understanding the quantitative relationship of fluid p-V-T, it is necessary to qualitatively understand the p-V-T three-dimensional phase diagram of pure substance, which intuitively describes the basic law of fluid state change. However, in practical use, three-dimensional drawings are difficult to form and understand, and two-dimensional projection drawings are often used, such as T-V diagram, p-T diagram and p-V diagram. That is, by fixing other variables, focus on the impact of the emphasized variables.

2.4.1 The T-V Diagram

The T-V diagram can be obtained by repeating the phase transition process of water under different pressures. Increase the weight on the top of the piston until the pressure inside the cylinder reaches 1 MPa. The specific volume of water at this pressure is going to be a little bit smaller than at 1 atm. When the feed water is heated at a pressure of 1 MPa, the process follows a path that looks very similar to the process path at a pressure of 1 atm, as shown in Fig. 2-5, but with some notable differences. First, at this pressure, water boils at a higher temperature (179.9 ℃). Second, at the pressure of 1 atm, the specific volume of saturated liquid is larger, and that of saturated vapor is less than the corresponding value. The horizontal line connecting the saturated liquid to the saturated vapor state is much shorter. The saturation line continues to shrink with the further increase of the pressure, and becomes a point when the pressure reaches the saturation line of 22.06 MPa. This point is called the critical point, and it is defined as the point at which the saturated liquid and saturated vapor are in the same state. The temperature, pressure, and volume of a substance at the critical point are called critical temperature T_{cr}, critical pressure p_{cr}, and critical specific volume V_{cr}, respectively. The critical point characteristics of water were p_{cr} = 22.06 MPa, T_{cr} = 373.95 ℃, and V_{cr} = 0.003106 m^3/kg.

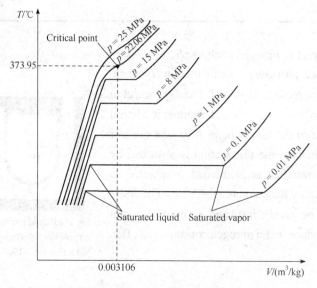

Fig.2-5 T-V diagram of constant-pressure phase-change processes of a pure substance at various pressures
(numerical values are for water)

When the pressure is higher than the critical pressure, the substance has no obvious phase transformation process (Fig. 2-6). In this state, the specific volume of substance is increasing, and only one phase exists at any one time. Above the critical state, no line separates the compressed liquid zone from the superheated vapor zone. The substance is customarily referred to as superheated vapor above the critical temperature and compressed liquid below it. The saturated liquid state in Fig. 2-7 can be connected by one line, called the saturated liquid line, and the saturated vapor state in the same figure can be connected by another line, called the saturated vapor line. The two lines intersect at the critical point to form a dome as shown in Fig. 2-7. All the compressed liquid states are located to the left of the saturated liquid line, called the compressed liquid region. All superheated vapor states are located to the right of the saturated vapor line, called the superheated vapor region. In both regions, substance exists as a single phase, a liquid or a vapor. All the states involved in two-phase equilibrium are located under the dome, called the saturated liquid-vapor mixing zone, or wet zone.

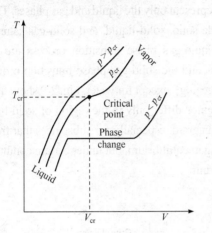
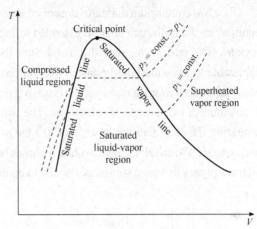

Fig.2-6 At supercritical pressures ($p > p_{cr}$), there is no distinct phase-change (boiling) process

Fig.2-7 *T-V* diagram of a pure substance

2.4.2 The *p-V* Diagram

As shown in Fig. 2-8, the general shape of the *p-V* diagram of the pure substance is very similar to that of the *T-V* diagram. In a piston-cylinder arrangement, there is liquid water at the pressure of 1 MPa and temperature of 150℃. Water exists in this state as a compressed liquid. Now, gradually reduce the pressure in the cylinder and keep the temperature of the system constant. As the pressure decrease gradually, the volume of water increases slightly (Wang et al., 2020). When the pressure reaches the saturation pressure value (0.4762 MPa) at the specified temperature, the water begins to boil. During this boiling process, the temperature and pressure remain the same, but the specific

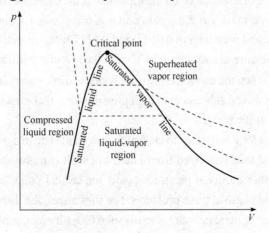

Fig.2-8 *p-V* diagram of a pure substance

15

volume increases. After the last drop of water in the system has evaporated, a further decrease in pressure results in a further increase in specific volume. When this process is repeated for other temperatures, a similar path for the phase transition process is obtained. The saturated liquid state and the saturated vapor state are connected by curves to obtain the p-V diagram of the pure substance.

The two equilibrium diagrams described above represent only the liquid and gas phases. These equilibrium diagrams can also be extended to include solid, solid-liquid, and solid-gas saturated regions. The basic principles discussed with the liquid-gas phase transition process are also applicable to the solid-liquid phase transition process and the solid-gas phase transition process. Most substances contract during solidification (i.e., freezing), except for water. Fig. 2-9 show the p-V diagrams of two groups of substances. The two figures differ only in the region of solid-liquid saturation. The T-V diagram looks a lot like the p-V diagram, especially for substances that freeze and contract. We are all familiar with two phases being in equilibrium, but under certain conditions all three phases of a pure substance coexist in equilibrium.

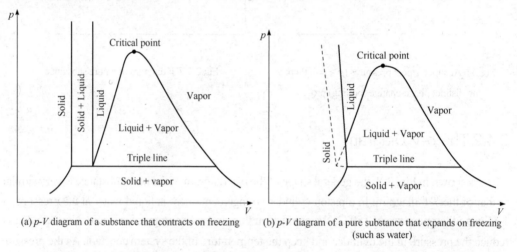

(a) p-V diagram of a substance that contracts on freezing (b) p-V diagram of a pure substance that expands on freezing (such as water)

Fig.2-9 p-V diagram of a pure substance (extending the diagrams to include the solid phase)

On a p-V or T-V diagram, these triple-phases form a line which is called a triple-line. The pressure and temperature of a substance on the triple-line is the same but have different volumes. A triple-line appears as a point on a p-T diagram and it is often called a triple point. For water, the triple point temperature and pressure are 0.01 ℃ and 0.6117 kPa, respectively. That is, only when the temperature and pressure are exactly 0.01 ℃ and 0.6117 kPa, all triple-phases of water can coexist in equilibrium. When the condition pressure is below triple point, there is no liquid phase in a stable equilibrium. However, substances at high pressures can also exist in the liquid phase when temperatures are below triple point.

There are two ways for a substance to change from a solid state to a gas state: it can be melted from solid to liquid, and then vaporized from liquid to gas. It can also sublimate directly from a solid into a gas. The latter occurs at pressures below the critical point, because pure substances cannot exist in the liquid phase at these pressures. For substances, the three-phase pressure of it is certainly higher than atmospheric pressure, such as solid CO_2 (dry ice), sublimation is the only way to change from solid to vapor phase under atmospheric conditions.

2.4.3 The p-T Diagram

Fig. 2-10 shows the p-T diagram of the pure substance. Because all three phases are separated by three lines from each other, this diagram is often also called phase diagram. The solid zone is separated from the vapor zone by the sublimation line, the liquid zone is separated from the vapor zone by the evaporation line, and the solid zone is separated from the liquid zone by the melting (or fusion) line. These three lines intersect at a point known as the triple point. At this point the three phases of substance coexist in equilibrium. Because there is no distinction can be made between liquid and vapor phases above the critical point, the evaporation line ends at the critical point. Substances expand and contract on freezing differ only in the melting line on the p-T diagram.

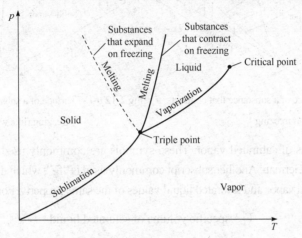

Fig.2-10 p-T diagram of pure substances

2.4.4 The p-V-T Surface

Two independents intensive properties can determine the state of a simple compressible substance. All the other properties will become dependent once the two appropriate properties are fixed. A spatial surface can be represented by any equation $z = z(x, y)$ with two independent variables, and the p-V-T behavior of substance can be represented as spatial surface, as shown in Fig. 2-11 and Fig. 2-12. Here T and V can be regarded as independent variables (the base) and p as dependent variables (the height).

All points on the surface represent equilibrium states of substance under certain conditions. It is important to note that because the path of the quasi-equilibrium must pass through equilibrium states, all states on the quasi-equilibrium path are on the p-V-T surface. The single-phase regions appear as curved surface on the p-V-T surface, while the two-phase region is perpendicular to the p-T plane. This is expected because the projections of two-phase region on the p-T plane are straight lines (Zhang et al., 2020).

Saturated Liquid and Saturated Vapor States

The subscript "f" is used to denote properties of a saturated liquid and the subscript "g" to

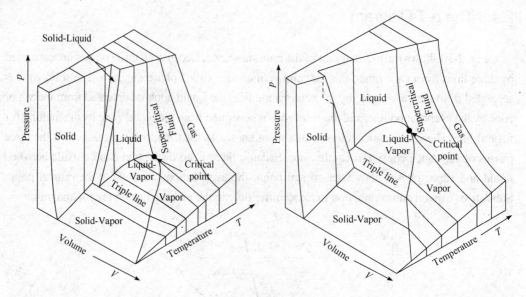

Fig.2-11 p-V-T surface of a substance that contracts on freezing

Fig.2-12 p-V-T surface of a substance that expands on freezing (like water)

denote the properties of saturated vapor. These symbols are commonly used in thermodynamics and originated from German. Another subscript commonly used is "fg", which denotes the difference between the saturated vapor and saturated liquid values of the same property. For example,

V_f = specific volume of saturated liquid

V_g = specific volume of saturated vapor

V_{fg} = difference between V_g and V_f, (that is $V_{fg} = V_g - V_f$)

The H_{fg} is called the enthalpy of vaporization (or latent heat of vaporization). It represents the amount of energy needed to vaporize a unit mass of saturated liquid at a given temperature or pressure. It decreases as the temperature or pressure increases and becomes zero at the critical point.

EXAMPLE 2-1

A rigid tank contains 100 kg saturated liquid water at 90℃. Determine the pressure in the tank and the volume of the tank.

SOLUTION

The state of the saturated liquid water is shown on the T-V diagram in Fig. 2-13. Since saturation conditions exist in the tank, the pressure must be the saturation pressure at 90℃:

$$p = p_{\text{sat, 90℃}} = 70.183 \text{ kPa}$$

The specific volume of the saturated liquid at 90℃ is,

$$V_{f,\,90℃} = 0.001036 \text{ m}^3/\text{kg}$$

Then the total volume of the tank becomes,

$$V = mV_{f,\,90°C} = 100 \text{ kg} \times 0.001036 \text{ m}^3/\text{kg} = 0.1036 \text{ m}^3$$

EXAMPLE 2-2

Saturated liquid water (m=500 g) is completely vaporized at a constant pressure of 100 kPa, the enthalpy H_{fg} of water gasification at 100 kPa is 2257.5 kJ/kg. Determine (a) the volume change and (b) the amount of energy transferred to the water.

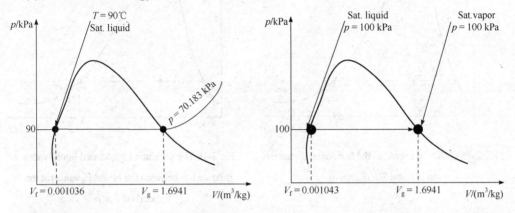

Fig.2-13 Schematic and T-V diagram for Example 2-1   Fig.2-14 Schematic and p-V diagram for Example 2-2

SOLUTION

(a) The process described is illustrated on the p-V diagram in Fig. 2-14. The volume change per unit mass during a vaporization process is V_{fg}, which is the difference between V_g and V_f. Reading these values from saturated water——pressure table at 100 kPa and substituting,

$$V_{fg} = V_g - V_f = (1.6941 - 0.001043) \text{m}^3/\text{kg} = 1.6931 \text{ m}^3/\text{kg}$$

Thus,

$$\Delta V = mV_{fg} = 0.5 \text{ kg} \times 1.6931 \text{ m}^3/\text{kg} = 0.84655 \text{ m}^3$$

(b) The amount of energy needed to vaporize a unit mass of a substance at a given pressure is the enthalpy of vaporization at that pressure, which is H_{fg} = 2257.5 kJ/kg for water at 100 kPa. Thus, the amount of energy transferred is,

$$mH_{fg} = 0.5 \text{ kg} \times 2257.5 \text{ kJ/kg} = 1128.75 \text{ kJ}$$

Saturated Liquid-Vapor Mixture

During a vaporization process, a substance exists as part liquid and part vapor. That is, it is a mixture of saturated liquid and saturated vapor (Fig. 2-15). To analyze this mixture properly, we need to know the proportions of the liquid and vapor phases in the mixture. This is done by defining a new property called the quality x as the ratio of the mass of vapor to the total mass of the mixture: A saturated mixture can be treated as a combination of two subsystems: the saturated liquid and the saturated vapor(Fig.2-16).

$$x = \frac{m_{vapor}}{m_{total}}$$

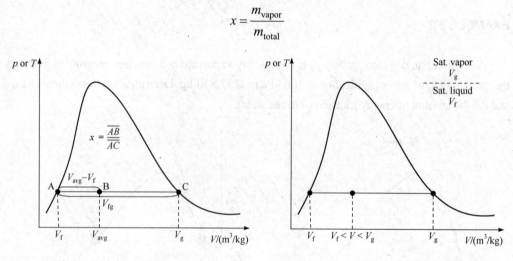

Fig.2-15 Quality is related to the horizontal distances on p-V and T-V diagrams

Fig.2-16 The V value of a saturated liquid-vapor mixture lies between the V_f and V_g values at the specified T or p

EXAMPLE 2-3

A rigid tank contains water (m=10 kg, T=90℃). If water (m=8 kg) is in the liquid phase and the rest is in the vapor phase, determine (a) the pressure in the tank and (b) the volume of the tank.

SOLUTION

(a) The state of the saturated liquid-vapor mixture is shown in Fig. 2-17. Since the two phases coexist in equilibrium, we have a saturated mixture, and the pressure must be the saturation pressure at the given temperature:

$$p = p_{sat,\ 90℃} = 70.183 \text{ kPa}$$

(b) At 90℃, we have V_f = 0.001036 m³/kg and V_g = 2.3593 m³/kg. One way of finding the volume of the tank is to determine the volume occupied by each phase and then add them up:

$$V = m_f V_f + m_g V_g$$
$$= 8 \text{ kg} \times 0.001036 \text{ m}^3/\text{kg} + 2 \text{ kg} \times 2.3593 \text{ m}^3/\text{kg}$$
$$= 4.73 \text{ m}^3$$

EXAMPLE 2-4

An 0.06 m³ vessel contains refrigerant-134a (m=3 kg, p=160 kPa). Determine (a) the temperature, (b) x, (c) the enthalpy of the refrigerant, and (d) the volume occupied by the vapor phase.

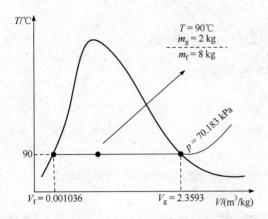
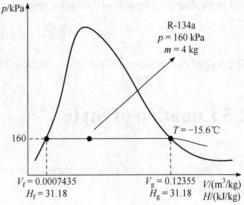

Fig.2-17 Schematic and *T-V* diagram for Example 2-3   Fig.2-18 Schematic and *p-V* diagram for Example 2-4

SOLUTION

(a) The state of the saturated liquid-vapor mixture is shown in Fig. 2-18. At this point we do not know whether the refrigerant is in the compressed liquid, superheated vapor, or saturated mixture region. This can be determined by comparing a suitable property to the saturated liquid and saturated vapor values. From the information given, we can determine the volume:

$$\hat{V} = \frac{V}{m} = \frac{0.060 \text{ m}^3}{3 \text{ kg}} = 0.02 \text{ m}^3/\text{kg}$$

At 160 kPa,

$$V_f = 0.0007435 \text{ m}^3/\text{kg}$$

$$V_g = 0.12355 \text{ m}^3/\text{kg}$$

Obviously, $V_f < \hat{V} < V_g$, and the refrigerant is in the saturated mixture region. Thus, the temperature must be the saturation temperature at the specified pressure:

$$T = T_{\text{sat, 160 kPa}} = -15.60 \text{°C}$$

(b) *x* can be determined from

$$x = \frac{\hat{V} - V_f}{V_{fg}} = \frac{0.02 - 0.0007435}{0.12355 - 0.0007435} = 0.157$$

(c) At 160 kPa, $H_f = 31.18$ kJ/kg and $H_{fg} = 209.96$ kJ/kg. Then,

$$H = H_f + xH_{fg}$$

$$= 31.18 \text{ kJ/kg} + 0.157 \times 209.96 \text{ kJ/kg}$$

$$= 64.1 \text{ kJ/kg}$$

(d) The mass of the vapor is,

$$m_g = xm_t = 0.157 \times 3 \text{ kg} = 0.471 \text{ kg}$$

and the volume occupied by the vapor phase is,

$$V_g = m_g V_g = 0.471 \text{ kg} \times 0.12355 \text{ m}^3/\text{kg} = 0.0582 \text{ m}^3$$

The rest of the volume (0.0018 m³) is occupied by the liquid.

2.5 Equation of State

Equations that relate the pressure, temperature, and specific volume of a substance are called equations of state.

$$f(p, \underline{V}, T) = 0 \tag{2-1}$$

Formally, an equation of state is a relationship among temperature, pressure, and molar volume. Therefore, if any two of the p, V and T of pure fluid are determined, the state of the system will be determined.

Eq. (2-1) is a function that describes the p-V-T relationship of fluids. It is also a thermodynamic equation related to state variables, which describes the state of substance under a given set of physical conditions, such as pressure, volume, temperature (p-V-T) or internal energy. There is a contradiction between the accuracy of equation of state and the simplicity of equation form. There are more than 150 equations of state. This book will introduce several common equations of state in the chemical industry.

2.5.1 The Ideal-Gas Equation of State

Research significance

The ideal-gas equation of state is applicable to the calculation of lower pressure and higher temperature condition. It provides the initial value for the calculation of the real-gas equation of state. In order to judge the correctness of the limit case of the real-gas equation of state, any equation of state will be simplified to the ideal-gas equation when the pressure is close to zero or the volume is infinite.

Both gas and vapor are usually referred to as the gas phase. Gas usually refers to substance above the critical temperature. Vapor usually means a substance not far from the condensed state. Avogadro's law states that "equal volumes of all gases, at the same temperature and pressure, have the same number of molecules." For a given mass of an ideal gas, the volume and amount (moles) of the gas are directly proportional if the temperature and pressure are constant.

$$pV = nRT \tag{2-2}$$

$$p\underline{V} = RT \tag{2-3}$$

$$p\hat{V} = \frac{RT}{M} \tag{2-4}$$

$$R = \begin{cases} 8.31447 \text{ J/(mol} \cdot \text{K)} \\ 1.987 \text{ cal/(mol} \cdot \text{K)} \\ 82.05 \text{ atm} \cdot \text{cm}^3/(\text{mol} \cdot \text{K}) \\ 1.98588 \text{ Btu/(l bmol} \cdot \text{R)} \\ 10.7316 \text{ psia} \cdot \text{ft}^3/(\text{l bmol} \cdot \text{R)} \\ 1545.37 \text{ ft} \cdot \text{lbf/(l bmol} \cdot \text{R)} \end{cases}$$

Where R is the gas constant [8.31447 J/(mol $\cdot$ K)]

An ideal gas is an imaginary substance that obeys the relation $p\underline{V} = RT$. Gases follow the ideal-gas equation of state closely at low pressures and high temperatures.

2.5.2 Nonideality of Gases

The ideal-gas equation is very simple and thus very convenient to use. However, gases deviate from ideal-gas behavior significantly at states near the saturation region and the critical point. This deviation from ideal-gas behavior at a given temperature and pressure can accurately be accounted for by the introduction of a correction factor called the compressibility factor Z.

The deviation of the real gas from the ideal gas can be expressed by the compressibility factor Z. The compressibility factor Z is a correction factor that must be considered when the ideal-gas equation of state is applied to the actual gas, which is used to represent the deviation in volume between the compressed real gas and the compressed ideal gas under the same pressure. When $Z=1$, it means that the V_m of the real gas is smaller than that of the ideal gas under the same conditions. At this time, the real gas is easier to compress than the ideal gas. This is because the actual molecular cohesion reduces the pressure generated by the collision of the gas molecules against the gas wall, so the measured pressure is smaller than that of the ideal state(Fig.2-19).

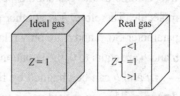

Fig.2-19 The compressibility factor is unity for ideal gases

Applying the concept of a compressibility factor to a critical point, a similar "critical compressibility factor" can be obtained. The measured critical compressibility factors of most gases are close, between 0.25 and 0.31. The critical state is real, so the critical parameters can be measured, and the calculated Z_c can also be checked in the corresponding list (Nan et al., 2021).

2.6 Other Equations of State

The ideal-gas equation of state is very simple, but its range of applicability is limited. However, we need the equations of state to accurately represent the p-V-T behavior of substance in a larger region without any limit. So, such equations are naturally more complicated. Several equations have been proposed for this purpose.

2.6.1 The van der Waals Equation of State

The first practical cubic equation of state was proposed by van der Waals of Leiden University in the Netherlands in 1873.

$$p = \frac{RT}{V-b} - \frac{a}{V^2} \tag{2-5}$$

Where, p is the gas pressure, Pa; T is the absolute temperature, K; R is the universal gas constant; V is the molar volume, Note that V is not the total volume, m^3. It is important to note that the units of R must correspond to the units of p and V. Please convert all physical quantities into SI units before substituting them into the equation to avoid calculation errors.

The vdW equation of state is actually derived by modifying the ideal gas model. It takes into account the volume of the molecules themselves and the gravitation between them, i.e., it treats the gas molecules as hard balls with mutual gravitation. Due to intermolecular gravitation, gas molecules hit the vessel wall when the momentum is reduced, therefore the pressure decreases, so the ideal state of pressure p_{ideal} should be equal to the actual pressure p_{actual} plus reduced pressure due to gravitation. $\frac{a}{V^2}$ is the result of pressure reducing, known as the internal pressure of the gas, which is a physical constant, is called the parameters of gravity. The volume of the molecule will make the movement space of the gas molecule smaller than the ideal state. Therefore, the actual free volume of the gas should be the container volume V minus the occupied volume b of the molecule. Therefore, b is another physical constant in the equation, which is called the repulsive parameter. So, a and b are only related to physical properties, and independent of p, V and T. When a and b are equal to 0, the equation of state is simplified as the ideal gas equation of state.

a and b can be obtained by fitting the p-V-T experimental data of the fluid, and can also be determined by the critical parameters T_c, p_c and V_c that reflect the properties of the substance. The specific method is to use the characteristics of horizontal inflection point of critical isotherms at critical points, i.e., $\left(\frac{\partial p}{\partial V}\right)_{T=T_c} = 0$ and $\left(\frac{\partial p^2}{\partial V^2}\right)_{T=T_c} = 0$, take the first and second partial derivatives of Eq. (2-5) with respect to the molar volume V, and order them to be equal to 0 when $T = T_c$ and $p = p_c$, thus we can accurately measure the parameters a and b expressed in terms of T_c and p_c.

$$a = \frac{27}{64} \times \frac{R^2 T_c^2}{p_c} \tag{2-6}$$

$$b = \frac{1}{8} \times \frac{RT_c}{p_c} \tag{2-7}$$

vdW equation of state has a great deviation in numerical calculation, but its contribution is the first to propose the inference of the influence of molecular gravity, repulsion, molecular volume and other factors on the properties of p-V-T fluid, providing very important enlightenment to the later, many subsequent equations are derived from it.

2.6.2 Redlich-Kwong (RK) Equation of State

After the vdW equation of state, there are hundreds of cubic equations of state improved on the basis of vdW equation of state. Among these equations, the Redlich-Kwong equation of state (RK equation) proposed by Redlich and Kwong in 1949 is the most accurate two parameter gas equation of state.

$$p = \frac{RT}{V-b} - \frac{a}{T^{1/2}V(V+b)} \tag{2-8}$$

with

$$a = 0.42748 R^2 \frac{T_c^2}{p_c} \quad b = 0.08664 R \frac{T_c}{p_c} \tag{2-9}$$

Compared with vdW equation of state, the calculation accuracy of RK equation of state is greatly improved. RK equation of state can be successfully used to calculate p-V-T in gas phase. The error for nonpolar and weak polar substances is about 2%, which can meet the needs of engineering, but the error for strong polar substances and liquid phase is large, ranging from 10% to 20%.

2.6.3 The Soave-Redlich-Kwong (SRK) Equation of State

In order to improve the accuracy, many researchers have modified the RK equation of state. Among these people, Soave's correction of RK equation of state is the most successful, which is called SRK (or RKS) equation of state. Soave believes that the lack of accuracy of RK equation of state in "expressing temperature effect" leads to difficulties in calculating multivariate vapor-liquid equilibrium, so he proposes to use $a(T)$ to replace $a/T^{1/2}$.

$$p = \frac{RT}{V-b} - \frac{a(T)}{V(V+b)} \tag{2-10}$$

With

$$a(T) = a_c \alpha(T_r) = 0.42748 R^2 \frac{T_c^2}{p_c} \alpha(T_r), \quad b = 0.08664 R \frac{T_c}{p_c}$$

$$\alpha(T_r) = \left[1 + m \times \left(1 - T_r^{0.5}\right)\right]^2 \tag{2-11}$$

$$m = 0.480 + 1.574\omega - 0.176\omega^2 \tag{2-12}$$

Where, ω is the eccentricity factor.

In the SRK equation of state, a is not only a physical property (critical property) but also a function of temperature. Due to the introduction of a more precise temperature function, the SRK equation of state greatly improves the calculation accuracy and application range (it can accurately calculate the p-V-T data of polar substances and substances containing hydrogen bonds and the density of saturated liquid), which has become the first cubic equation of state widely accepted and used by the engineering community. This equation has even been used as the standard method for phase equilibrium calculation. It can be said that SRK equation of state is an important milestone in the history of equation of state.

2.6.4 Peng-Robinson (PR) Equation of State

Peng and Robinson equation of state found that SRK equation of state was still not accurate enough in calculating the density of saturated liquid. So, Peng-Robinson equation of state, abbreviated as PR equation of state, was proposed in 1976.

$$p = \frac{RT}{\underline{V}-b} - \frac{a}{\underline{V}(\underline{V}+b)+b(\underline{V}-b)} \tag{2-13}$$

with

$$a = a_c \alpha \tag{2-14}$$

$$a_c = 0.45724 R^2 \frac{T_c^2}{p_c} \tag{2-15}$$

$$\alpha = \left[1 + \kappa\left(1 - T_r^{0.5}\right)\right]^2 \tag{2-16}$$

$$\kappa = 0.37464 + 1.54226\omega - 0.269932\omega^2 \tag{2-17}$$

$$b = 0.07780 R \frac{T_c}{p_c} \tag{2-18}$$

In PR equation of state, a is still a function of temperature, and the expression of volume is more precise, which makes it more accurate than SRK equation of state in predicting liquid molar volume. Moreover, PR equation of state is applicable to both polar substances and vapor-liquid phases. It is one of the most commonly used methods in engineering phase equilibrium calculation.

2.6.5 Virial Equation of State

The cubic equation of state introduced above belongs to the semi-empirical and semi-theoretical equation of state. However, the virial equation of state has a strict theoretical basis. In 1901, H. K. Onnes of Leiden University in the Netherlands proposed the virial equation of state, which is an equation of state expressed in the form of power series. There are two forms: density type and pressure type.

Density type:

$$Z = \frac{pV}{RT} = 1 + \frac{B}{V} + \frac{C}{V^2} + \frac{D}{V^3} \cdots \tag{2-19}$$

$$Z = \frac{pV}{RT} = 1 + B'p + C'p^2 + D'p^3 \cdots \tag{2-20}$$

In the case of taking infinite terms, the two are equivalent. There is the following relationship between different virial coefficients.

$$B' = \frac{B}{RT}, \quad C' = \frac{C-B^2}{R^2T^2}, \quad D' = \frac{D-3BC+2B^3}{R^3T^3} \tag{2-21}$$

Microscopically, the virial coefficient reflects the interaction between molecules. Macroscopically, the virial coefficient of pure substance is only a function of temperature. Generally, the virial coefficient is determined by experiments. The data of the second virial coefficient not only have rich measured values, but also can be estimated by comparing the temperature T_r. However, we know little about the virial coefficient after the third virial coefficient, so the virial truncation formula is often used in practical application.

Binomial virial truncation:

$$Z = \frac{pV}{RT} = 1 + \frac{B}{V} = 1 + B'p \tag{2-22}$$

Trinomial virial truncation:

$$Z = \frac{pV}{RT} = 1 + \frac{B}{V} + \frac{C}{V^2} = 1 + B'p + C'p^2 \tag{2-23}$$

The smaller the number of intercepted items are, the lower the accuracy and the applicable pressure are. Binomial truncation formula is only applicable to the gas with $T < T_c$ and $p < 1.5$ MPa; The binomial truncation formula is suitable for gases with $p < 5$ MPa. For gases with p greater than 5 MPa, other equations of state are usually used, such as SRK, PR equation of state, etc.

However, the virial equation of state can only be used for gas calculation, not for liquid calculation like the cubic equation of state, and the virial truncated formula is not suitable for calculation under high pressure.

Virial equation of state has more theoretical significance than practical application value. It can not only be used in the calculation of p-V-T relation, but also to relate the viscosity, sound velocity, heat capacity and other properties of gas based on molecular thermodynamics. Other multi-parameter equations of state such as BWR equation of state and MH equation of state are improved on this basis.

2.6.6 Multiparameter Equation of State

Compared with the simple equation of state, the multi-parameter equation of state can accurately describe the p-V-T relationship of different substances in a wider range of T and p, but its disadvantages are complicated equation form, difficulty in calculation and large workload.

Beattie-Bridgeman Equation of State

The Beattie-bridgeman equation of state based on five experimentally determined constants (Table 2-3), was proposed in 1928.

$$p = \frac{RT}{\underline{V}^2}\left(1 - \frac{c}{\underline{V}T^3}\right)(\underline{V} + B) - \frac{A}{\underline{V}^2} \tag{2-24}$$

$$A = A_0 \left(1 - \frac{a}{\underline{V}}\right) \tag{2-25}$$

$$B = B_0 \left(1 - \frac{b}{\underline{V}}\right) \tag{2-26}$$

Table 2-3 Constants that appear in the Beattie-Bridgeman equations of state (When p is in kPa, $\underline{V}$ is in m^3/kmol, T is in K, and R_u= 8.314 kPa · m^3/(kmol · K))

Gas	A_0	a	B_0	b	c
Air	131.8441	0.01931	0.04611	−0.001101	4.34×10^4
Ar	130.7802	0.02328	0.03931	0.0	5.99×10^4
CO$_2$	507.2836	0.07132	0.10476	0.07235	6.60×10^5
He	2.1886	0.05984	0.01400	0.0	40
H	20.0117	−0.00506	0.02096	−0.04359	504
N$_2$	136.2315	0.02617	0.05046	−0.00691	4.20×10^4
O$_2$	151.0857	0.02562	0.04624	0.004208	4.80×10^4

Benedict-Webb-Rubin Equation of State

BWR equation of state, named after Manson Benedict, G. B. Webb, and L. C. Rubin, is the first multi-parameter equation which can express p-V-T relation of fluid and calculate vapor-liquid equilibrium in high density region (Table 2-4).

$$p = \frac{RT}{\underline{V}} + \left(B_0 RT - A_0 - \frac{C_0}{T^2}\right)\frac{1}{\underline{V}^2} + \frac{bRT - a}{\underline{V}^3} + \frac{a\alpha}{\underline{V}^6} + \frac{c}{\underline{V}^3 T^2}\left(1 + \frac{\gamma}{\underline{V}^2}\right)e^{-\frac{\gamma}{\underline{V}^2}} \tag{2-27}$$

Table 2-4 Constants that appears in the Benedict-Webb-Rubin equation of state (When p is in kPa, $\underline{V}$ is in m^3/kmol, T is in K, and R_u= 8.314 kPa · m^3/(kmol · K))

Gas	a	A_0	b	B_0	c	C_0	α	γ
C$_4$H$_{10}$	190.68	1021.6	0.039998	0.12436	3.205×10^7	1.006×10^8	1.101×10^{-3}	0.0340
CO$_2$	13.86	277.30	0.007210	0.04991	1.511×10^6	1.404×10^7	8.470×10^{-5}	0.00539
CO	3.71	135.87	0.002632	0.05454	1.054×10^5	8.673×10^5	1.350×10^{-4}	0.0060
CH$_4$	5.00	187.91	0.003380	0.04260	2.578×10^5	2.286×10	1.244×10^{-4}	0.0060
N$_2$	2.54	106.73	0.002328	0.04074	7.379×10^4	8.164×10^5	1.272×10^{-4}	0.0053

In 1962, Strobridge further extended this equation by raising the number of constants to 16.

$$p = \frac{RT}{\underline{V}} + \left(B_0 RT - A_0 - \frac{C_0}{T^2} + \frac{D_0}{T^3} - \frac{E_0}{T^4}\right)\frac{1}{\underline{V}^2} + \left(bRT - a - \frac{d}{T}\right)\frac{1}{\underline{V}^3}$$
$$+ \left(a + \frac{d}{T}\right)\frac{\alpha}{\underline{V}^6} + \frac{c}{\underline{V}^3 T^2} \times \left(1 + \frac{\gamma}{\underline{V}^2}\right)e^{-\frac{\gamma}{\underline{V}^2}} \tag{2-28}$$

$$A_0 = \sum_i\sum_j x_i x_j A_{0i}^{1/2} A_{0j}^{1/2}\left(1-k_{ij}\right) \tag{2-29}$$

$$B_0 = \sum_i x_i B_{0i} \tag{2-30}$$

$$C_0 = \sum_i\sum_j x_i x_j C_{0i}^{1/2} C_{0j}^{1/2}\left(1-k_{ij}\right)^3 \tag{2-31}$$

$$D_0 = \sum_i\sum_j x_i x_j D_{0i}^{1/2} D_{0j}^{1/2}\left(1-k_{ij}\right)^4 \tag{2-32}$$

$$E_0 = \sum_i\sum_j x_i x_j E_{0i}^{1/2} E_{0j}^{1/2}\left(1-k_{ij}\right)^5 \tag{2-33}$$

$$\alpha = \left[\sum_i x_i \alpha_i^{1/3}\right]^3 \tag{2-34}$$

$$\gamma = \left[\sum_i x_i \gamma_i^{1/2}\right]^2 \tag{2-35}$$

$$a = \left[\sum_i x_i a_i^{1/3}\right]^3 \tag{2-36}$$

$$b = \left[\sum_i x_i b_i^{1/3}\right]^3 \tag{2-37}$$

$$c = \left[\sum_i x_i c_i^{1/3}\right]^3 \tag{2-38}$$

$$d = \left[\sum_i x_i d_i^{1/3}\right]^3 \tag{2-39}$$

Martin-Hou Equation of State

MH equation of state was put forward by Professor Martin and Chinese scholar Hou Yujun in 1955. In order to improve the accuracy of the equation in the high-density region, Marin further improved the equation in 1959. In 1981, Professor Hou Yujun extended the application scope of the equation to the liquid region, and the improved equation was called MH-81 equation.

The general formula of MH equation of state is:

$$p = \sum_{i=1}^{5} \frac{f_i(T)}{(V-b)^i} \tag{2-40}$$

$$f_i(T) = RT \qquad (i=1) \tag{2-41}$$

$$f_i(T) = A_i + B_i T + C_i \exp(-5.475T/T_c) \tag{2-42}$$

Where, A_i, B_i, C_i and b are constants of the equation, which can be obtained from critical parameters of pure substance and the data of a point on the saturated vapor pressure curve. Among them, in MH-55 equation of state, constant $B_4=C_4=A_5=C_5=0$, and in MH-81 type equation of state, constant $C_4=A_5=C_5=0$.

MH-81 equation of state can be used in two phases of gas and liquid at the same time, with high accuracy and wide application range. It can be used in non-polar to strong polar substances (such as NH_3 and H_2O), as well as quantum gases such as H_2 and He. It is widely used in engineering design of ammonia synthesis.

EXAMPLE 2-5

Predict the pressure of N_2 at $T = 175$ K and $\hat{V} = 0.00375$ m³/kg on the basis of
(a) The ideal-gas equation of state,
(b) The van der Waals equation of state,
(c) The Beattie-Bridgeman equation of state,
(d) The Benedict-Webb-Rubin equation of state.

SOLUTION

(a) Ideal-gas equation of state

$$p = \frac{RT}{V} = \frac{0.2968 \text{ kPa/(kg·K)} \times 175 \text{K}}{0.00375 \text{ m}^3/\text{kg}} = 1.3851 \times 10^4 \text{ kPa}$$

The error is 38.5%.

(b) Van der Waals equation of state

$$a = 0.175 \text{ m}^6 \cdot \text{kPa/kg}^2$$

$$b = 0.00138 \text{ m}^3/\text{kg}$$

$$p = \frac{RT}{V-b} - \frac{a}{V^2} = 9.471 \times 10^3 \text{ kPa}$$

The error is 5.3%.

(c) Beattie-Bridgeman equation of state

$$A = 102.29$$

$$B = 0.05378$$

$$c = 4.2 \times 10^4$$

$$\overline{V} = M V = 28.013 \text{ kg/kmol} \times 0.00375 \text{ m}^3/\text{kg} = 0.10505 \text{ m}^3/\text{kmol}$$

$$p = \frac{RT}{\overline{V}^2}\left(1 - \frac{c}{\overline{V}T^3}\right)(\overline{V} + B) - \frac{A}{\overline{V}^2} = 1.011 \times 10^3 \text{ kPa}$$

The error is 1.1%.

(d) Benedict-Webb-Rubin equation

$$a = 2.54$$
$$A_0 = 106.73$$
$$b = 0.002328$$
$$B_0 = 0.04074$$
$$c = 7.379 \times 10^4$$
$$C_0 = 8.164 \times 10^5$$
$$\alpha = 1.272 \times 10^{-4}$$
$$\gamma = 0.0053$$

$$p = \frac{RT}{\overline{V}} + \left(B_0 RT - A_0 - \frac{C_0}{T^2}\right)\frac{1}{\overline{V}^2} + \frac{bRT - a}{\overline{V}^3}$$
$$+ \frac{a\alpha}{\overline{V}^6} + \frac{c}{\overline{V}^3 T^2}\left(1 + \frac{\gamma}{\overline{V}^2}\right)e^{-\frac{\gamma}{\overline{V}^2}} = 1.0009 \times 10^4 \text{ kPa}$$

Which is in error by only 0.09%. Thus, the accuracy of the Benedict-Webb Rubin equation of state is rather impressive in this case(Fig.2-20).

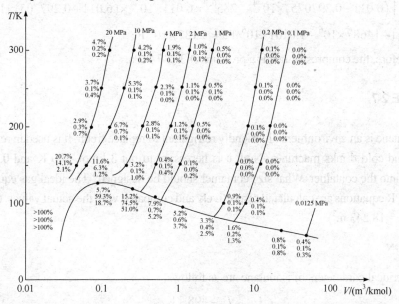

Fig.2-20 Percentage of error involved in various equations of state for nitrogen beyond the cubic equations of state

EXAMPLE 2-6

5 mol and 285 K CO_2 gas is compressed into medium pressure gas through the compressor, and then loaded into a steel cylinder with a volume of 3005.5 cm³ for use in the food industry. How much pressure does the compressor need to add to achieve this purpose? Try to calculate using the RK equation of state.

SOLUTION

The molar volume of CO_2 gas is as follows,

$$V = \frac{V_{total}}{n} = \left(\frac{3005.5}{5}\right) \text{cm}^3/\text{mol} = 601.1 \text{ cm}^3/\text{mol} = 6.011 \times 10^{-4} \text{ m}^3/\text{mol}$$

The critical parameters of CO_2 are as follows,

$$T_c = 304.2 \text{ K}$$

$$p_c = 7.376 \text{ MPa}$$

$$a = 0.42748 \frac{R^2 T_c^{2.5}}{p_c} = \left(0.42748 \times \frac{8.314^2 \times 304.2^{2.5}}{7.376 \times 10^6}\right) \text{Pa} \cdot \text{m}^3 \cdot \text{K}^{0.5}/\text{mol}^2$$

$$= 6.4657 \text{ Pa} \cdot \text{m}^3 \cdot \text{K}^{0.5}/\text{mol}^2$$

$$b = 0.08664 \frac{RT_c}{p_c} = \left(0.08664 \times \frac{8.314 \times 304.2}{7.376 \times 10^6}\right) \text{m}^3/\text{mol} = 2.97075 \times 10^{-5} \text{ m}^3/\text{mol}$$

$$p = \frac{RT}{V-b} - \frac{a}{T^{0.5} V(V+b)}$$

$$= \left[\frac{8.314 \times 285}{(6.011 - 0.297075) \times 10^{-4}} - \frac{6.4657}{285^{0.5} \times 6.011 \times 10^{-4} \times (6.011 + 0.297705) \times 10^{-4}}\right] \text{Pa}$$

$$= (4.14687 \times 10^6 - 1.0101 \times 10^6) \text{Pa} = 3.137 \times 10^6 \text{ Pa}$$

Therefore, the compressor needs a pressure of 3.137 MPa.

EXAMPLE 2-7

Isobutane is an environmentally friendly refrigerant to replace freon. It is used in refrigerators, freezers and cold drinks machines. Now it is necessary to put 3 kmol, 300 K and 0.3704 MPa isobutane into the container. What size container should be designed? The ideal gas equation, RK, SRK and PR equations are calculated respectively and compared with the actual values. (The actual value $V_{total} = 18.243 \text{ m}^3$).

SOLUTION

The critical parameters of isobutane are as follows,

$$T_c = 408.1 \text{ K}$$

$$p_c = 3.648 \text{ MPa}$$

$$\omega = 0.176$$

(1) Ideal gas equation

$$V = \frac{RT}{p} = \left(\frac{8.314 \times 300}{0.3704 \times 10^6}\right) \text{m}^3/\text{mol} = 0.006734 \text{ m}^3/\text{mol}$$

$$V_{total} = (3000 \times 0.006734) \text{m}^3 = 20.201 \text{ m}^3$$

(2) RK equation of state

$$a = 0.42748\frac{R^2 T_c^{2.5}}{p_c} = 27.25 \text{ Pa} \cdot \text{m}^6 \cdot \text{K}^{0.5}/\text{mol}^2$$

$$b = 0.08664\frac{RT_c}{p_c} = 8.058 \times 10^{-5} \text{ m}^3/\text{mol}$$

Given that p and T solve V iteratively, the unit of volume is m^3/mol.

$$V_{n+1} = \frac{RT}{p} + b - \frac{a(V_n - b)}{pT^{0.5}V_n(V_n + b)}$$

$$V_{n+1} = \left[\frac{8.314 \times 300}{0.3704 \times 10^6} + 8.058 \times 10^{-5}\right] \text{m}^3/\text{mol} - \frac{27.25(V_n - 8.058 \times 10^{-5})}{0.3704 \times 10^6 \times 300^{0.5} \times V_n(V_n + 8.058 \times 10^{-5})}$$

$$= 6.814 \times 10^{-3} \text{ m}^3/\text{mol} - \frac{4.248 \times 10^{-6}(V_n - 8.058 \times 10^{-5})}{V_n(V_n + 8.058 \times 10^{-5})}$$

Substitute $V_0 = \frac{RT}{p} = \left(\frac{8.314 \times 300}{0.3704 \times 10^6}\right) \text{m}^3/\text{mol} = 0.006734 \text{ m}^3/\text{mol}$ as the initial value into the above equation,

$$V_1 = 0.006198 \text{ m}^3/\text{mol}$$

And then substitute that into the equation above,

$$V_2 = 0.006146 \text{ m}^3/\text{mol}$$
$$V_3 = 0.006146 \text{ m}^3/\text{mol}$$
$$V_4 = 0.0061408 \text{ m}^3/\text{mol}$$

$$Z_4 = \frac{pV_4}{RT} = \frac{0.3704 \times 10^6 \times 6.1408 \times 10^{-3}}{8.314 \times 300} = 0.9119$$

$$V_5 = 0.006140 \text{ m}^3/\text{mol}$$

$$Z_5 = \frac{pV_5}{RT} = \frac{0.3704 \times 10^6 \times 6.140 \times 10^{-3}}{8.314 \times 300} = 0.91182$$

$$|Z_5 - Z_4| = 8.2 \times 10^{-5} \leq 10^{-4}$$

$$V = 0.006140 \text{ m}^3/\text{mol}$$

2.7 Principle of Corresponding States and Generalized Association

At the same temperature and pressure, the compressibility factor Z of different real gases is not equal. This indicates that the degree of deviation of real gas from ideal gas does not only depend on temperature and pressure. Through a large number of experimental studies, it is found that for different fluids, when they have the same reduced temperature and pressure, they have roughly the same Z. That is, the degree of deviation from the ideal gas is roughly the same. This is the famous principle of corresponding states. It found commonalities for various substances with different properties and a key to open the generalized equation of state.

2.7.1 Principle of Corresponding States

The reduced temperature, pressure and molar volume are defined as:

$$T_r = \frac{T}{T_c}, \quad p_r = \frac{p}{p_c}, \quad V_r = \frac{V}{V_c} \tag{2-43}$$

According to the principle of corresponding state, the reduced molar volumes of different gases are approximately equal under the same reduced temperature and pressure. It can also be seen from Fig. 2-21.

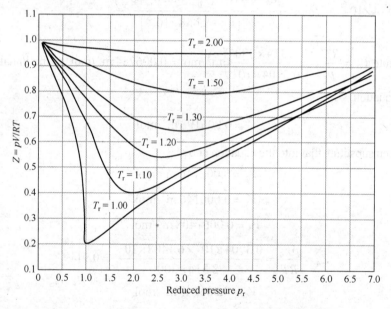

Fig.2-21 Relationship between real gas compressibility factor Z, T_r and p_r

2.7.2 Principle of Corresponding States with Two Parameters

The equation of state of real gas can be generalized by means of the principle of corresponding states. Eq. (2-44) can be obtained by substituting $T_r = \frac{T}{T_c}, p_r = \frac{p}{p_c}, V_r = \frac{V}{V_c}$ into vdW equation of state. Eq. (2-44) is the principle of corresponding states with two parameters first proposed by van der Waals.

$$\left(p_r + \frac{3}{V_r^2}\right)(3V_r - 1) = 8T_r \tag{2-44}$$

Experiments show that the principle of corresponding state with two parameters is not strictly correct. Eq. (2-44) is very accurate in calculating spherical nonpolar simple molecules (such as Ar, Kr and Xe). For non-spherical weakly polar molecules, the error is generally small, but sometimes the error is considerable. For some complex gases with non-spherical strong polar molecules, there is an obvious deviation. When the third parameter reflecting the characteristics of molecular structure is introduced, this situation has been significantly improved.

2.7.3 Principle of Corresponding States with Three Parameters

In order to improve the accuracy of calculating the compressibility factor of complex molecules, the third parameter is introduced. There are generally two methods:
(1) Critical compressibility factor Z_c, $Z = f(T_r, p_r, Z_c)$
(2) Acentric factor ω, $Z = f(T_r, p_r, \omega)$

The acentric factor is

$$\omega = -\log_{10}\left(\frac{p^{sat}}{p_c}\right)\bigg|_{T_r=0.7} - 1 \tag{2-45}$$

The acentric factor is said to be a measure of the non-sphericity (centricity) of molecules. The physical meaning of ω is: the eccentricity of general fluid and spherical nonpolar simple fluid (Ar, Kr and Xe) in shape and polarity. $0 < \omega < 1$, the greater the ω, the greater the deviation. According to the definition of ω, ω of simple spherical fluid is 0, and ω of polar molecule is larger, such as ω of ethanol is 0.635.

Pitzer's principle of corresponding state with three parameters can be expressed as follows: under the same T_r and p_r, all fluids with the same ω have the same compressibility factor Z, so they deviate from the ideal gas to the same extent. This is a great improvement over the original principle of corresponding state with two parameters. Such a concept can be obtained from this principle. The behavior of gas deviating from ideal gas is not determined by temperature and pressure alone, but by T_r, p_r and ω.

According to the above conclusions, Pitzer puts forward two very useful generalization relations: ① generalization relations expressed by polynomials of compressibility factors (generalized compressibility factor graph method); ② The generalized second virial coefficient relation expressed by two virial equations (generalized virial coefficient method).

2.7.4 Generalized Compressibility Factor Graph Method

Pitzer proposed that the relationship of compressibility factor Z was:

$$Z = Z^0 + \omega Z^1 \tag{2-46}$$

Where Z^0 is the compressibility factor of simple fluid; Z^1 is the deviation of the fluid from the simple fluid. They are all complex functions of p and T, which are difficult to be accurately described by simple equations. In order to facilitate manual calculation, predecessors have made these complex functions into charts, as shown in the Fig. 2-22, which provides convenience for engineering application.

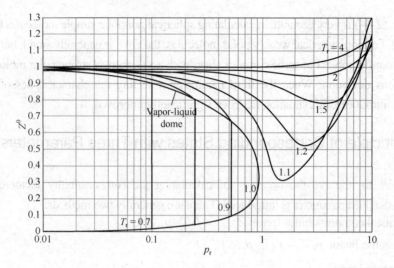

Fig 2-22 The generalized relation of Z^0

2.7.5 Generalized Virial Coefficient Method

The virial equation of state can also be expressed as a generalized equation of state, when substitute $T_r = \dfrac{T}{T_c}$, $p_r = \dfrac{p}{p_c}$ into Eq. (2-47),

$$Z = 1 + \frac{Bp}{RT} = 1 + \frac{Bp_c}{RT_c} \times \frac{p_r}{T_r} = 1 + \hat{B} \frac{p_r}{T_r} \tag{2-47}$$

Where, $\hat{B}$ is the comparative second virial coefficient, a dimensionless variable, which can be written as,

$$\hat{B} = \frac{Bp_c}{RT_c} \tag{2-48}$$

For the value of $\hat{B}$, Pitzer gives a form similar to Eq. (2-46),

$$\hat{B} = \frac{Bp_c}{RT_c} = B^0 + \omega B^1 \tag{2-49}$$

Combining Eq. (2-47) and Eq. (2-49),

$$Z = 1 + B^0 \frac{p_r}{T_r} + \omega B^1 \frac{p_r}{T_r} \tag{2-50}$$

Compared with Eq. (2-46),

$$Z^0 = 1 + B^0 \frac{p_r}{T_r}, \quad Z^1 = B^1 \frac{p_r}{T_r} \tag{2-51}$$

The second virial coefficient is only a function of temperature. Similarly, B^0 and B^1 are only functions of reduced temperature, which can be reasonably expressed by the following formula:

$$B^0 = 0.083 - \frac{0.422}{T_r^{1.6}} \tag{2-52}$$

$$B^1 = 0.139 - \frac{0.172}{T_r^{4.2}} \tag{2-53}$$

Tsonopoulos improved the generalized second virial coefficient with higher accuracy.

$$B^0 = 0.1445 - \frac{0.33}{T_r} - \frac{0.1385}{T_r^2} - \frac{0.0121}{T_r^3} - \frac{0.000607}{T_r^8} \tag{2-54}$$

$$B^1 = 0.0637 + \frac{0.331}{T_r^2} - \frac{0.423}{T_r^3} - \frac{0.008}{T_r^8} \tag{2-55}$$

Although the second virial coefficient can be estimated by T and ω, the determination of the second virial coefficient needs to be determined experimentally at first, which is the advantage of generalization. The generalized second virial coefficient equation has the same limitations as the binomial virial truncation in the calculation of fluid p-V-T, i.e., it is only suitable for low-pressure non-polar gases and has a narrow application range. Large scale chemical engineering software usually adopts PR or SRK equation instead of generalizing the second virial coefficient equation. But the generalized second virial coefficient equation is more suitable for manual calculation.

The generalized equation of state established by the generalized compressibility factor graph method and the generalized second virial coefficient method proposed by Pitzer express the compressibility factor Z as a function of T_r, p_r and ω, but the application scope of the two equations is different, as shown in the Fig. 2-23. If T_r, p_r is above the slash, the generalized virial coefficient method is applicable, otherwise the generalized compressibility factor graph method is applicable.

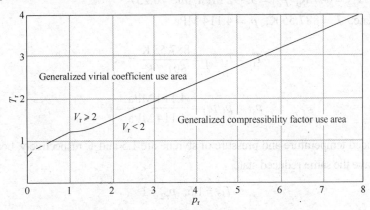

Fig.2-23 The applicable region of the generalized relational equation

EXAMPLE 2-8

Calculate the volume of ethane (n=1 kmol) at 382 K and 21.5 MPa. (Ethane: $T_c = 305.4$ K, $p_c = 4.884$ MPa, $\omega = 0.098$)

SOLUTION

$$T_r = \frac{382}{305.4} = 1.25, \quad p_r = \frac{21.5}{4.884} = 4.40$$

According to Fig. 2-23, the generalized compressibility factor graph method is used for calculation. According to Fig. 2-22, $Z^0 = 0.67, Z^1 = 0.06$.

$$Z = Z^0 + \omega Z^1 = 0.67 + 0.098 \times 0.06 = 0.675$$

$$V = \frac{ZRT}{p} = \left(\frac{0.675 \times 8.314 \times 382}{21.5 \times 10^6}\right) \text{m}^3/\text{mol} = 0.0000998 \text{ m}^3/\text{mol}$$

$$V_{\text{Total}} = 1000 \times 0.0000998 = 0.0998 \text{ m}^3$$

Therefore, the volume of ethane (n=1 kmol) at 382 K and 21.5 MPa is 0.0998 m^3.

EXAMPLE 2-9

The acentric factor of toluene and styrene are equal (i.e., $\omega = 0.257$). It is known that the pressure of toluene is 4.114 MPa and the temperature is 887.55 K. Toluene and styrene have the same reduced state. What is the temperature, pressure and volume of styrene at this time? What is the volume of toluene? (Please use the generalized relation to calculate.)

SOLUTION

Known: the critical parameters of toluene and styrene,
Toluene. $T_{c1} = 591.7$ K, $p_{c1} = 4.114$ MPa, $\omega_1 = 0.257$
Styrene. $T_{c2} = 647$ K, $p_{c2} = 3.992$ MPa, $\omega_2 = 0.257$
(1) Toluene: $T_1 = 887.55$ K, $p_1 = 4.114$ MPa

$$T_{r1} = T_1/T_{c1} = \frac{887.55 \text{ K}}{591.7 \text{ K}} = 1.5$$

$$p_{r1} = p_1/p_{c1} = \frac{4.114 \text{ MPa}}{4.114 \text{ MPa}} = 1$$

The reduced temperature and pressure of styrene are 1.5 and 1, respectively, because styrene and toluene have the same reduced state.

$$T_{r2} = 1.5, \quad p_{r2} = 1$$

$$T_2 = T_{r2}T_{c2} = (1.5 \times 647)\text{K} = 970.5 \text{ K}$$

$$p_2 = p_{r2}p_c = 3.992 \text{ MPa}$$

(2) Volume of styrene: According to Fig. 2-23, the generalized second virial coefficient method should be used when $T_{r2} = 1.5$, $p_{r2} = 1$.
According to Eq. (2-52) and Eq. (2-53),

$$B_2^0 = 0.083 - \frac{0.442}{T_{r2}^{1.6}} = 0.083 - \frac{0.422}{1.5^{1.6}} = -0.13758$$

$$B_2^1 = 0.139 - \frac{0.172}{T_{r2}^{4.2}} = 0.139 - \frac{0.172}{1.5^{4.2}} = 0.10767$$

According to Eq. (2-49),

$$\hat{B}_2 = B_2^0 + \omega B_2^1 = -0.13758 + 0.257 \times 0.10767 = -0.1099$$

According to Eq. (2-47),

$$Z_2 = 1 + \hat{B}_2 \frac{p_{r2}}{T_{r2}} = 1 - 0.1099 \times \frac{1}{1.5} = 0.9267$$

$$V_2 = \frac{Z_2 RT_2}{p_2} = \left(\frac{0.9267 \times 8.314 \times 970.5}{3.992 \times 10^6}\right) \text{m}^3/\text{mol} = 1.873 \times 10^{-3} \text{ m}^3/\text{mol}$$

(3) Volume of toluene: According to the principle of corresponding state with three parameters, all fluids with the same value of Z have the same compressibility factor at the same reduced temperature and pressure. Therefore, the compressibility factor of toluene is:

$$V_1 = \frac{Z_1 RT_1}{p_1} = \left(\frac{0.9267 \times 8.314 \times 887.55}{4.114 \times 10^6}\right) \text{m}^3/\text{mol} = 1.662 \times 10^{-3} \text{ m}^3/\text{mol}$$

2.8 Application of Aspen Plus in Calculation of Thermodynamic Equation of State

Please use SRK and PR equations of state to calculate the molar volume of saturated vapor of isobutane at 300 K and 370.4 kPa respectively. The known experimental value is $V = 6.081 \times 10^{-3} \text{ m}^3/\text{mol}$.

Calculation with manual calculation process

The critical parameter of isobutane is known as,

$$T_c = 408.1 \text{ K}, \quad p_c = 3.648 \times 10^6 \text{ Pa}, \quad \omega = 0.176$$

(1) SRK equation of state

$$T_r = \frac{300}{408.1} = 0.7351$$

$$m = 0.480 + 1.574 \times 0.176 - 0.176^3 = 0.7516$$

$$\alpha(T) = \left[1 + 0.7516 \times (1 - 0.7351^{0.5})\right]^2 = 1.2259$$

$$a = \left[0.42748 \times \frac{(8.314)^2 \times (408.1)^2}{3.648 \times 10^6} \times 1.2259\right] \text{Pa} \cdot \text{m}^6/\text{mol}^2 = 1.6537 \text{ Pa} \cdot \text{m}^6/\text{mol}^2$$

$$b = \left(0.08664 \times \frac{8.314 \times 408.1}{3.648 \times 10^6}\right) \text{m}^3/\text{mol} = 8.0582 \times 10^{-5} \text{ m}^3/\text{mol}$$

$$A = \frac{1.6537 \times 3.704 \times 10^5}{8.314^2 \times 300^2} = 0.09846$$

$$B = \frac{8.0582 \times 10^{-5} \times 3.704 \times 10^5}{8.314 \times 300} = 0.01197$$

so

$$Z = \frac{1}{1-h} - \frac{A}{B}\left(\frac{h}{1+h}\right) = \frac{1}{1-h} - 8.2256 \times \left(\frac{h}{1+h}\right)$$

and

$$h = \frac{B}{Z} = \frac{0.01197}{Z}$$

Iterative calculation.

(2) PR equation of state

$$k = 0.37464 + 1.54226 \times 0.176 - 0.26992 \times 0.176^2 = 0.6377$$

$$\alpha(T) = \left[1 + 0.6377 \times (1 - 0.7351^{0.5})\right]^2 = 1.1902$$

$$a = \left[0.45727 \times \frac{(8.314)^2 \times (408.1)^2}{3.648 \times 10^6} \times 1.1902\right] \text{Pa} \cdot \text{m}^6/\text{mol}^2 = 1.7175 \text{ Pa} \cdot \text{m}^6/\text{mol}^2$$

$$b = \left(0.07780 \times \frac{8.314 \times 408.1}{3.648 \times 10^6}\right) \text{m}^3/\text{mol} = 7.2360 \times 10^{-5} \text{ m}^3/\text{mol}$$

$$A = \frac{1.7175 \times 3.704 \times 10^5}{8.314^2 \times 300^2} = 0.1023$$

$$B = \frac{7.2360 \times 10^{-5} \times 3.704 \times 10^5}{8.314 \times 300} = 0.01075$$

so

$$Z = \frac{1}{1-h} - \frac{A}{B}\left(\frac{h}{1+2h-h^2}\right) = \frac{1}{1-h} - 9.5163 \times \left(\frac{h}{1+2h-h^2}\right)$$

and

$$h = \frac{B}{Z} = \frac{0.01075}{Z}$$

Iterative calculation.

(3) Take the initial value $Z=1$, and see the following table for the iterative calculation results.

The number of iterations	SRK equation of state		PR equation of state	
	Z	h	Z	h
0	1	0.01197	1	0.01075
1	0.9148	0.01308	0.9107	0.01184
2	0.9071	0.01320	0.9010	0.01192
3	0.9062	0.01321	0.9013	0.01193
4	0.9061		0.9012	

SRK equation of state:

$$V = \frac{ZRT}{p} = \left(\frac{0.9061 \times 8.314 \times 300}{3.704 \times 10^5}\right) \text{m}^3/\text{mol} = 6.1015 \times 10^{-3} \text{ m}^3/\text{mol}$$

Error

$$\frac{(6.081 - 6.1015) \times 10^{-3}}{6.081 \times 10^{-3}} = -0.34\%$$

PR equation of state:

$$V = \frac{ZRT}{p} = \left(\frac{0.9012 \times 8.314 \times 300}{3.704 \times 10^5}\right) \text{m}^3/\text{mol} = 6.0685 \times 10^{-3} \text{ m}^3/\text{mol}$$

Error

$$\frac{(6.081 - 6.0685) \times 10^{-3}}{6.081 \times 10^{-3}} = 0.21\%$$

It can be seen from the above calculation that the iteration convergence speed is very fast, and the end point can be obtained only after 4 iterations. SRK and PR equations have good accuracy, and their results are acceptable to engineering circles.

Although equation of state can be applied to the calculation under different conditions, the manual calculation process has many problems such as large amount of calculation, complex calculation and time-consuming. So, the commercial chemical process simulation software (Aspen Plus) is considered to solve these problems.

Estimate the specific molar volume of isobutane at 101.325 kPa and 298.15 K. Compare the results of manual calculation with those calculated by Aspen Plus.

Calculation with Aspen Plus

The SRK equation of state

Step 1. Input component isobutene(Fig.2-24).
Step 2. Select the thermodynamic physical property method(Fig.2-25).
Step 3. Click Next, enter the simulation environment(Fig.2-26).
Step 4. Establish a simple process as shown in Figure 2-27.

Fig.2-24 Component | Specification | Selection

Step 5. Enter the stream information for isobutane (S1)(Fig.2-28).
Step 6. Click Run and view the results(Fig.2-29).

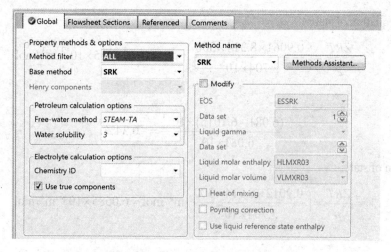

Fig.2-25 Methods | Specification | Global

Fig.2-26 Enter the simulation environment

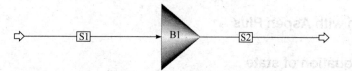

Fig.2-27 Simple flowsheet of isobutane

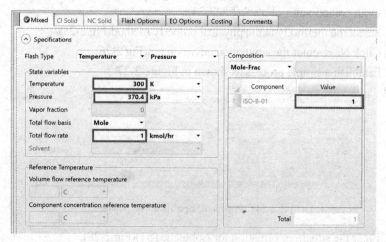

Fig.2-28 Streams | Input | Mixed

Chapter 2 The Physical Properties of Pure Substances

Material	Heat	Load	Work	Vol.% Curves	Wt.% Curves	Petroleum	Polymers	Solids

	Units	S1	S2	
Temperature	C	26.85	26.85	
Pressure	bar	3.704	3.704	
Molar Vapor Fraction		1	1	
Molar Liquid Fraction		0	0	
Molar Solid Fraction		0	0	
Mass Vapor Fraction		1	1	
Mass Liquid Fraction		0	0	
Mass Solid Fraction		0	0	
Molar Enthalpy	kcal/mol	-32.3579	-32.3579	
Mass Enthalpy	kcal/kg	-556.71	-556.71	
Molar Entropy	cal/mol-K	-93.7454	-93.7454	
Mass Entropy	cal/gm-K	-1.61287	-1.61287	
Molar Density	kmol/cum	0.163955	0.163955	
Mass Density	kg/cum	9.52961	9.52961	
Enthalpy Flow	Gcal/hr	-0.0323579	-0.0323579	
Average MW		58.1234	58.1234	

Fig.2-29 Results Summary | Streams | Material

Step 7. The reciprocal of the result of molar density is the result calculated by Aspen Plus using SRK ($6.0992 \text{ m}^3/\text{kmol}$). Compared with the experimental results, the error is:

$$\frac{(6.081-6.0992) \times 10^{-3}}{6.081 \times 10^{-3}} = 0.30\%$$

When the PR equation of state is adopted, only the selection of equation of state in step 2 needs to be changed to PR, and other steps are the same as above.

EXERCISES

1. Calculate the saturated vapor pressure and saturated liquid molar volume of isobutane at 273.15 K (experimental values were 152561 Pa and 100.1 cm$^3 \cdot$ mol^{-1}), and estimate the saturated vapor phase molar volume.

2. At 323.15 K, 6.08×10^7 KPa, a mixture of 0.401 (mole fraction) nitrogen and 0.5999 (mole fraction) ethylene, try to find the volume of the mixture by:
   (1) Ideal gas equation of state;
   (2) Kay rules. $Z=1.40$ from the experiment.

3. Ammonia gas with a mass of 500 g is stored in a steel shell with a volume of 0.03 m^3, and the steel shell is immersed in a constant temperature bath with a temperature of 65 ℃. Calculate the ammonia pressure by using the second Virial coefficient and compare with the reference value p =2.382 MPa (ammonia: T_c=405.6 K, p_c=11.28 MPa, V_c=72.6$\times 10^{-6}$ m^3/mol, ω=0.250).

4. 1kmol nitrogen is compressed and stored in a cylinder with a volume of 0.04636 m^3 and a temperature of 273.15 K. What is the pressure of nitrogen? The ideal gas equation of state, RK equation and RKS equation are used to calculate. (The experimental value is 101.33 MPa) (nitrogen: T_c=126.1 K, p_c=3.394 MPa, ω=0.176)

5. Try to rewrite the RK equation of the following form into a generalized form. The molar volume of ethylene at 277.6 K and 4.513$\times 10^6$ Pa is calculated(ethylene: T_c=282.4, p_c=5.036 MPa)

$$Z = \frac{1}{1-h} - \frac{A}{B} \times \left(\frac{h}{1+h}\right)$$

and

$$h=B/Z,\ B=bp/RT,\ A/B=a/bRT^{1.5}$$

a and b are constants of the RK equation.

6. Try to calculate the molar volume of methane gas at 400℃ and 4.053 MPa by the following three methods. (a) Using ideal gas equation; (b) Using RK equation; (c) Using virial truncation. B is calculated by Pitzer's generalized correlation method.

7. The gas phase molar volume of n-butane at 510 K and 2.5 MPa is calculated by ideal gas equation and generalized correlation method. The known experimental value is 1.4807 m³/kmol.

8. The ratio of feed gas in a synthetic ammonia plant is $N_2 : H_2 = 1 : 3$ (molar ratio). Before entering the synthetic tower, the mixed gas pressure is first reduced to 40.532 MPa (400 atm) and heated to 300℃. Since the molar volume of mixed gas is the necessary data for the specification design of synthesis tower, try the following method to calculate. Known literature value $Z_m =$ 1.1155. (1) Using ideal gas equation; (2) Combination of AMAGAT classification and generalized z-graph; (3) It is calculated with virtual critical parameters (Kay rule).

9. The compressibility factor of butane vapor at 350 K and 1.2 MPa is calculated by generalized RK equation and SRK equation respectively. The known experimental value is 0.7731.

10. Apply PR equation and SRK equation to calculate pressure of perchloride fluoride at 404.42 K, $V=2.7925\times10^{-4}$ m³/mol and 404.42 K, $V=2.0382\times10^{-2}$ m³/mol. The measured values are 68.32 atm and 1.488 atm respectively (from the former Laboratory of chemical thermodynamics, Department of chemical engineering, Zhejiang University).

REFERENCES

[1] Chang H Y, Ma P S. On the core role of chemical thermodynamics in the curriculum system of chemical engineering[J]. Beijing: Higher Education in Chemical Engineering, 2005(4): 28-30.

[2] Chen X Z, Zhao Q, Chao Q. Teaching of chemical thermodynamics based on Aspen Plus (Ⅰ) calculation of homogeneous properties[J]. Higher education in Chemical Engineering, 2011, 28(5): 5.

[3] Jiang N, Wang D L, Zhou H R, et al. Solution of cubic equation of state based on improved dichotomy[J]. Contemporary Chemical Research, 2021(9): 133-135.

[4] Ma P S. Chemical thermodynamics course[M]. Beijing: Higher Education Press, 2011.

[5] Sun X Y, Xia L, Wang Y L, et al. Thermodynamic models suitable for industrial applications, from classical and advanced mixing rules to association theory[M]. Beijing: Chemical Industry Press, 2020.

[6] Tian W D, Wang H, Wang Y L. Fundamentals of computer aided design of chemical process[M]. Beijing: Chemical Industry Press, 2012.

[7] Tian W D, Wang Y L, Li Z Y. Computer applications in chemical engineering[M]. Beijing: Chemical Industry Press, 2020.

[8] Wang K L, Li J, Li L, et al. Application of Aspen Plus in regression analysis of vapor-liquid equilibrium data of binary system in chemical thermodynamics[J]. Shandong Chemical Industry, 2020, 49(18): 200-204.

[9] Wang Y L, Cui P Z, Tian W D. Chemical Process Simulation: Green, Energy Saving and Precise Control[M]. Beijing: Chemical Industry Press, 2019.

[10] Zhang X, Huo Y T, Li Y, et al. A near ideal gas equation of state and calculation of its thermodynamic properties[J]. Contemporary Chemical Research, 2020(13): 19-22.

Chapter 3

Thermodynamic Properties of Pure Fluids

Learning Objectives

- Learn the relationships between thermodynamic properties.
- Learn how to express thermodynamic matter (mainly entropy) that cannot be directly measured as a function of temperature, pressure, and volume that can be directly measured.
- Learn the method of calculating process exact change and entropy change by p, V, T and heat capacity, and lay the foundation for thermodynamic analysis of chemical process.

In the previous chapters, property tables are widely used. This chapter focuses on how the property tables are prepared and how to identify some unknown attributes from the limited available data. Temperature, pressure, volume and mass can be measured directly. Other properties can be determined by some simple relationships, such as density and specific volume. However, properties such as internal energy, enthalpy and entropy are not easy to determine because they cannot be measured directly or associated with measured properties easily through some simple relationships. Therefore, some basic relationships must be established between common thermodynamic properties, and properties that cannot be measured directly must be expressed by properties.

3.1 Mathematical Relationship between Functions

3.1.1 Partial Differentials

Firstly, consider a function f that depends on a single variable x, $f = f(x)$. The derivative of the function at that point is defined as:

$$\frac{df}{dx} = \lim_{\Delta x \to 0} \frac{\Delta f}{\Delta x} = \lim_{\Delta x \to 0} \frac{f(x+\Delta x)-f(x)}{\Delta x} \tag{3-1}$$

As shown in Fig. 3-1, the derivative of a function $f(x)$ about x represents the rate of change of f with x.

Under the condition of certain pressure, the heat absorbed by a substance per unit mass to

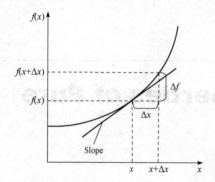

Fig.3-1 The derivative of a function at certain point

increase its temperature by 1K is called the specific heat capacity, represented by the symbol C_p. The international system unit is J/(kg · K). For the same gas, the specific heat capacity at certain pressure is generally greater than the specific heat capacity at certain volume. The C_p is determined by differentiating function $H(T)$ about T. So, the C_p of air at 300 K can be evaluated. The $\hat{C}_p$ of ideal gas depends on temperature only.

When a function that depends on two variables, such as:

$$z = z(x, y)$$

The z is determined by x and y. If one variable changes, the z is to change. When y is constant, the variation of $z(x, y)$ with x is called the partial derivative of z with respect to x. As is shown in Fig.3-2 and Fig.3-3:

$$\left(\frac{\partial z}{\partial x}\right)_y = \lim_{\Delta x \to 0}\left(\frac{\Delta z}{\Delta x}\right)_y = \lim_{\Delta x \to 0}\frac{z(x+\Delta x, y) - z(x, y)}{\Delta x} \tag{3-2}$$

They differ in the symbol that d represents the total differential change of a function and reflects the influence of all variables, whereas ∂ represents the partial differential change due to the variation of a single variable (Ma et al., 2015).

The function is expressed by an explicit function, x and y are independent variable, the derivative can be obtained:

$$dz = \left(\frac{\partial z}{\partial x}\right)_y dx + \left(\frac{\partial z}{\partial y}\right)_x dy \tag{3-3}$$

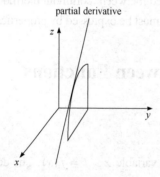

Fig.3-2 Total differential change in $z(x, y)$ for simultaneous changes in x and y

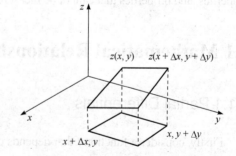

Fig.3-3 Geometric representation of total derivative dz for a function $z(x, y)$

EXAMPLE 3-1

The air state changes from 300 K and 0.86 m³/kg to 302 K and 0.87 m³/kg due to some interference. Estimate the change in the pressure of air.

SOLUTION

Air is assumed to be an ideal gas.

According to Eq. (3-3) The change of T and V can be expressed as:
$$dT \cong \Delta T = (302 - 300)\text{K} = 2\text{ K}$$

and

$$dV \cong \Delta V = (0.87 - 0.86)\text{m}^3/\text{kg} = 0.01\text{ m}^3/\text{kg}$$

Since the relation of ideal gas is $pV=RT$. So

$$p = \frac{RT}{V}$$

According to Eq. (3-3) and using average values for T and V,

$$dp = \left(\frac{\partial p}{\partial T}\right)_V dT + \left(\frac{\partial p}{\partial V}\right)_T dV = \frac{RdT}{V} - \frac{RTdV}{V^2} = 0.664\text{ kPa} - 1.155\text{ kPa}$$
$$= -0.491\text{ kPa}$$

3.1.2 Partial Differential Relations

According to Eq. (3-3),

$$dz = \left(\frac{\partial z}{\partial x}\right)_y dx + \left(\frac{\partial z}{\partial y}\right)_x dy$$

Where:

$$M = \left(\frac{\partial z}{\partial x}\right)_y \quad N = \left(\frac{\partial z}{\partial y}\right)_x$$

According to Eq. (3-2),

$$dz = Mdx + Ndy \quad (3\text{-}4)$$

Derivation of M respect to y, derivation of N respect to y,

$$\left(\frac{\partial M}{\partial y}\right)_x = \frac{\partial^2 z}{\partial x \partial y}$$

$$\left(\frac{\partial N}{\partial x}\right)_y = \frac{\partial^2 z}{\partial y \partial x}$$

According to mathematics knowledge, derivation of continuous function is independent of order, and can be obtained:

$$\left(\frac{\partial M}{\partial y}\right)_x = \left(\frac{\partial N}{\partial x}\right)_y$$

This is an important relation for partial derivatives. In thermodynamics, this relation forms the basis for the development of the Maxwell relations discussed in the next section.

The partial derivative is independent of the expression of letters, so

$$\mathrm{d}x = \left(\frac{\partial x}{\partial y}\right)_z \mathrm{d}y + \left(\frac{\partial x}{\partial z}\right)_y \mathrm{d}z \tag{3-5}$$

Combining Eq. (3-3) to Eq. (3-5). Eliminating $\mathrm{d}x$

$$\left[\left(\frac{\partial z}{\partial x}\right)_y\left(\frac{\partial x}{\partial y}\right)_z + \left(\frac{\partial z}{\partial y}\right)_x\right]\mathrm{d}y = \left[1 - \left(\frac{\partial x}{\partial z}\right)_y\left(\frac{\partial z}{\partial x}\right)_y\right]\mathrm{d}z \tag{3-6}$$

Finally, two important relations of partial derivatives are developed —— the reciprocity and the cyclic relations.

According to triple product rule

$$\left(\frac{\partial x}{\partial y}\right)_z\left(\frac{\partial y}{\partial z}\right)_x\left(\frac{\partial z}{\partial x}\right)_y = -1 \tag{3-7}$$

The variables y and z are independent of each other and thus can be varied independently.

3.1.3 Fundamental Thermodynamic Relation

The definition of H as strictly limited to enthalpy was formally proposed by Alfred W. Porter in 1922. The term enthalpy first appeared in print in 1909. In thermodynamics, an important state parameter enthalpy representing the energy of matter system is equal to the sum of the products of internal energy, pressure and volume, which is often represented by symbol H (Solomonov et al., 2016).

Benoît Paul Émile Clapeyron and Rudolf Clausius (Clausius-Clapeyron relation, 1850) introduce the concept of "heat content". Helmholtz energy is the ability to do all work at certain temperature, and its size is equal to the product of internal energy subtract absolute temperature and entropy, also named work function, which is a thermodynamic parameter with extensive property. In physics, the letter A is often used to express Helmholtz energy (Johannes et al., 2014).

In thermodynamics, the fundamental thermodynamic relation is generally expressed as a microscopic change in internal energy.

Josiah Willard Gibbs, American scientist, has made great theoretical contributions in physics, chemistry and mathematics. Gibbs provided a microscopic interpretation of the laws of thermodynamics through ensemble theory, thus became one of the founders of statistical mechanics. He also introduced the term "statistical mechanics". At the same time, Gibbs also introduced Maxwell's equations into the study of physical optics, and developed the modern vector analysis theory with British scientist.

He introduced that Gibbs free energy (Jia et al., 2019) is the part of the reduced internal energy of a system that can be converted into external work in a certain thermodynamic process. A thermodynamic function introduced in chemical thermodynamics to judge the direction of a process.

$$\mathrm{d}U = T\mathrm{d}S - p\mathrm{d}V \tag{3-8}$$

$$\mathrm{d}H = T\mathrm{d}S + V\mathrm{d}p \tag{3-9}$$

$$\mathrm{d}A = -p\mathrm{d}V - S\mathrm{d}T \tag{3-10}$$

$$\mathrm{d}G = V\mathrm{d}p - S\mathrm{d}T \tag{3-11}$$

For multi-component systems. The U_i are the chemical potentials corresponding to particles of component i. $\sum_i U_i dn_i$ must be zero for a reversible process.

If the component can also change,

$$dU = TdS - pdV + \sum_i U_i dn_i \qquad (3\text{-}12)$$

$$dH = TdS + Vdp + \sum_i U_i dn_i \qquad (3\text{-}13)$$

$$dA = -SdT - pdV + \sum_i U_i dn_i \qquad (3\text{-}14)$$

$$dG = -SdT + Vdp + \sum_i U_i dn_i \qquad (3\text{-}15)$$

According to the first law of thermodynamics

$$dU = \delta Q + \delta W$$

According to the second law of thermodynamics, a reversible process can be written as:

$$dS = \frac{\delta Q}{T}$$

Hence: $\delta Q = TdS$

δW is the reversible work done by the system on its surroundings,

$$\delta W = -pdV$$

Hence:

$$dU = TdS - pdV$$

It can be known from $H = U + pV$

$$dH = dU + d(pV) = TdS - pdV + Vdp + pdV = TdS + Vdp$$

Eq. (3-8) to Eq. (3-9) can be obtained by the same method.

In the fundamental equations of thermodynamics, the change variable of U, H, A and G are all related to p, V, T and S. Although p, V, T are measurable, S is unmeasurable. To solve this problem, the relationship between unmeasurable properties S and measurable properties p, V, T must be established. Maxwell relation is the bridge, which links the unmeasurable properties with measurable properties(Fig.3-4).

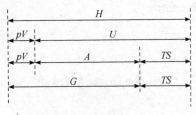

Fig.3-4 The relationship of H, U, A and G

3.2 The Maxwell Relations

$\left(\dfrac{\partial M}{\partial y}\right)_x = \dfrac{\partial^2 z}{\partial x \partial y}$, $\left(\dfrac{\partial N}{\partial x}\right)_y = \dfrac{\partial^2 z}{\partial y \partial x}$ is used in Eq. (3-8) to Eq. (3-11). Then get the following equations.

$$\left(\frac{\partial T}{\partial V}\right)_S = -\left(\frac{\partial p}{\partial S}\right)_V \tag{3-16}$$

$$\left(\frac{\partial T}{\partial p}\right)_S = \left(\frac{\partial V}{\partial S}\right)_p \tag{3-17}$$

$$\left(\frac{\partial p}{\partial T}\right)_V = \left(\frac{\partial S}{\partial V}\right)_T \tag{3-18}$$

$$\left(\frac{\partial V}{\partial T}\right)_p = -\left(\frac{\partial S}{\partial p}\right)_T \tag{3-19}$$

Eq. (3-16) to Eq. (3-19) are important Maxwell relations (José et al., 2017), of which Eq. (3-18) to Eq. (3-19) play important roles. It is important because it links the unmeasurable S to the measurable p-V-T. Eq. (3-18) using $\left(\frac{\partial p}{\partial T}\right)_V$ replaces $\left(\frac{\partial S}{\partial V}\right)_T$.

EXAMPLE 3-2

Derive the expressions of dS and dH with T and V as parameters.

SOLUTION

When S is a function of T and V, there is

$$dS = \left(\frac{\partial S}{\partial T}\right)_V dT + \left(\frac{\partial S}{\partial V}\right)_T dV$$

$$TdS = T\left(\frac{\partial S}{\partial T}\right)_V dT + T\left(\frac{\partial S}{\partial V}\right)_T dV$$

Because

$$C_V = \left(\frac{\partial Q}{\partial T}\right)_V = \left(\frac{T\partial S}{\partial T}\right)_V = T\left(\frac{\partial S}{\partial T}\right)_V$$

According to the Maxwell relationship, get:

$$\left(\frac{\partial S}{\partial V}\right)_T = \left(\frac{\partial p}{\partial T}\right)_V$$

So

$$TdS = C_V dT + T\left(\frac{\partial p}{\partial T}\right)_V dV$$

or

$$dS = \frac{C_V}{T} dT + \left(\frac{\partial p}{\partial T}\right)_V dV$$

According to

$$dH = TdS + Vdp$$

Bring $dS = \dfrac{C_V}{T}dT + \left(\dfrac{\partial p}{\partial T}\right)_V dV$ into the above equation and $dp = \left(\dfrac{\partial p}{\partial T}\right)_V dT + \left(\dfrac{\partial p}{\partial V}\right)_T dV$

$$dH = T\left[C_V \dfrac{dT}{T} + \left(\dfrac{\partial p}{\partial T}\right)_V dV\right] + V\left[\left(\dfrac{\partial p}{\partial T}\right)_V dT + \left(\dfrac{\partial p}{\partial V}\right)_T dV\right]$$

$$dH = \left[C_V + V\left(\dfrac{\partial p}{\partial T}\right)_V\right]dT + \left[T\left(\dfrac{\partial p}{\partial T}\right)_V + V\left(\dfrac{\partial p}{\partial V}\right)_T\right]dV$$

3.3 The Clapeyron Equation

The Maxwell relations have far reached implications in thermodynamics and are frequently used to derive useful thermodynamic relations. The Clapeyron equation is such a relation, and it enables us to determine the enthalpy change associated with a phase change (such as the enthalpy of vaporization H_{fg}) from a knowledge of p, V and T data.

Consider the third Maxwell relation, Eq. (3-18):

$$\left(\dfrac{\partial p}{\partial T}\right)_V = \left(\dfrac{\partial S}{\partial V}\right)_T$$

In the process of phase transition, pressure is a kind of saturation pressure, which only depends on temperature and is independent of volume.

$$S_g - S_f = \left(\dfrac{dp}{dT}\right)_{sat}(V_g - V_f) \tag{3-20}$$

$$\left(\dfrac{dp}{dT}\right)_{sat} = \dfrac{S_{fg}}{V_{fg}} \tag{3-21}$$

During this process, the pressure also remains constant. Therefore, Substitute Eq. (3-9) into Eq. (3-11) to obtain

$$\left(\dfrac{dp}{dT}\right)_{sat} = \dfrac{H_{fg}}{TV_{fg}} \tag{3-22}$$

It is based on a French engineer and physicist E. Clapeyron named as Clapeyron equation (Ma et al., 2015) (1799—1864). It is an important thermodynamic relationship that the evaporation enthalpy at a certain temperature can be determined by simply measuring the inclination of the saturation curve on the p-T diagram and the specific volume of saturated liquid and saturated vapor at a certain temperature.

The Clapeyron equation can be used phase-change process that occurs at certain temperature and pressure. So, Eq. (3-22) can be expressed as:

$$\left(\dfrac{dp}{dT}\right)_{sat} = \dfrac{H_{12}}{TV_{f12}} \tag{3-23}$$

Where 1 and 2 indicate the two phases.

Several approximations can be used to simplify the Clapeyron equation of gas-liquid phase change and solid-gas phase change. At low pressure $V_g \gg V_f$, so $V_{fg} \cong V_g$. If steam is the ideal gas, there is $V_g = RT/p$. These approximations are substituted for Eq. (3-22).

$$\left(\frac{dp}{dT}\right)_{sat} = \frac{pH_{fg}}{RT^2}$$

For small temperature intervals, H_{fg} can be seen as a constant with several averages. And this equation is obtained by integrally accumulating between the two saturation states.

$$\ln\left(\frac{p_2}{p_1}\right)_{sat} \cong \frac{H_{fg}}{R}\left(\frac{1}{T_1} - \frac{1}{T_2}\right)_{sat}$$

This equation is called the Clapeyron-Clauseus equation and can be used to determine the change in saturation pressure with temperature. It can also be used in the solid-gas region, using H_{ig} instead of H_{fg}.

EXAMPLE 3-3

1 kg and 2×10^7 Pa of high-pressure air flows through the heat exchanger under constant pressure, heat exchanger under constant pressure is cooled from 27°C to 133°C. What is the heat that needs to be removed from the cooling medium?

SOLUTION

It can be seen that $T_1 = 300$ K and $p_1 = 20$ MPa,

$$H_1 = 114 \text{ kcal/kg} = 476.976 \text{ kJ/kg}$$

$$S_1 = 0.52 \text{ kcal/(kg} \cdot \text{K)} = 2.176 \text{ kJ/(kg} \cdot \text{K)}$$

$T_2 = 140$ K and $p_2 = p_1 = 20$ MPa,

$$H_2 = 51.5 \text{ kcal/kg} = 215.476 \text{ kJ/kg}$$

$$S_2 = 0.23 \text{ kcal/(kg} \cdot \text{K)} = 0.9623 \text{ kJ/(kg} \cdot \text{K)}$$

$$q_p = \Delta H = H_2 - H_1 = 215.476 - 476.976 = -261.500 \text{ kJ/kg}$$

3.4 General Relations for dU, dH, dA, and dG

In chemical calculation, the calculation of internal energy, entropy and enthalpy plays an important role.

Thermodynamic properties of fluids can be divided into direct measurement and non-direct measurement. The following can be directly measured: temperature, volume, pressure, the heat capacity, enthalpy (ΔH can be measured by calorimeter determination) and so on; The following cannot be directly measured: thermodynamics energy U (internal energy), entropy S, Helmholtz free energy A (Paulechka et al., 2010), Gibbs free energy G, etc. And the thermodynamic properties that cannot be directly measured can be calculated by establishing a relationship with the measurable thermodynamic properties, which is the greatness and cleverness of thermodynamics. It is particularly important to establish the functional relation between the two thermodynamic properties which act as a bridge.

Internal Energy Change

$U=U(T, V)$, The total derivation for U

$$dU = \left(\frac{\partial U}{\partial T}\right)_V dT + \left(\frac{\partial U}{\partial V}\right)_T dV$$

Where $C_V = \left(\dfrac{\partial U}{\partial T}\right)_V$

$$dU = C_V dT + \left(\frac{\partial U}{\partial V}\right)_T dV \tag{3-24}$$

$S=S(T, V)$, and the total derivation for S

$$dS = \left(\frac{\partial S}{\partial T}\right)_V dT + \left(\frac{\partial S}{\partial V}\right)_T dV \tag{3-25}$$

Substituting Eq. (3-25) into the TdS relation

$$dU = T\left(\frac{\partial S}{\partial T}\right)_V dT + \left[T\left(\frac{\partial S}{\partial V}\right)_T - p\right] dV \tag{3-26}$$

Equating the coefficients of dT and dV in Eq. (3-24) and Eq. (3-26) gives

$$\left(\frac{\partial S}{\partial T}\right)_V = \frac{C_V}{T}$$

$$\left(\frac{\partial U}{\partial T}\right)_V = T\left(\frac{\partial S}{\partial V}\right)_T - p \tag{3-27}$$

According to Maxwell relation get

$$\left(\frac{\partial U}{\partial T}\right)_V = T\left(\frac{\partial p}{\partial T}\right)_V - p$$

Integration Eq. (3-24)

$$dU = C_V dT + \left[T\left(\frac{\partial p}{\partial T}\right)_V - p\right] dV$$

State from (T_1, V_1) to (T_2, V_2) obtain

$$U_2 - U_1 = \int_{T_1}^{T_2} C_V dT + \int_{V_1}^{V_2} \left[T\left(\frac{\partial p}{\partial T}\right)_V - p\right] dV \tag{3-28}$$

Enthalpy Change

Under certain conditions, the change in enthalpy of a system can be measured from the transfer of heat between the system and environment.

Enthalpy is a function of T and p, that is, $H=H(T, p)$, and take its total differential,

$$dH = \left(\frac{\partial H}{\partial T}\right)_p dT + \left(\frac{\partial H}{\partial p}\right)_T dp$$

Using the definition of C_p,

$$dH = C_p dT + \left(\frac{\partial H}{\partial p}\right)_T dp \tag{3-29}$$

take $S=S(T, p)$ and take its total differential,

$$dS = \left(\frac{\partial S}{\partial T}\right)_p dT + \left(\frac{\partial S}{\partial p}\right)_T dp \tag{3-30}$$

Substituting this into the TdS relation $dH = TdS + Vdp$ gives

$$\left(\frac{\partial S}{\partial T}\right)_p = \frac{C_p}{T}$$

$$\left(\frac{\partial S}{\partial T}\right)_V = \frac{C_V}{T}$$

$$\left(\frac{\partial H}{\partial p}\right)_T = V + T\left(\frac{\partial S}{\partial p}\right)_T \tag{3-31}$$

According to Maxwell relation

$$\left(\frac{\partial H}{\partial p}\right)_T = V - T\left(\frac{\partial S}{\partial T}\right)_p$$

Substituting this into Eq. (3-29), obtain the desired relation for dH:

$$dH = C_p dT + \left[V - T\left(\frac{\partial V}{\partial T}\right)_p\right] dp$$

Enthalpy change of state from (T_1, V_1) to (T_2, V_2) (Radmanović et al., 2014).

$$H_2 - H_1 = \int_{T_1}^{T_2} C_p dT + \int_{p_1}^{p_2}\left[V - T\left(\frac{\partial V}{\partial T}\right)_p\right] dp \tag{3-32}$$

When the system is in a certain condition, according to the definition formula, it can be obtained

$$H_2 - H_1 = U_2 - U_1 + (p_2 V_2 - p_1 V_1)$$

Entropy Change

The general relationship of entropy change (Sharma et al., 2010) of simple compressible system is as follows. The first relation is obtained by replacing the first partial derivative in the total differential dS Eq. (3-24) by Eq. (3-27) and the second partial derivative by Maxwell relation, so

$$dS = \frac{C_V}{T}dT + \left(\frac{\partial p}{\partial T}\right)_V dV \qquad (3\text{-}33)$$

and

$$S_2 - S_1 = \int_{T_1}^{T_2} \frac{C_V}{T}dT + \int_{V_1}^{V_2} \left(\frac{\partial p}{\partial T}\right)_V dV \qquad (3\text{-}34)$$

The second relation is obtained by replacing the first partial derivative in the total differential of dS in Eq. (3-35) by Eq. (3-36), and the second partial derivative by Maxwell relation, so

$$dS = \frac{C_p}{T}dT + \left(\frac{\partial V}{\partial T}\right)_p dp \qquad (3\text{-}35)$$

and

$$S_2 - S_1 = \int_{T_1}^{T_2} \frac{C_p}{T}dT - \int_{V_1}^{V_2} \left(\frac{\partial V}{\partial T}\right)_p dp \qquad (3\text{-}36)$$

Constant-pressure Heat Capacity

The heat capacity in the isobaric process is named as the constant-pressure heat capacity (Zaitsev et al., 2010).

$$C_p = \left(\frac{\partial H}{\partial T}\right)_p$$

The heat capacity in the constant volume process is named as the constant-volume heat capacity (Abbas et al., 2011).

$$C_V = \left(\frac{\partial U}{\partial T}\right)_V$$

For a general pure substance, the specific heats depend on specific volume or pressure as well as the temperature. The general relationship between the specific heat, pressure, volume and temperature is derived. At low pressure, gases can be seen as ideal gases, and their specific heats depend on temperature only. These specific heats are called constant pressure specific heats or constant volume specific heats (denoted C_{V0} and C_{p0}), and they are relatively easier to determine. Thus, it is desirable to have some general relations that enable us to calculate the specific heats at higher pressures (or lower specific volumes) from knowledge of C_{V0} and C_{p0} and the p-V-T behavior of the substance. Such relations are obtained from Eq. (3-25) and Eq. (3-33).

$$\left(\frac{\partial C_V}{\partial V}\right)_T = T\left(\frac{\partial^2 p}{\partial T^2}\right)_V \qquad (3\text{-}37)$$

and

$$\left(\frac{\partial C_p}{\partial p}\right)_T = -T\left(\frac{\partial^2 V}{\partial T^2}\right)_p \qquad (3\text{-}38)$$

For example, the deviation of C_p from C_p to C_{p0} with increasing pressure is determined by the integral of Eq. (3-38) from zero pressure to any pressure p along the isothermal path:

$$(C_p - C_{p0}) = -T\int_0^p \left(\frac{\partial^2 V}{\partial T^2}\right)_p dp \qquad (3\text{-}39)$$

The integration on the right-hand side requires knowledge of the p-V-T behavior of the substance alone. Volume (V) should be differentiated twice with respect to T while p is constant. The resulting expression should be integrated about p while T is constant.

Another general relation involving specific heats is one that relates the two specific heats C_p, and C_V. It is convenient to determine one specific heat (usually C_p) and calculate another specific heat (usually C_p) using this relationship and the p-V-T data of the substance. By equating the two dS relations in Eq. (3-33) and Eq. (3-35) and solving for dT:

$$dT = \frac{T\left(\frac{\partial p}{\partial T}\right)_V}{C_p - C_V} dV + \frac{T\left(\frac{\partial T}{\partial p}\right)_T}{C_p - C_V} dp$$

Choosing $T = T(V, p)$ and differentiating, get

$$dT = \left(\frac{\partial T}{\partial V}\right)_p dV + \left(\frac{\partial T}{\partial p}\right)_V dp$$

Equating the coefficient of either dV or dp of the above two equations give the result:

$$C_p - C_V = -T\left(\frac{\partial V}{\partial T}\right)_p \left(\frac{\partial p}{\partial T}\right)_V \qquad (3\text{-}40)$$

According to Maxwell relation:

$$C_p - C_V = -T\left(\frac{\partial V}{\partial T}\right)_p^2 \left(\frac{\partial p}{\partial V}\right)_T \qquad (3\text{-}41)$$

This relation can be expressed by two other thermodynamic properties, which is named as the volume expansivity β (Acree et al., 2010) and the isothermal compressibility α (Achour et al., 2020).

$$\beta = \frac{1}{V}\left(\frac{\partial V}{\partial T}\right)_p \qquad (3\text{-}42)$$

and

$$\alpha = -\frac{1}{V}\left(\frac{\partial V}{\partial p}\right)_T \qquad (3\text{-}43)$$

Substitute β and α into Eq. (3-41):

$$C_p - C_V = \frac{VT\beta^2}{\alpha} \qquad (3\text{-}44)$$

We can draw several conclusions from these equations:

(1) The isothermal compressibility α is a positive quantity for all substances in all phases. The volume expansivity could be negative for some substances (such as liquid water below 4℃), but its square is always positive or zero. The temperature T in this relation is thermodynamic temperature, which is also positive. Therefore, we conclude that the constant-pressure specific heat is always greater than or equal to the constant-volume specific heat.

(2) The difference between C_p and C_V approaches zero as the absolute temperature approaches zero.

(3) The two specific heats are identical for truly incompressible substances since volume is constant. The difference between the two specific heats is very small and is usually disregarded for substances for liquids and solids.

EXAMPLE 3-4

The relationship between the change of internal energy of gas is deduced, which obeys van der Waals equation of state. It is assumed that $C_V = C_1 + C_2 T$, and C_1 and C_2 are constants.

SOLUTION

A relation is to be obtained for the internal energy change of gas which obeys a van der Waals.

$$U_2 - U_1 = \int_{T_1}^{T_2} C_V \, dT + \int_{V_1}^{V_2} \left[T \left(\frac{\partial p}{\partial T} \right)_V - p \right] dV$$

The van der Waals equation of state is

$$\left(p + \frac{a}{V^2} \right)(V - b) = RT$$

Then

$$\left(\frac{\partial p}{\partial T} \right)_V = \frac{R}{V - b}$$

Thus

$$T \left(\frac{\partial p}{\partial T} \right)_V - p = \frac{RT}{V - b} - \frac{RT}{V - b} + \frac{a}{V^2} = \frac{a}{V^2}$$

Substituting

$$U_2 - U_1 = \int_{T_1}^{T_2} (C_1 + C_2 T) \, dT + \int_{V_1}^{V_2} \frac{a}{V^2} \, dV$$

Integrating

$$U_2 - U_1 = C_1 (T_2 - T_1) + \frac{C_2}{2} (T_2^2 - T_1^2) + a \left(\frac{1}{V_1} - \frac{1}{V_2} \right)$$

EXAMPLE 3-5

Prove that $C_p - C_V = R$ for an ideal gas.

SOLUTION

It is to be shown that the difference of specific heat for an ideal gas is equal to its gas constant R.

ANALYSIS

This relation is easily proved by showing that the right-hand side of Eq. (3-36) is equal to the gas constant R of the ideal gas:

$$C_p - C_V = -T\left(\frac{\partial V}{\partial T}\right)_p^2 \left(\frac{\partial p}{\partial T}\right)_V$$

$$p = \frac{RT}{V} \rightarrow \left(\frac{\partial p}{\partial V}\right)_T = -\frac{RT}{V^2} = \frac{p}{V}$$

$$V = \frac{RT}{p} \rightarrow \left(\frac{\partial p}{\partial T}\right)_p^2 = \left(\frac{R}{p}\right)^2$$

Substituting

$$-T\left(\frac{\partial V}{\partial T}\right)_p^2 \left(\frac{\partial p}{\partial T}\right)_V = -T\left(\frac{R}{p}\right)^2 \left(-\frac{p}{V}\right) = R$$

Therefore

$$C_p - C_V = R$$

3.5 Joule-Thomson Coefficient

When the fluid passes through restrictions such as porous plugs, capillary tubes, and common valves, its pressure decreases. In the process of passing through the aperture, the temperature of the fluid may be greatly reduced. It is the basic operating principle of refrigerators and air conditioners. But in the diaphragm, the temperature of the fluid may keep constant or rise.

The temperature of a fluid is described by the Joule-Thomson coefficient (Mammana et al., 2021) during a throttling (H = constant) process, defined as

$$\mu_{JT} = \left(\frac{\partial T}{\partial p}\right)_H \tag{3-45}$$

Thus, the Joule-Thomson coefficient is a measure of the change in temperature with pressure during a constant-enthalpy process. Notice that if

$$\mu_{JT} \begin{cases} < 0 & \text{temperature increases} \\ = 0 & \text{temperature remains constant} \\ > 0 & \text{temperature decreases} \end{cases}$$

The Jules-Thomson coefficient represents the slope of H=constant line on the T-p diagram. Such graphs can be easily plotted simply by measuring the temperature and pressure during the aperture. Fix the temperature T_1 and pressure p_1 flowing through the inlet of the porous plug, and the temperature and pressure downstream in T_2 and p_2 are measured. Experiments are repeated on porous plugs of different sizes, different T_2 and p_2 are given. As shown in Fig. 3-5, plotting the relationship between temperature and pressure on a T-p diagram, H=constant line is obtained. As shown in Fig. 3-6, the experimental results of different inlet pressures and temperatures are plotted, and T-p diagrams with several H=constant lines are obtained.

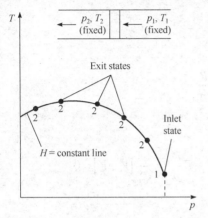 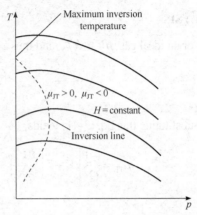

Fig.3-5 The development of an H=constant line on a T-p diagram

Fig.3-6 Constant-enthalpy lines of a substance on a T-p diagram

Some equal enthalpy lines on the T-p diagram pass through a point with a slope of zero or a Joule-Thomson coefficient of zero. The line through these points is named as a reverse temperature line. The point of enthalpy line cross is named as a reverse temperature line. The temperature at the intersection of the $p = 0$ line (vertical axis) and the upper part of the inverted line is called the maximum reversal temperature.

It is clear from this figure that a cooling effect cannot be achieved by throttling unless the fluid is below its maximum inversion temperature. This presents a problem for substances whose maximum inversion temperature is well below room temperature.

For example, the maximum hydrogen conversion temperature is $-68\,^\circ\text{C}$. This requires further cooling by throttling, and the hydrogen must be cooled below this temperature. The general relationship of Joule-Thomson coefficient is derived in terms of specific heat, specific pressure, specific volume and temperature. It can be achieved by modifying the enthalpy relationship. As shown in the following equation:

$$dH = C_p dT + \left[V - T\left(\frac{\partial V}{\partial T}\right)_p\right]dp$$

When H=constant and $dH = 0$, equation can be gotten:

$$-\frac{1}{C_p}\left[V - T\left(\frac{\partial V}{\partial T}\right)_p\right] = \left(\frac{\partial T}{\partial p}\right)_H = \mu_{JT} \qquad (3\text{-}46)$$

Therefore, the Joule-Thomson coefficient can be determined according to the theory of constant-pressure specific heat and p-V-T data. Of course, the Joule-Thomson coefficient and the p-V-T data also can be used to predict the constant-pressure specific heat.

EXAMPLE 3-6

Verify that the Joule-Thomson coefficient of ideal gas is zero.

SOLUTION

It is to be shown that $\mu_{JT} = 0$ for an ideal gas.

ANALYSIS

For an ideal gas $pV = RT$, and thus

$$\left(\frac{\partial V}{\partial T}\right)_p = \frac{R}{p}$$

Substituting this Eq. (3-41) yields,

$$\mu_{JT} = \frac{-1}{C_p}\left[V - T\left(\frac{\partial V}{\partial T}\right)_p\right] = \frac{-1}{C_p}\left[V - T\frac{R}{p}\right] = \frac{-1}{C_p}[V - V] = 0$$

3.6 The ΔH, ΔU, and ΔS of Real Gas

The ideal gas is the limit state of the real gas $p \to 0, V \to \infty$. It follows the relationship $pV = RT$. The properties of ideal gases are easy to evaluate since the properties U, H, C_V and C_p depend on temperature only. However, the behavior of gas under high pressure is quite different from that of ideal gas, and it becomes necessary to account for this deviation. we accounted for the deviation in properties p, V, and T by either using more complex equations of state or evaluating the compressibility factor Z from the compressibility graph. Now we evaluate the changes in the enthalpy, internal energy, and entropy of nonideal (real) gases by using the general relations for dU, dH, and dS obtained.

Enthalpy Change of Real Gas

The enthalpy of the real gas depends on pressure and temperature. Thus, the enthalpy change of real gas can be evaluated from the general relation for dH in Eq. (3-32)

$$H_2 - H_1 = \int_{T_1}^{T_2} C_p dT$$

where p_1, T_1 and p_2, T_2 are the pressures and temperatures of the gas at the initial and the final states, respectively.

For an isothermal process, $dT=0$,

$$H_2 - H_1 = \int_{p_1}^{p_2}\left[V - T\left(\frac{\partial V}{\partial T}\right)_p\right]dp$$

For a constant-pressure process, $dp=0$,

$$H_2 - H_1 = \int_{T_1}^{T_2} C_p \, dT$$

The enthalpy change between two specified states is the same no matter which process path is followed. This fact can be exploited to greatly simplify the integration of Eq. (3-32). H_2–H_1 can be determined by Eq. (3-32) along a path that consists of two isothermal (T_1 = constant and T_2 = constant) lines and one isobaric (p_0=constant) line instead of the actual process path.

The pressure p_0 can be chosen to be very low or zero, so that the gas can be treated as an ideal gas during the p_0 = constant process. Using a superscript asterisk (*) to represent the ideal gas state, and the enthalpy change of the real gas in process 1-2 is expressed as:

$$H_2 - H_1 = \left(H_2 - H_2^*\right) + \left(H_2^* - H_1^*\right) + \left(H_1^* - H_1\right) \tag{3-47}$$

The difference between H and H^* is called the enthalpy departure, and it represents the variation of the enthalpy of a gas with pressure at certain temperature. The calculation of enthalpy departure requires knowledge of the p-V-T behavior of the gas. In the absence of such data, we can use the relation $pV=ZRT$, where Z is the compressibility factor. The enthalpy difference at any temperature T and pressure p is written as:

$$\left(H^* - H\right)_T = -RT^2 \int_0^p \left(\frac{\partial Z}{\partial T}\right)_p \frac{dp}{p}$$

By substituting $T = T_{cr}T_R$ and $p = p_{cr}p_R$ into the above equation, the final enthalpy deviation can be expressed as:

$$Z_2 = \frac{\left(H^* - H\right)_T}{R_U T_{cr}} = T_R^2 \int_0^{p_R} \left(\frac{\partial Z}{\partial T_R}\right)_{p_R} d(\ln p_R) \tag{3-48}$$

The difference of S^* and S is called the entropy departure (Z_s), which is also named as the entropy departure factor.

Internal Energy Change of Real Gas

The change in internal energy of the real gas is determined by the change in enthalpy.

$$\bar{H} = \bar{U} + p\bar{V} = \bar{U} + ZR_U T$$

$$\bar{U}_2 - \bar{U}_1 = (\bar{H}_2 - \bar{H}_1) - R_U(Z_2 T_2 - Z_1 T_1) \tag{3-49}$$

Entropy Chang of Real Gas

The entropy change of real gas is determined by following an approach similar to that used above for the enthalpy change. There is some difference in derivation, however, owing to the dependence of the ideal gas entropy on pressure as well as the temperature.

The correlation equation of dS is expressed as:

$$S_2 - S_1 = \int_{T_1}^{T_2} \frac{C_p}{T} dT + \int_{p_1}^{p_2} \left(\frac{\partial V}{\partial T}\right)_V dp$$

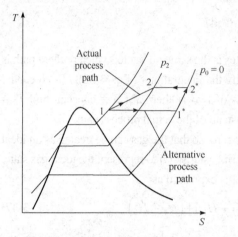

Fig.3-7 An alternative process path to evaluate the entropy changes of real gases during process 1-2.

Where p_1, T_1 and p_2, T_2 are the pressures and temperatures of the gas at the initial and the final states, respectively. In this process, it is similar to enthalpy. First along the T_1 = constant line to zero pressure, then along the $p = 0$ line to T_2, and finally along the T_2 = constant line to p_2. However, this approach is not suitable for entropy change calculations. Because it involves the value of entropy at zero pressure, which is infinity value. The difficulty can be avoided by selecting different (but more complex) paths between the two states, as shown in Fig. 3-7.

State 1 and State 1^* are same ($T_1 = T_1^*$ and $p_1 = P_1^*$) and State 2 and State 2^* are also the same. The gas is assumed to behave as an ideal gas at the imaginary State 1^* and State 2^* as well as at the states between the State 1^* and State 2^*. Therefore, the entropy changes during process $1^* - 2^*$ can be determined from the entropy-change relations for ideal gas. However, the calculation of entropy change between a real state and the corresponding imaginary ideal gas state is more complexed.

An isothermal process from the actual state p, T to zero (or close to zero) pressure and then back to the imaginary ideal-gas state p^*, T^* (denoted by superscript*), as shown in Fig. 3-6. The entropy change can be expressed as during this isothermal process:

$$S_2 - S_1 = \left(S_2 - S_b^*\right) + \left(S_b^* - S_2^*\right) + \left(S_2^* - S_1^*\right) + \left(S_1^* - S_a^*\right) + \left(S_b^* - S_1\right) \tag{3-50}$$

The entropy difference between ideal gas and real gas at the same temperature and pressure. The isothermal process from the real state p, T to zero (or close to zero) pressure and then back to the imaginary ideal-gas state p^*, T^* (denoted by superscript*).

3.7 Application of Aspen in Thermodynamic Properties

At present, ionic liquids are widely used as green solvents. It is very important to master the thermodynamic properties of ionic liquids. Aspen makes it easy to calculate thermodynamic properties. For example, Aspen Plus software is used to estimate the physical properties of 1-n-butyl-3-methylimidazole trifluoromethane sulfonate ([BMIM][OTF]) according to the structure and molecular weight found in literature.

First choose Estimate all missing parameters.
Enter the name of the component to be estimated.
Mapping molecular structure.
Computational molecular skeleton.
Establish the physical parameters of pure components, input the normal boiling point and molecular weight, select the type of physical parameters of pure components, and calculate.

Chapter 3 Thermodynamic Properties of Pure Fluids

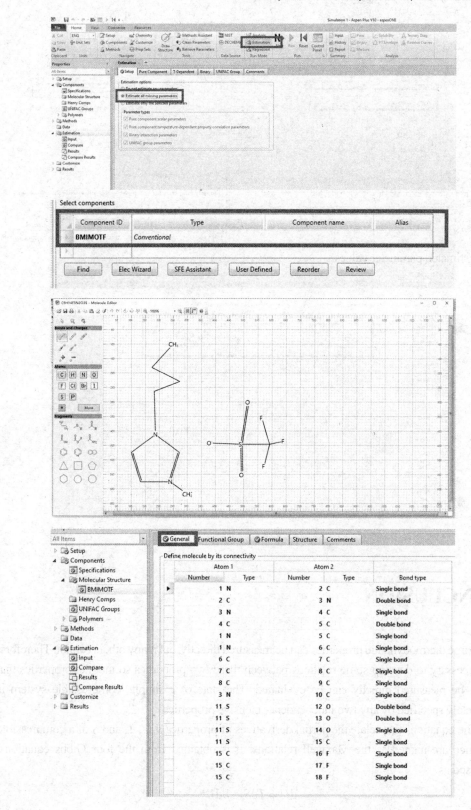

63

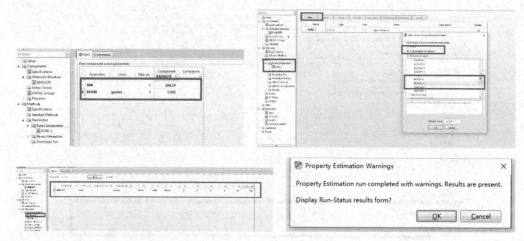

View estimated parameter values.

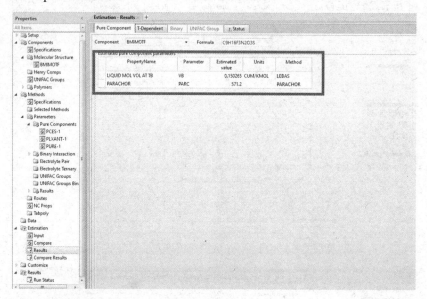

CONCLUSION

Some thermodynamic properties can be measured directly, but many others cannot. Therefore, it is necessary to develop some relations between these two properties so that the properties that cannot be measured directly can be evaluated. The state of a simple, compressible system is completely specified by any two independent, intensive properties.

The equations that relate the partial derivatives of properties p, V, T, and S of a compressible substance are named as the Maxwell relations. It is obtained from the four Gibbs equations, expressed as

$$dU = TdS - pdV$$

$$dH = TdS + Vdp$$

$$dA = -pdV - SdT$$

$$dG = Vdp - SdT$$

The Maxwell equations are

$$\left(\frac{\partial T}{\partial V}\right)_S = -\left(\frac{\partial p}{\partial S}\right)_V$$

$$\left(\frac{\partial T}{\partial p}\right)_S = \left(\frac{\partial V}{\partial S}\right)_p$$

$$\left(\frac{\partial p}{\partial T}\right)_V = \left(\frac{\partial S}{\partial V}\right)_T$$

$$\left(\frac{\partial V}{\partial T}\right)_p = -\left(\frac{\partial S}{\partial p}\right)_T$$

For liquid-vapor and solid-vapor phase-change processes at low pressures, it can be approximated as

$$\ln\left(\frac{p_2}{p_1}\right)_{sat} \cong \frac{H_{fg}}{R}\left(\frac{1}{T_1} - \frac{1}{T_2}\right)_{sat}$$

The changes in internal energy, enthalpy, and entropy of a compressible substance can be expressed pressure, specific volume, temperature, and specific heats as

$$dU = C_V dT + \left[T\left(\frac{\partial p}{\partial T}\right)_V - p\right]dV$$

$$dH = C_p dT + \left[V - T\left(\frac{\partial V}{\partial T}\right)_p\right]dp$$

$$dS = \frac{C_V}{T}dT + \left(\frac{\partial p}{\partial T}\right)_V dV$$

For specific heats, the following equation can be written as:

$$\left(\frac{\partial C_V}{\partial V}\right)_T = T\left(\frac{\partial^2 p}{\partial T^2}\right)_V$$

$$\left(\frac{\partial C_p}{\partial p}\right)_T = -T\left(\frac{\partial^2 V}{\partial T^2}\right)_p$$

$$C_p - C_V = -T\left(\frac{\partial V}{\partial T}\right)_V^2$$

β is the volume expansivity and α is the isothermal compressibility, defined as:

$$\beta = \frac{1}{V}\left(\frac{\partial V}{\partial T}\right)_p$$

$$\alpha = -\frac{1}{V}\left(\frac{\partial V}{\partial p}\right)_T$$

The difference $C_p - C_V$ is equal to R for ideal gas and to zero for incompressible substances.

The temperature of a fluid during a throttling (H = constant) process is described by the Joule-Thomson coefficient as:

$$\mu_{JT} = \left(\frac{\partial T}{\partial p}\right)_H$$

The Joule-Thomson coefficient is a measure of the change in temperature with pressure during a constant enthalpy process, and it can also be expressed as:

$$\mu_{JT} = \frac{-1}{C_p}\left[V - T\left(\frac{\partial V}{\partial T}\right)_p\right]$$

The enthalpy, internal energy, and entropy changes of real gas can be determined accurately by enthalpy or entropy departure graph to account for the deviation from the ideal-gas behavior by using the following equation:

$$\bar{U}_2 - \bar{U}_1 = (\bar{H}_2 - \bar{H}_1) - R_U(Z_2 T_2 - Z_1 T_1)$$

This chapter mainly discusses the calculation of the thermodynamic properties of fluids. In a homogeneous closed system, given any two independent variables of the eight common variables, the other variables can be calculated by equations or read out by common thermodynamic property diagrams. However, when we don't know the properties of the fluid, we can't get the other properties through calculation. Therefore, we need the software Aspen Plus with a powerful database. The database in Aspen Plus system includes constant parameters (absolute temperature and pressure), property parameters of phase transition (boiling point and three-phase point), property parameters of reference state (standard formation energy and standard formation Gibbs free energy), thermodynamic property parameters varying with temperature (saturated vapor pressure), parameters in equation of state and other parameters, Therefore, we can use Aspen Plus software to realize physical property analysis and physical property estimation.

EXERCISES

1. Deduce the equation $\left(\frac{\partial U}{\partial V}\right)_T = T\left(\frac{\partial p}{\partial T}\right)_V - p$ where T, V is an independent variable.

2. Prove the fluid expressed by equation of state $p(V-b) = RT$:

(1) C_p is independent of pressure:

(2) In the constant enthalpy process, the temperature rises as the pressure decreases.

3. Using $\left(\dfrac{\partial U}{\partial V}\right)_T = T\left(\dfrac{\partial p}{\partial T}\right)_V - p$ proof $\mu = \dfrac{1}{C_p}\left[T\left(\dfrac{\partial V}{\partial T}\right)_p - V\right]$

4. The laboratory needs to prepare 1500 cm³ antifreeze containing 30% (mol%) methanol (1) and 70% H₂O (2). Try to figure out what volume of methanol you need to mix with water at 25℃. Partial molar volumes of methanol and water at 25℃ and 30% (mol%) methanol are known:

$$\overline{V_1} = 38.632 \text{ cm}^3/\text{mol}$$

$$\overline{V_2} = 17.765 \text{ cm}^3/\text{mol}$$

The volume of the pure substance at 25 degrees Celsius:

$$V_1 = 40.727 \text{ cm}^3/\text{mol}$$

$$V_2 = 18.068 \text{ cm}^3/\text{mol}$$

5. Assuming that nitrogen obeys the ideal gas law, calculate the internal energy, enthalpy, entropy, C_p, C_V of 1 kmol N₂ at 500℃ and 10.13 MPa.
It is known that:
(1) The relationship between C_p and temperature at 0.1013 MPa is as follows:

$$C_p = (27.22 + 0.004187T)\text{J}/(\text{mol}\cdot\text{K})$$

(2) Assume that the enthalpy value of N₂ is zero at 0℃ and 0.1013 MPa;
(3) The entropy of N₂ at 25℃ and 0.1013 MPa is 191.76 J/(mol · K).
6. Try to derive the expressions dS and dH with T and V as parameters.
7. Fill a closed container with liquid water at 25℃ and 0.1 MPa. If the water is heated to 60℃, what is the pressure? It is known that the specific volume of water at 25℃ is 1.003 cm³/g, and the average value of volume expansion coefficient β between 25℃ and 60℃ is 36.2×10^{-5} K⁻¹, the compressibility coefficient K is 4.42×10^{-4} MPa⁻¹ at 0.1 MPa and 60℃ and assumed to be independent of pressure.
8. Try to prove the following relationship

$$\left(\dfrac{\partial \beta}{\partial p}\right)_T = -\left(\dfrac{\partial \alpha}{\partial T}\right)_p$$

Where β and α are volume expansion coefficient and isothermal compression coefficient respectively,

$$\beta = \dfrac{1}{V}\left(\dfrac{\partial V}{\partial T}\right)_p$$

$$\alpha = -\dfrac{1}{V}\left(\dfrac{\partial V}{\partial p}\right)_T$$

9. An ideal gas is loaded into a steel cylinder with the help of a piston. The pressure is 34.45 MPa and the temperature is 93℃. It resists a constant external pressure of 3.45 MPa and expands isothermal until it is twice the initial volume. Try to calculate the process of ΔU, ΔH, ΔS,

$\Delta A, \Delta G, \int T\mathrm{d}S, \int p\mathrm{d}V, Q, W$.

10. The mixture of saturated steam and water at 230℃ is in equilibrium. If the specific volume of the mixed phase is 0.04166 m³/kg, try to calculate: (1) the content of steam in the mixed phase; (2) enthalpy of mixed phase; (3) entropy of mixed phase.

REFERENCES

[1] Abbas R, Ihmels C, Enders S, Gmehling J, et al. Joule-Thomson coefficients and Joule-Thomson inversion curves for pure compounds and binary systems predicted with the group contribution equation of state VTPR[J]. Fluid Phase Equilibria, 2011, 306(2): 181-189.

[2] Achour S H, Okuno R. Phase stability analysis for tight porous media by minimization of the Helmholtz free energy[J]. Fluid Phase Equilibria, 2020.

[3] Acree W, Chickos J S. Phase transition enthalpy measurements of organic and organometallic compounds[J]. Journal of Physical and Chemical Reference Data, 2010, 39(4): 73-2484.

[4] Jia C S, Zhang L H, Peng X L, et al. Prediction of entropy and Gibbs free energy for nitrogen[J]. Chemical Engineering Science, 2019, 202: 70-74.

[5] Johannes G, Andreas J, Roland S, et al. Calculation of phase equilibria for multi-component mixtures using highly accurate Helmholtz energy equations of state[J]. Fluid Phase Equilibria, 2014, 375.

[6] José W, Wen C. Generalized Maxwell relations in thermodynamics with metric dserivatives[J]. Entropy, 2017, 19(8):407.

[7] Ma H C, Yao D D, Peng X F. Exact solutions of non-linear fractional partial differential equations by fractional sub-equation method[J]. Thermal Science, 2015, 19(4): 1239-1244.

[8] Ma W, Zhang L, Yang C. Discussion of the applicability of the generalized Clausius-Clapeyron equation and the frozen fringe process[J]. Earth-Science Reviews, 2015, 142: 47-59.

[9] Mammana C Z. The Relativistic Nature of Entropy Changes[J]. Brazilian Journal of Physics, 2021, 51(6): 1833-1843.

[10] Paulechka Y U, Kabo A G, Blokhin A V, et al. Heat capacity of ionic liquids: experimental determination and correlations with molar volume[J]. Journal of Chemical & Engineering Data, 2010, 55(8):2719-2724.

[11] Radmanović K, Dukić I, Pervan S. Specific heat capacity of wood[J]. Drvna Industrija, 2014, 65(2):151-157.

[12] Sharma S K, Sharma B K. Volume dependence of thermal expansivity for hcp iron[J]. Physica B: Condensed Matter, 2010, 405(15): 3145-3148.

[13] Solomonov B N, Nagrimanov R N, Mukhametzyanov T A. Additive scheme for calculation of solvation enthalpies of heterocyclic aromatic compounds sublimation/vaporization enthalpy at 298.15 K[J]. Thermochemical Acta, 2016, 633: 37-47.

Chapter 4

The Thermodynamics of Multicomponent Mixtures

Learning Objectives

- Understand the difference between partial molar properties and pure component properties.
- Compute partial molar properties from experimental data. Use the Gibbs-Duhem equation to simplify the equation.
- Estimate mixture composition and fugacity.

This chapter mainly studies the thermodynamic properties of multicomponent mixtures. The effect of composition on properties is of the most importance when applying thermodynamic principles to describe the mixtures. It is a complicated process to study the thermodynamic properties of multicomponent mixtures because the no linear changes arise when relating the thermodynamic properties of a mixture to its temperature, pressure and composition.

The thermodynamic state of a pure, single-phase system is completely specified by two of its intensive state variables. The size and thermodynamic state of a pure system can be specified by its mass or number of moles and two of its intensive state variables.

$$\underline{M} = f(x, y) \tag{4-1}$$

$$n\underline{M} = f(x, y, n) \tag{4-2}$$

For a c-component single-phase mixture, the size and state of the system are determined by the values of two state variables and the mole numbers of all species present.

$$n\underline{M} = f(x, y, n_1, n_2, \cdots, n_C) \tag{4-3}$$

The state of the c-component system is also specified by specifying the values of two state variables and the mole fractions of $C-1$ components present.

$$\underline{M} = f(x, y, x_1, x_2, \cdots, x_{C-1}) \tag{4-4}$$

For a one-phase mixture of c-component, we can predict the properties of the system by specifying the number of moles for all components and the values of the two state variables. In other words, for a mixture of c-components:

$$U = U(T, p, n_1, n_2, \cdots, n_C) \tag{4-5}$$

$$V = V(T, p, n_1, n_2, \cdots, n_C) \tag{4-6}$$

$$\underline{U} = \underline{U}(T, p, x_1, x_2, \cdots, x_{C-1}) \tag{4-7}$$

$$\underline{V} = \underline{V}(T, p, x_1, x_2, \cdots, x_{C-1}) \tag{4-8}$$

Our main research objective is to explore the relationship between the thermodynamic properties of the mixture and the concentration of components. i.e. the concentration dependence of the mixture properties.

A naive ideal is that the properties of a thermodynamic mixture are the sum of the related properties of the pure components weighted at the same temperature and pressure. That is, if $\underline{H}_i$. is the enthalpy for per mole of pure i component at same T and p, it is convenient to calculate the mixture of c-component. In other words, if $\underline{H}_i$ is the enthalpy per unit mass of component i at temperature T and pressure p, then it would be easy for components. If, for a mixture:

$$\underline{H}(T, p, x_1, x_2, \cdots, x_{C-1}) = \sum_{i=1}^{C} x_i \underline{U}_i(T, p) \tag{4-9}$$

$$\hat{U}(T, p, \omega_1, \omega_2, \cdots, \omega_{C-1}) = \sum_{i=1}^{C} \omega_i \hat{U}_i(T, p) \tag{4-10}$$

4.1 Excess Property

The real thermodynamic behavior of the mixture is quite different. The ideal solution is a mixture with similar size and similar intermolecular interaction. The bigger the difference, the larger the magnitude of value. Therefore, it is important to understand the key properties of the ideal solution. For an ideal solution,

$$\underline{H} = x_1 \underline{H}_1 + x_2 \underline{H}_2 \tag{4-11}$$

The molar enthalpy of the ideal solution can be used as a function of the composition.

$$\underline{H} = x_1 \underline{H}_1 + (1 - x_1) \underline{H}_2 \tag{4-12}$$

$$\underline{H} = x_1 (\underline{H}_1 - \underline{H}_2) + \underline{H}_2 \tag{4-13}$$

Under the same temperature, pressure and composition conditions, the difference between the molar properties of the real solution and the ideal solution is defined the excess property. $\underline{H}^E$ is the excess molar enthalpy of the mixture. It changes as a function of the composition the solution when its temperature and pressure are fixed. Its value at the endpoints (that represent pure solutions of either ethanol or benzene) can be specified (Abello et al., 1973; Smith et al., 1970).

It describes the molar enthalpy that exceeds the calculation result of the ideal solution model.

In general, not only the excess properties of molar enthalpy can be defined, but the excess properties of any solution properties can also be defined.

$$\underline{M}^E = \underline{M} - \underline{M}^{im} \tag{4-14}$$

im is the ideal solution.

$$\underline{U}^E = \underline{U} - \underline{U}^{im} = \underline{U} - (x_1 \underline{U}_1 + x_2 \underline{U}_2) \tag{4-15}$$

$$\underline{V}^E = \underline{V} - \underline{V}^{im} = \underline{V} - (x_1\underline{V}_1 + x_2\underline{V}_2) \tag{4-16}$$

Now, does the above methods apply to entropy? The molar entropy of an ideal solution is not the weighted sum of the molar entropy of the pure components, but is equal to:

$$\underline{S}^{im} = x_1\underline{S}_1 + x_2\underline{S}_2 - Rx_1\ln(x_1) - Rx_2\ln(x_2) \tag{4-17}$$

So, the result of mixing for ideal solutions is an increase in molar entropy.

The excess molar entropy:

$$\underline{S}^E = \underline{S} - \underline{S}^{im} = \underline{S} - \left[x_1\underline{S}_1 + x_2\underline{S}_2 - Rx_1\ln(x_1) - Rx_2\ln(x_2)\right] \tag{4-18}$$

The molar Gibbs free energy of ideal solution:

$$\begin{aligned}\underline{G}^{im} &= \underline{H}^{im} - T\underline{S}^{im} \\ &= x_1\underline{H}_1 + x_2\underline{H}_2 - T\left[x_1\underline{S}_1 + x_2\underline{S}_2 - Rx_1\ln(x_1) - Rx_2\ln(x_2)\right] \\ &= x_1\underline{G}_1 + x_2\underline{G}_2 + RTx_1\ln(x_1) + RTx_2\ln(x_2)\end{aligned} \tag{4-19}$$

The excess molar Gibbs free energy:

$$\underline{G}^E = \underline{G} - \underline{G}^{im} = \underline{G} - \left[x_1\underline{G}_1 + x_2\underline{G}_2 + RTx_1\ln(x_1) + RTx_1\ln(x_1)\right] \tag{4-20}$$

4.2 Properties Change on Mixing

Compared with the macroscopic properties of the mixing process, the intrinsic change of the mixing process seems to be significant, and the change of enthalpy is the main content of the chapter in the mixing process.

The change of molar enthalpy of the mixing can be written as:

$$\Delta\underline{H} = \underline{H} - (x_1\underline{H}_1 + x_2\underline{H}_2) \tag{4-21}$$

The enthalpy of mixture is linear combination of the enthalpy of two pure components. The enthalpies of the sulfuric acid-water mixtures at different temperatures are shown in Fig. 4-1. Fig. 4-2 plots the volume changes of methyl formate + methanol and methyl formate + ethanol mixtures as well as the enthalpy changes of benzene and aromatic fluorocarbons. The enthalpy changes of mixing, and the volume change of mixing (Treszczanowicz et al., 2010) for mixtures can be expressed as:

$$\Delta_{mix}\underline{H} = \underline{H}(T,p,x_1) - x_1\underline{H}_1(T,p) - x_2\underline{H}_2(T,p) \tag{4-22}$$

$$\Delta_{mix}\underline{V} = \underline{V}(T,p,x_1) - x_1\underline{V}_1(T,p) - x_2\underline{V}_2(T,p) \tag{4-23}$$

If the equations are valid for these mixing as Eq. (4-10), then both $\Delta_{mix}\underline{V}$ and $\Delta_{mix}\underline{H}$ are to be equal to zero at all components compositions. But the data in the figures show clearly that this is not the case.

In general, we can define properties change on mixing as:

$$\Delta_{mix}M = \underline{M} - \sum_{i=1}^{C}(x_i - M_i) \tag{4-24}$$

The basic mixture equation cannot be used for the calculation of real mixtures. The interaction between molecules affects the internal energy of fluid and other thermodynamic properties. As shown in the previous chapter, only the interaction between molecules and other similar molecules

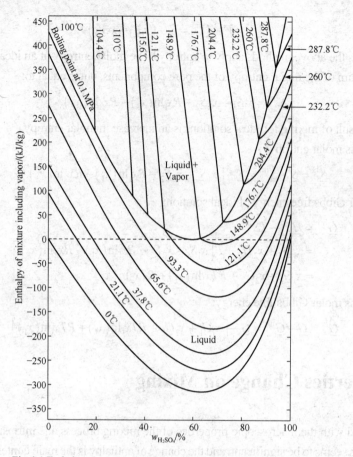

Fig.4-1 Enthalpy-concentration figure for aqueous sulfuric acid at 0.1 MPa

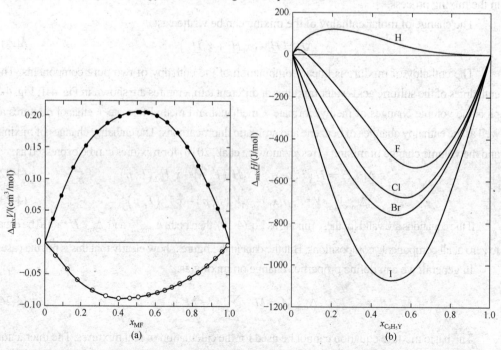

Fig.4-2 (a) Volume change on mixing at 298.15 K: methyl formate + methanol, methyl formate + ethanol.
(b) Enthalpy change on mixing of benzene (C_6H_6) and aromatic fluorocarbons (C_6F_5Y) at 298.15 K

Chapter 4 The Thermodynamics of Multicomponent Mixtures

affects the results of thermodynamics for pure fluids. However, the total interaction energy is the result of molecular interactions in binary fluid mixtures of component 1 and component 2. Although the 1-1 and 2-2 molecular interactions that occur in the mixture are approximately taken into account through the pure-component properties, the effects of the 1-2 interactions are not. Therefore, these fundamental equations cannot approximate the properties of most real fluid mixtures (Williamson et al., 1960).

EXAMPLE 4-1

Calculating the energy of an exothermic mixing process. 5 mol water and 2 mol sulfuric acid are mixed at 0℃. How much heat must be absorbed or released to keep the mixture at 0℃? ($M_{H_2O} = 18.015 \times 10^{-3}$ kg/mol, $M_{H_2SO_4} = 98.078 \times 10^{-3}$ kg/mol)

SOLUTION

Mass fraction

$$\omega_A = \frac{n_A M_A}{n_A M_A + n_B M_B}$$

$$= \frac{2 \times 98.078 \times 10^{-3}}{2 \times 98.078 \times 10^{-3} + 5 \times 18.015 \times 10^{-3}} = 68.53\%$$

It can be seen from Fig. 4-1 that the enthalpy of the mixture is about -315 kJ/kg.

$$m = n_A M_A + n_B M_B = \left(2 \times 98.078 \times 10^{-3} + 5 \times 18.015 \times 10^{-3}\right) \text{kg} = 0.286 \text{ kg}$$

$$\Delta_{mix} \hat{H} = \hat{H}(T,p,x_1) - x_1 \hat{H}_1(T,p) - x_2 \hat{H}_2(T,p) = -315 \text{ kJ/kg}$$

$$Q = m \Delta_{mix} \hat{H} = \left[0.286 \times (-315)\right] \text{kJ} = -90.09 \text{ kJ}$$

In general, the molar change of mixing for any thermodynamic property is

$$\Delta \underline{M} = \underline{M} - (x_1 \underline{M}_1 + x_2 \underline{M}_2) \tag{4-25}$$

The definition of change of mixing is that mixture properties subtract pure component properties.

$$\Delta \underline{M} = \underline{M} - (x_1 \underline{M}_1 + x_2 \underline{M}_2) \tag{4-26}$$

The excess property is the property of the real mixture subtract the property of the ideal solution.

$$\underline{M}^E = \underline{M} - \underline{M}^{im} \tag{4-27}$$

It is clear that molar enthalpy of mixing and the excess molar enthalpy are the same.

$$\Delta \underline{H} = \underline{H}^E \tag{4-28}$$

It is also the case for the molar internal energy and the molar volume. But molar entropy and the molar Gibbs free energy are not so. Because the molar entropy and the molar Gibbs free energy of the ideal solution are not as not the same as weighting the average of pure component properties by their mole fraction. So, it cannot apply to molar entropy and the molar Gibbs free energy. The above as shown in Table 4-1.

Table 4-1 Thermodynamic properties of multi-component mixtures

Real mixture	Ideal Solution	Property of mixing (ΔM)	Excess Property ($\underline{M}^E$)
Enthalpy $\underline{H}_1$	$x_1\underline{H}_1 + x_2\underline{H}_2$	$\underline{H}_1 - (x_1\underline{H}_1 + x_2\underline{H}_2)$	$\underline{H} - (x_1\underline{H}_1 + x_2\underline{H}_2)$
Internal Energy $\underline{U}_1$	$x_1\underline{U}_1 + x_2\underline{U}_2$	$\underline{U}_1 - (x_1\underline{U}_1 + x_2\underline{U}_2)$	$\underline{U} - (x_1\underline{U}_1 + x_2\underline{U}_2)$
Volume $\underline{V}_1$	$x_1\underline{V}_1 + x_2\underline{V}_2$	$\underline{V}_1 - (x_1\underline{V}_1 + x_2\underline{V}_2)$	$\underline{V} - (x_1\underline{V}_1 + x_2\underline{V}_2)$
Entropy $\underline{S}_1$	$x_1\underline{S}_1 + x_2\underline{S}_2 - Rx_1\ln(x_1) - Rx_2\ln(x_2)$	$\underline{S}_1 - (x_1\underline{S}_1 + x_2\underline{S}_2)$	$\underline{S} - [x_1\underline{S}_1 + x_2\underline{S}_2 - Rx_1\ln(x_1) - Rx_2\ln(x_2)]$
Gibbs free energy $\underline{G}_1$	$x_1\underline{G}_1 + x_2\underline{G}_2 + RTx_1\ln(x_1) + RTx_2\ln(x_2)$	$\underline{G}_1 - (x_1\underline{G}_1 + x_2\underline{G}_2)$	$\underline{G} - [x_1\underline{G}_1 + x_2\underline{G}_2 + RTx_1\ln(x_1) + RTx_2\ln(x_2)]$

At present, the research mainly focuses on the variation of $\underline{M}$ under certain temperature and pressure. It is necessary to relate various thermodynamic properties of mixtures to their component variables. so

$$M = n\underline{M} = \varphi(n_1, n_2, \cdots, n_C) \tag{4-29}$$

As long as the relative amounts of each component are constant, φ is a linear function of the total number of moles. It suggests that φ is a function of the mole ratio. Generally, the relationship between M and concentration is shown by the following equation:

$$M = n\underline{M} = n_1\varphi_1\left(\frac{n_2}{n_1}, \frac{n_3}{n_1}, \cdots, \frac{n_C}{n_1}\right) \tag{4-30}$$

The change in the value of $n\underline{M}$ as the number of moles of component 1 changes is

$$\begin{aligned}
\left[\frac{\partial(n\underline{M})}{\partial n_1}\right]_{T,p,n_2,n_3,\cdots} &= \left\{\frac{\partial\left[n_1\varphi_1\left(\frac{n_2}{n_1}, \frac{n_3}{n_1}, \cdots, \frac{n_C}{n_1}\right)\right]}{\partial n_1}\right\}_{T,p,n_2,n_3,\cdots} \\
&= \varphi_1\left(\frac{n_2}{n_1}, \frac{n_3}{n_1}, \cdots, \frac{n_C}{n_1}\right) + n_1\left[\frac{\partial\varphi_1\left(\frac{n_2}{n_1}, \frac{n_3}{n_1}, \cdots, \frac{n_C}{n_1}\right)}{\partial n_1}\right]_{T,p,n_2,n_3,\cdots} \\
&= \varphi_1\left(\frac{n_2}{n_1}, \frac{n_3}{n_1}, \cdots, \frac{n_C}{n_1}\right) + n_1\sum_{i=2}^{C}\frac{\partial\varphi_1}{\partial\left(\frac{n_i}{n_1}\right)} \times \frac{\partial\left(\frac{n_i}{n_1}\right)}{\partial n_1}
\end{aligned} \tag{4-31}$$

Where the last term arises from the chain rule of partial differentiation and the fact that φ_1 is a function of the mole ratios. Now multiplying by n_1, and using the fact that

$$\left[\frac{\partial\left(\frac{n_i}{n_1}\right)}{\partial n_1}\right]_{i\neq 1} = -\frac{n_i}{n_1^2}$$

gives

$$n_1\left[\frac{\partial(n\underline{M})}{\partial n_1}\right]_{T,p,n_{j\neq 1}} = n_1\left[\varphi_1\left(\frac{n_2}{n_1},\cdots\right) - \frac{1}{n_1}\sum_{i=2}^{C}n_i\frac{\partial\varphi_1}{\partial\left(\frac{n_i}{n_1}\right)}\right]$$

$$= n\underline{M} - \sum_{i=2}^{C}n_i\frac{\partial\varphi_1}{\partial\left(\frac{n_i}{n_1}\right)}$$

(4-32)

or

$$n\underline{M} = n_1\left[\frac{\partial(n\underline{M})}{\partial n_1}\right]_{T,p,n_{j\neq 1}} + \sum_{i=2}^{C}n_i\frac{\partial\varphi_1}{\partial\left(\frac{n_i}{n_1}\right)}$$

(4-33)

For the derivative with respect to $n_{i,i\neq 1}$,

$$n\underline{M} = n_1\varphi_1\left(\frac{n_2}{n_1},\frac{n_3}{n_1},\cdots,\frac{n_C}{n_1}\right)$$

$$\left[\frac{\partial(n\underline{M})}{\partial n_i}\right]_{T,p,n_{j\neq i}} = n_1\sum_{j=2}^{C}\frac{\partial\varphi_1}{\partial\left(\frac{n_j}{n_1}\right)}\frac{\partial\left(\frac{n_j}{n_1}\right)}{\partial n_i} = \frac{\partial\varphi_1}{\partial\left(\frac{n_i}{n_1}\right)}$$

(4-34)

since $\frac{\partial n_j}{\partial n_i}$ is equal to 1 for $j=i$, and 0 for $j\neq i$

$$n\underline{M} = n_1\left[\frac{\partial(n\underline{M})}{\partial n_1}\right]_{T,p,n_{j\neq 1}} + \sum_{i=2}^{C}n_i\left[\frac{\partial(n\underline{M})}{\partial n_i}\right]_{T,p,n_{j\neq i}}$$

(4-35)

or

$$n\underline{M} = \sum_{i=1}^{C}n_i\left[\frac{\partial(n\underline{M})}{\partial n_i}\right]_{T,p,n_{j\neq i}}$$

(4-36)

$$\underline{M} = \sum_{i=1}^{C}x_i\left[\frac{\partial(n\underline{M})}{\partial n_i}\right]_{T,p,n_{j\neq i}}$$

(4-37)

Define the partial molar property of component i as

$$\bar{M}_i = \left[\frac{\partial(n\underline{M})}{\partial n_i}\right]_{T,p,n_{j\neq i}} = \bar{M}_i(T,p,x_1,\cdots,x_{C-1})$$

(4-38)

Here, $\underline{M}$ is molar enthalpy, molar internal energy, molar volume, etc., while n is the number of moles in the system.

Although Eq. (4-38) are formally similar to Eq. (4-38) and Eq. (4-39). There is difference between the two equations, which is that $\bar{M}_i$ in Eq. (4-30) is the property of real mixtures, and each mixture is usually evaluated experimentally. Eq. (4-38) and Eq. (4-39) are usually used to express the properties of pure components. It should be noted that in general, $\underline{M} \neq \bar{M}_i$, that is, partial properties of mixtures are not equal to the thermodynamic properties of the pure components. Table 4-2 lists some common thermodynamic properties. The measurement and estimation of the partial molar properties is the most parts of the thermodynamics of mixtures. It is shown in this chapter and the next chapter.

Table 4-2 Partial molar thermodynamic properties

	Partial Molar Property	Molar Property of Mixture
$\bar{U}_i(T,p)$	$\bar{U}_i = \left[\dfrac{\partial(n\underline{U})}{\partial n_i}\right]_{T,p,n_{j\neq i}}$	$\underline{U} = \sum_{i=1}^{C} x_i \bar{U}_i$
$\bar{V}_i(T,p)$	$\bar{V}_i = \left[\dfrac{\partial(n\underline{V})}{\partial n_i}\right]_{T,p,n_{j\neq i}}$	$\underline{V} = \sum_{i=1}^{C} x_i \bar{V}_i$
$\bar{H}_i(T,p)$	$\bar{H}_i = \left[\dfrac{\partial(n\underline{H})}{\partial n_i}\right]_{T,p,n_{j\neq i}}$	$\underline{H} = \sum_{i=1}^{C} x_i \bar{H}_i$
$\bar{S}_i(T,p)$	$\bar{S}_i = \left[\dfrac{\partial(n\underline{S})}{\partial n_i}\right]_{T,p,n_{j\neq i}}$	$\underline{S} = \sum_{i=1}^{C} x_i \bar{S}_i$
$\bar{G}_i(T,p)$	$\bar{G}_i = \left[\dfrac{\partial(n\underline{G})}{\partial n_i}\right]_{T,p,n_{j\neq i}}$	$\underline{G} = \sum_{i=1}^{C} x_i \bar{G}_i$
$\bar{A}_i(T,p)$	$\bar{A}_i = \left[\dfrac{\partial(n\underline{A})}{\partial n_i}\right]_{T,p,n_{j\neq i}}$	$\underline{A} = \sum_{i=1}^{C} x_i \bar{A}_i$

$$\underline{M} = \sum_{i=1}^{C} x_i \bar{M}_i \tag{4-39}$$

The partial molar property is the change rate of the certain properties of the system with the molar number of a component under isothermal and isostatic conditions. $\bar{M}_i$ is a real mixture property which is evaluated experimentally for each mixture. The partial molar thermodynamic properties are not equal to pure component thermodynamic properties. An important content in the thermodynamics properties of mixtures is the measurement or estimation of partial molar properties.

The partial molar properties of the components in the mixture are different from those of the pure components. Therefore, the thermodynamic properties of pure components change during the mixing.

The total volume and enthalpy of the pure components are

$$V = \sum_{i=1}^{C} n_i \underline{V}_i(T,p) \tag{4-40}$$

$$H = \sum_{i=1}^{C} n_i \underline{H}_i(T,p) \tag{4-41}$$

Chapter 4 The Thermodynamics of Multicomponent Mixtures

The volume and enthalpy of the mixture at the same temperature and pressure are

$$V(T,p,n_1,n_2,\cdots) = \sum_{i=1}^{C} n_i \overline{V}_i(T,p,x_1,\cdots,x_{C-1}) \qquad (4\text{-}42)$$

$$H(T,p,n_1,n_2,\cdots) = \sum_{i=1}^{C} n_i \overline{H}_i(T,p,x_1,\cdots,x_{C-1}) \qquad (4\text{-}43)$$

Volume change of mixing,

$$\Delta V_{mix} = \sum_{i=1}^{C} n_i \left[\overline{V}_i(T,p,x_1,\cdots,x_{C-1}) - \underline{V}_i(T,p) \right] \qquad (4\text{-}44)$$

Enthalpy change of mixing,

$$\Delta H_{mix} = \sum_{i=1}^{C} n_i \left[\overline{H}_i(T,p,x_1,\cdots,x_{C-1}) - \underline{H}_i(T,p) \right] \qquad (4\text{-}45)$$

Other thermodynamic property changes of mixing can also be calculated. The relationship between Gibbs free energy, internal energy, enthalpy and entropy of pure components is shown in Eq. (4-46), which is also available to mixtures.

$$\underline{A} = \underline{U} - T\underline{S} \qquad (4\text{-}46)$$

$$n\underline{A} = n\underline{U} - nT\underline{S} \qquad (4\text{-}47)$$

When $\underline{U}$, $\underline{H}$ and $\underline{S}$ are values of the mixture, Eq. (4-47) can be used for mixtures.

$$\left[\frac{\partial(n\underline{A})}{\partial n_i} \right]_{T,p,n_{j\neq i}} = \left[\frac{\partial(n\underline{U})}{\partial n_i} \right]_{T,p,n_{j\neq i}} - T \left[\frac{\partial(n\underline{S})}{\partial n_i} \right]_{T,p,n_{j\neq i}} \qquad (4\text{-}48)$$

$$\overline{A}_i = \overline{U}_i - T\overline{S}_i \qquad (4\text{-}49)$$

Similarly,

$$\overline{G}_i = \overline{H}_i - T\overline{S}_i \qquad (4\text{-}50)$$

The relationship between partial molar thermodynamic properties in a mixture is the same as the relationship between thermodynamic properties in a pure fluid.

EXAMPLE 4-2

It is known that at 25℃, 0.1013 MPa, the relationship between the volume V_t and n_B of the solution formed by n_B mol of NaCl (B) dissolved in 55.5 mol of H$_2$O (A) is

$$V_t = (1001.38 + 16.6253 n_B + 1.7738 n_B^{1.5} + 0.1194 n_B^2) \text{cm}^3/\text{mol}$$

When $n_B = 0.4$, solve the partial molar volume $\overline{V}_A$ and $\overline{V}_B$ of H$_2$O and NaCl.

SOLUTION

By the definition of partial mole property, there is

77

$$\overline{V_B} = \left[\frac{\partial(nV)}{\partial n_B}\right]_{T,p,n_A} = (16.6253 + 1.5 \times 1.7738 n_B^{1.5} + 2 \times 0.11994 n_B) \text{cm}^3/\text{mol}$$

$$= (16.6253 + 2.6607 n_B^{1.5} + 0.2388 n_B) \text{cm}^3/\text{mol}$$

When $n_B = 0.4$,

$$V_B = (16.6253 + 2.6607 \times 0.4^{0.5} + 0.2388 \times 0.4) \text{cm}^3/\text{mol} = 18.4036 \text{ cm}^3/\text{mol}$$

According to the relationship between partial molar properties and solution properties, $nV = n_1 \overline{V_1} + n_2 \overline{V_2}$ for volume so:

$$\overline{V_A} = \frac{(nV) - n_B \overline{V_B}}{n_A}$$

$$\overline{V_A} = \frac{1001.38 + 16.6253 n_B + 1.7738 n_B^{1.5} + 2 \times 0.11994 n_B^2}{n_A}$$

$$- \frac{n_B \left(16.6253 + 2.6607 n_B^{0.5} + 0.2388 n_B\right)}{n_A}$$

So,

$$\overline{V_A} = \left[18.043 - 0.01598 \times n_B^{1.5} + 0.002151 n_B^2\right] \text{cm}^3/\text{mol}$$

When $n_B = 0.4$,

$$\overline{V_A} = (18.043 - 0.01598 \times 0.4^{1.5} + 0.002151 \times 0.4^2) \text{cm}^3/\text{mol} = 18.039 \text{ cm}^3/\text{mol}$$

4.3 Partial Molar Gibbs Free Energy

Since the Gibbs free energy of a multi-component mixture is a function of temperature, pressure, and each component the mole number, the total differential of the Gibbs free energy function can be written as

$$dG = \left(\frac{\partial G}{\partial T}\right)_{p,n_1,n_2,\cdots} dT + \left(\frac{\partial G}{\partial p}\right)_{T,n_1,n_2,\cdots} dp + \sum_{i=1}^{C} \left(\frac{\partial G}{\partial n_i}\right)_{p,T,n_{j \neq i}} dn_i \quad (4-51)$$

$$dG = -SdT + Vdp + \sum_{i=1}^{C} \overline{G_i} dn_i \quad (4-52)$$

The partial molar Gibbs free energy has been named the chemical potential (μ_i). Chemical potential is the partial derivative of Gibbs free energy for components, also known as partial molar potential energy. Partial moles are the intensive properties of the system, which is also written in the form of partial derivative in physical chemistry.

$$\mu_i = \overline{G_i} = \left[\frac{\partial(nG)}{\partial n_i}\right]_{T,p,n_{j \neq i}} \quad (4-53)$$

Chapter 4 The Thermodynamics of Multicomponent Mixtures

Since enthalpy can be expressed as a function of entropy and pressure, Eq. (4-54) is as following.

$$dH = \left(\frac{\partial H}{\partial p}\right)_{S,n_1,n_2,\cdots} dp + \left(\frac{\partial H}{\partial S}\right)_{p,n_1,n_2,\cdots} dS + \sum_{i=1}^{C}\left(\frac{\partial H}{\partial n_i}\right)_{p,S,n_{j\ne i}} dn_i \qquad (4\text{-}54)$$

$$dH = Vdp + TdS + \sum_{i=1}^{C}\left(\frac{\partial H}{\partial n_i}\right)_{p,S,n_{j\ne i}} dn_i \qquad (4\text{-}55)$$

from $H = G + TS$ and Eq. (4-55),

$$\begin{aligned} dH &= dG + TdS + SdT = -SdT + Vdp + \sum_{i=1}^{C}\bar{G}_i dn_i + TdS + SdT \\ &= Vdp + TdS + \sum_{i=1}^{C}\bar{G}_i dn_i \end{aligned} \qquad (4\text{-}56)$$

Comparing Eq. (4-55) and Eq. (4-56),

$$\left(\frac{\partial H}{\partial n_i}\right)_{p,S,n_{j\ne i}} = \bar{G}_i = \left(\frac{\partial G}{\partial n_i}\right)_{T,p,n_{j\ne i}} \qquad (4\text{-}57)$$

also, can get:

$$dU = TdS - pdV + \sum_{i=1}^{C}\bar{G}_i dn_i \qquad (4\text{-}58)$$

$$dA = -pdV - SdT + \sum_{i=1}^{C}\bar{G}_i dn_i \qquad (4\text{-}59)$$

with

$$\bar{G}_i = \left(\frac{\partial U}{\partial n_i}\right)_{S,V,n_{j\ne i}} = \left(\frac{\partial A}{\partial n_i}\right)_{T,V,n_{j\ne i}} \qquad (4\text{-}60)$$

As can be seen from the above equations, partial molar Gibbs free energy assumes special importance in mixtures, as the molar Gibbs free energy does in pure liquids.

4.4 Gibbs-Duhem Equation

According to the partial molar properties discussed in Section 4.2, thermodynamic function M can be written as:

$$n\underline{M} = \sum_{i=1}^{C} n_i \bar{M}_i \qquad (4\text{-}61)$$

so,

$$d(n\underline{M}) = \sum_{i=1}^{C} n_i d\bar{M}_i + \sum_{i=1}^{C} \bar{M}_i dn_i \qquad (4\text{-}62)$$

79

By the rule of differentiation, $n\underline{M}$ can also be considered to be a function of T, p, and the mole numbers, according to Eq. (4-66):

$$d(n\underline{M}) = \left[\frac{\partial(n\underline{M})}{\partial T}\right]_{p,n_1,n_2,\cdots} dT + \left[\frac{\partial(n\underline{M})}{\partial p}\right]_{T,n_1,n_2,\cdots} dp + \sum_{i=1}^{C}\left[\frac{\partial(n\underline{M})}{\partial n_i}\right]_{p,T,n_{j\neq i}} dn_i$$

$$= n\left(\frac{\partial \underline{M}}{\partial T}\right)_{p,n_1,n_2,\cdots} dT + n\left(\frac{\partial \underline{M}}{\partial p}\right)_{T,n_1,n_2,\cdots} dp + \sum_{i=1}^{C}\overline{M}_i dn_i \qquad (4\text{-}63)$$

Subtracting Eq. (4-62) from Eq. (4-63) gives the relation

$$-n\left(\frac{\partial \underline{M}}{\partial T}\right)_{p,n_1,n_2,\cdots} dT - n\left(\frac{\partial \underline{M}}{\partial p}\right)_{T,n_1,n_2,\cdots} dp + \sum_{i=1}^{C} n_i d\overline{M}_i = 0 \qquad (4\text{-}64)$$

and dividing by the total number of moles n yields

$$-\left(\frac{\partial \underline{M}}{\partial T}\right)_{p,n_i} dT - \left(\frac{\partial \underline{M}}{\partial p}\right)_{T,n_i} dp + \sum_{i=1}^{C} x_i d\overline{M}_i = 0 \qquad (4\text{-}65)$$

The Eq. (4-65) is form of the generalized Gibbs-Duhem equation. It describes the relation between the intensive properties T and p and the molar properties of each component in a homogeneous open system. It applies to any thermodynamic function M in a homogeneous system.

The Gibbs-Duhem equation at certain constant temperature and pressure is

$$\sum_{i=1}^{C} x_i d(\overline{M}_i)_{T,p} = 0 \qquad (4\text{-}66)$$

and

$$\sum_{i=1}^{C} n_i d(\overline{M}_i)_{T,p} = 0 \qquad (4\text{-}67)$$

So, for any thermodynamic property Y (except mole fraction) at certain temperature and pressure, Eq. (4-66) can be rewritten as

$$\sum_{i=1}^{C} n_i \left(\frac{\partial \overline{M}_i}{\partial Y}\right)_{T,p} = 0 \qquad (4\text{-}68)$$

For the change in the number of moles of component j at certain temperature and pressure:

$$\sum_{i=1}^{C} n_i \left(\frac{\partial \overline{M}_i}{\partial n_j}\right)_{T,p,n_{k\neq j}} = 0 \qquad (4\text{-}69)$$

The Gibbs-Duhem equations for the Gibbs free energy, obtained by setting $M = G$ in Eq. (4-64) and Eq. (4-65), are

$$0 = SdT - Vdp + \sum_{i=1}^{C} n_i d\overline{G}_i \qquad (4\text{-}70)$$

$$0 = \underline{S}dT - \underline{V}dp + \sum_{i=1}^{C} x_i d\overline{G}_i \qquad (4\text{-}71)$$

and the similar relationship with Eq. (4-66), Eq. (4-67), Eq. (4-68) and Eq. (4-69) are

$$\sum_{i=1}^{C} n_i \mathrm{d}(\overline{G}_i)_{T,p} = 0 \tag{4-72}$$

$$\sum_{i=1}^{C} x_i \mathrm{d}(\overline{G}_i)_{T,p} = 0 \tag{4-73}$$

$$\sum_{i=1}^{C} n_i \left(\frac{\partial \overline{G}_i}{\partial Y} \right)_{T,p} = 0 \tag{4-74}$$

and

$$\sum_{i=1}^{C} n_i \left(\frac{\partial \overline{G}_i}{\partial n_j} \right)_{T,p,n_{k \neq j}} = 0 \tag{4-75}$$

All of these equations are to be used.

The Gibbs-Duhem equation is a thermodynamic consistency relation, which it indicates that only $C+1$ is independent in the partial molar Gibbs energy sets of $C+2$ state variables, T, p and C in c-component system. For example, even though the temperature, pressure, $\overline{G}_1$, $\overline{G}_2$, $\cdots$, $\overline{G}_{C-1}$ can be independently varied, but $\overline{G}_C$ cannot be changed; its change $\mathrm{d}\overline{G}_C$ is related to $\mathrm{d}T$, $\mathrm{d}p$, $\mathrm{d}\overline{G}_1$ and $\mathrm{d}\overline{G}_{C-1}$ through the Eq. (4-70), so that

$$0 = S\mathrm{d}T - V\mathrm{d}p + \sum_{i=1}^{C} n_i \mathrm{d}\overline{G}_i \tag{4-76}$$

$$\mathrm{d}\overline{G}_C = \frac{1}{n_C} \times \left(-S\mathrm{d}T + V\mathrm{d}p - \sum_{i=1}^{C-1} n_i \mathrm{d}\overline{G}_i \right) \tag{4-77}$$

The correlation that the Eq. (4-65) to Eq. (4-75) provide is actually a constraint on the mixed equation of state. Therefore, these equations are important in minimizing the amount of necessary experimental data, in evaluating the thermodynamic properties of mixtures, in simplifying the description of multi-component systems, and in testing the consistency of certain types of experimental data. In this chapter, we are to describe in detail how to determine the molar properties of the experimental data based on variation equation of mixture and the Gibbs-Duhem equation. Although Eq. (4-65) to Eq. (4-68) are applicable to calculations with temperature, pressure and partial molar properties as independent variables, it is usually more convenient to use T, p, and the mole fraction x_i as independent variables. The change of variables is realized by the fact that the mixture of component C has only $C-1$ independent mole fraction. (Since $\sum_{i=1}^{C} x_i = 1$) written as

$$\mathrm{d}\overline{M}_i = \left(\frac{\partial \overline{M}_i}{\partial T} \right)_{p,x_1,x_2,\cdots} \mathrm{d}T + \left(\frac{\partial \overline{M}_i}{\partial p} \right)_{T,x_1,x_2,\cdots} \mathrm{d}p + \sum_{j=1}^{C-1} \left(\frac{\partial \overline{M}_i}{\partial x_j} \right)_{T,p} \mathrm{d}x_j \tag{4-78}$$

where $T, p, x_1, \cdots, x_{C-1}$ have been chosen as the independent variables. Substituting this expansion in Eq. (4-65) as

$$0 = -\left(\frac{\partial \underline{M}}{\partial T}\right)_{p,n_1,n_2,\cdots} dT - \left(\frac{\partial \underline{M}}{\partial p}\right)_{T,n_1,n_2,\cdots} dp$$
$$+ \sum_{i=1}^{C} x_i \left[\left(\frac{\partial \bar{M}_i}{\partial T}\right)_{p,x_1,x_2,\cdots} dT + \left(\frac{\partial \bar{M}_i}{\partial p}\right)_{T,x_1,x_2,\cdots} dp + \sum_{j=1}^{C-1}\left(\frac{\partial \bar{M}_i}{\partial x_j}\right)_{T,p} dx_j\right] \quad (4\text{-}79)$$

Since

$$\sum_{i=1}^{C} x_i \left(\frac{\partial \bar{M}_i}{\partial T}\right)_{p,x_1,x_2,\cdots} dT = \left[\frac{\partial}{\partial T}\left(\sum_{i=1}^{C} x_i \bar{M}_i\right)\right]_{p,x_1,x_2,\cdots} dT = \left(\frac{\partial \underline{M}}{\partial T}\right)_{p,x_1,x_2,\cdots}$$

Keeping all the mole numbers constant is keeping all the mole fractions constant, so the first and third terms in the Eq. (4-79) are cancelled. Similarly, the second and fourth are also cancelled.

$$\sum_{i=1}^{C} x_i \sum_{j=1}^{C-1}\left(\frac{\partial \bar{M}_i}{\partial x_j}\right)_{T,p} dx_j = 0 \quad (4\text{-}80)$$

For a binary mixture,

$$\sum_{i=1}^{2} x_i \left(\frac{\partial \bar{M}_i}{\partial x_1}\right)_{T,p} dx_1 = 0 \quad (4\text{-}81)$$

Equivalently,

$$x_1 \left(\frac{\partial \bar{M}_1}{\partial x_1}\right)_{T,p} + x_2 \left(\frac{\partial \bar{M}_2}{\partial x_1}\right)_{T,p} = 0 \quad (4\text{-}82)$$

For the special case where M is equal to Gibbs free energy,

$$x_1 \left(\frac{\partial \bar{G}_1}{\partial x_1}\right)_{T,p} + x_2 \left(\frac{\partial \bar{G}_2}{\partial x_1}\right)_{T,p} = 0 \quad (4\text{-}83)$$

4.5 The Experimental Measurement of Partial Molar Volume and Enthalpy

Some physical quantities that cannot be measured directly can be obtained through some measurable physical quantities. In particular, mixture density measurements can be used to obtain partial molar volumes, and heat-of-mixing data can be used to obtain partial molar enthalpy. Phase equilibrium measurements can be used to obtain the partial molar Gibbs free energy of a component in mixture.

Table 4-3 Density values for the water (1) + methanol (2) system at $T = 298.15\text{K}$ (Albuquerque et al., 1996)

x_1	$\rho/(\text{kg/m}^3)$	$\underline{V} \times 10^{6*}/(\text{m}^3/\text{mol})$	$\Delta_{mix}\underline{V} \times 10^6/(\text{m}^3/\text{mol})$
0.0	786.846	40.7221	0
0.1162	806.655	37.7015	−0.3883
0.2221	825.959	35.0219	−0.6688
0.2841	837.504	33.5007	−0.7855

Continued

x_1	ρ/(kg/m³)	$\underline{V} \times 10^{6*}$/(m³/mol)	$\Delta_{mix}\underline{V} \times 10^6$/(m³/mol)
0.3729	855.031	31.3572	−0.9174
0.4186	864.245	30.2812	−0.9581
0.5266	887.222	27.7895	−1.0032
0.6119	905.376	25.9108	−0.9496
0.7220	929.537	23.5759	−0.7904
0.8509	957.522	20.9986	−0.4476
0.9489	981.906	19.0772	−0.1490
1.0	997.047	18.0686	0

*Note that the notation $\underline{V} \times 10^6$/(m³/mol) means that the entries in the table have been multiplied by the factor 10^6. Therefore, for example, the volume of pure methanol $\underline{V}$ ($x_1 = 0$) is 40.7221×10^{-6} m³/mol = 40.7221 cm³/mol.

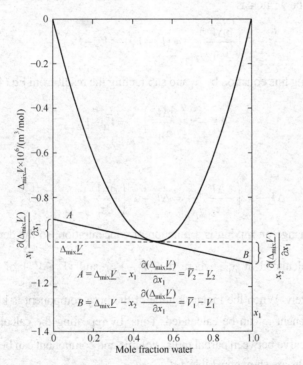

Fig.4-3 Isothermal volume change on mixing for the water(1)+methanol(2) system

The slope is related to the partial molar volume in Fig. 4-3. Table 4-3 shows the water (1) + methanol (2) system of mixture density at certain temperature and pressure. Line 4 is the change value of molar volume when different components are mixed (i.e. $\Delta\underline{V} = \left(x_1\overline{V}_1 + x_2\overline{V}_2\right) - x_1\underline{V}_1 - x_2\underline{V}_2$), The values in Table 4-3 are plotted in Fig. 4-5.

$$\Delta\underline{V} = \left(x_1\overline{V}_1 + x_2\overline{V}_2\right) - x_1\underline{V}_1 - x_2\underline{V}_2 = x_1\left(\overline{V}_1 - \underline{V}_1\right) + x_2\left(\overline{V}_2 - \underline{V}_2\right) \tag{4-84}$$

The pure component molar volumes are independent of mixture. So,

$$\left(\frac{\partial \underline{V}_i}{\partial x_1}\right)_{T,p} = 0 \tag{4-85}$$

$$x_2 = 1 - x_1 \tag{4-86}$$

so that

$$\left(\frac{\partial \Delta \underline{V}}{\partial x_1}\right)_{T,p} = \left(\overline{V}_1 - \underline{V}_1\right) + x_1\left(\frac{\partial \overline{V}_1}{\partial x_1}\right)_{T,p} - \left(\overline{V}_2 - \underline{V}_2\right) + x_2\left(\frac{\partial \overline{V}_2}{\partial x_1}\right)_{T,p} \tag{4-87}$$

Here, using the fact that the pure component molar volume is independent of the composition of the mixture (i.e., $(\partial \overline{V}_i / \partial x_1)_{T,p}=0$), and that for a binary mixture $x_2 = 1 - x_1$, so $\left(\dfrac{\partial x_2}{\partial x_1}\right) = -1$.

From the Gibbs-Duhem equation, Eq. (4-81) with $\theta_i = V_i$,

$$x_1\left(\frac{\partial \overline{V}_1}{\partial x_1}\right)_{T,p} + x_2\left(\frac{\partial \overline{V}_2}{\partial x_1}\right)_{T,p} = 0 \tag{4-88}$$

Eq. (4-86) can be written as

$$\left(\frac{\partial \Delta \underline{V}}{\partial x_1}\right)_{T,p} = \left(\overline{V}_1 - \underline{V}_1\right) - \left(\overline{V}_2 - \underline{V}_2\right) \tag{4-89}$$

Now multiplying this equation by x_1 and subtracting the result from Eq.(4-81) gives.

$$\Delta \underline{V} - x_1\left(\frac{\partial \Delta \underline{V}}{\partial x_1}\right)_{T,p} = \overline{V}_2 - \underline{V}_2 \tag{4-90}$$

Similarly,

$$\Delta \underline{V} - x_2\left(\frac{\partial \Delta \underline{V}}{\partial x_2}\right)_{T,p} = \Delta \underline{V} + x_2\left(\frac{\partial \Delta \underline{V}}{\partial x_1}\right)_{T,p} = \overline{V}_1 - \underline{V}_1 \tag{4-91}$$

The volume change on mixing is a function of concentration, so the derivative of $\partial \Delta \underline{V}$ and $\left(\dfrac{\partial \Delta \underline{V}}{\partial x_1}\right)$ can be calculated at x_1, where the $\left(\overline{V}_1 - \underline{V}_1\right)$ and $\left(\overline{V}_2 - \underline{V}_2\right)$ of component can be calculated immediately. When the molar volume of the pure component is known, $\overline{V}_1$ and $\overline{V}_2$ at the specified component x_1 can be calculated. Thus, by repeating the calculation at other mole fraction values, the curve between partial molar volume and component can be completely plotted. The calculation results are shown in Table 4-4.

Table 4-4 Molar volume of Pure Component

x_1	$(\overline{V}_1 - \underline{V}_1) \times 10^6 /(\text{m}^3/\text{mol})$	$\overline{V}_1 \times 10^6 /(\text{m}^3/\text{mol})$	$(\overline{V}_2 - \underline{V}_2) \times 10^6 /(\text{m}^3/\text{mol})$	$\overline{V}_2 \times 10^6 /(\text{m}^3/\text{mol})$
0.0	−3.8893	14.180*	0	40.722
0.1162	−2.9741	15.095	−0.0530	40.669
0.2221	−2.3833	15.686	−0.1727	40.549
0.2841	−2.0751	15.994	−0.2773	40.445
0.3729	−1.6452	16.424	−0.4884	40.234
0.4186	−1.4260	16.643	−0.6321	40.090
0.5266	−0.9260	17.143	−1.0822	39.640

Continued

x_1	$(\bar{V}_1-\underline{V}_1)\times10^6/(m^3/mol)$	$\bar{V}_1\times10^6/(m^3/mol)$	$(\bar{V}_2-\underline{V}_2)\times10^6/(m^3/mol)$	$\bar{V}_2\times10^6/(m^3/mol)$
0.6119	−0.5752	17.494	−1.5464	39.176
0.7220	−0.2294	17.840	−2.2363	38.486
0.8509	−0.0254	18.044	−2.9631	37.759
0.9489	−0.0026	18.072	−3.1689	37.553
1.0	0	18.069	−3.0348	37.687*

*Partial molar volume at infinite dilution.

The other way to determine partial molar property is graphic method. At a given composition, say x_1^*, a tangent line to the $\Delta\underline{V}$ curve is drawn, the intersection of the tangent and the coordinate axis is A and B.

The slope of the tangent line is $\left(\dfrac{\partial\Delta\underline{V}}{\partial x_1}\right)_{x_1^*}$, so that $-x_1^*\left(\dfrac{\partial\Delta\underline{V}}{\partial x_1}\right)_{x_1^*}$ is the distance shown in Fig. 4-4. Obviously, the value of point A on the vertical axis is equal to the value to the left of Eq. (4-93), so it is equal to the value of $\left(\bar{V}_2-\underline{V}_2\right)$ at x_1^*.

The third way to determine partial molar property is: in order to get accurate partial molar property, it is better to use analytical rather than graphical methods. The volume change on mixing ($\Delta\underline{V}$) is fitted with a polynomial of mole fraction, and then the derivative is obtained analytically. Because $\Delta\underline{V}$ is equal to zero at $x_1=0$ and $x_1=1$ ($x_2=0$), it is usually fit with a polynomial in the Redlich-Kister form:

Analytical process

$$\Delta\underline{V}=x_1 x_2\sum_{i=0}^{n}a_i(x_1-x_2)^i \qquad (4\text{-}92)$$

Then $\Delta\underline{V}$ is rewritten as:

$$\Delta\underline{V}=x_1(1-x_1)\sum_{i=0}^{n}a_i(2x_1-1)^i \qquad (4\text{-}93)$$

$$\left(\dfrac{\partial\Delta\underline{V}}{\partial x_1}\right)_{T,p}=-\sum_{i=0}^{n}a_i(2x_1-1)^{i+1}+2x_1(1-x_1)\sum_{i=0}^{n}a_i i(2x_1-1)^{i-1} \qquad (4\text{-}94)$$

and

$$\Delta\underline{V}-x_1\left(\dfrac{\partial\Delta\underline{V}}{\partial x_1}\right)_{T,p}=\bar{V}_2-\underline{V}_2=x_1^2\sum_{i=0}^{n}a_i\left[(x_1-x_2)^i-2ix_2(x_1-x_2)^{i-1}\right] \qquad (4\text{-}95)$$

Similarly,

$$\bar{V}_1-\underline{V}_1=\Delta\underline{V}+x_2\left(\dfrac{\partial\Delta\underline{V}}{\partial x_1}\right)_{T,p}=x_2^2\sum_{i=0}^{n}a_i\left[(x_1-x_2)^i+2ix_1(x_1-x_2)^{i-1}\right] \qquad (4\text{-}96)$$

The same procedure can be extended to $\Delta\underline{H}$, $\Delta\underline{U}$, and other thermodynamic properties.

An accurate representation of the water-methanol data is obtained using Eq. (4-91) (in units of m³/mol) $a_0 = -4.0034 \times 10^{-6}$, $a_1 = -0.17756 \times 10^{-6}$, $a_2 = 0.54139 \times 10^{-6}$, $a_3 = 0.60481 \times 10^{-6}$. The partial molar volumes in Table 4-4 are calculated using these constants and Eq. (4-95) and Eq. (4-96) (Arce et al., 1993).

The volume change on mixing of the water-methanol system is negative for the mixing, which is $(\overline{V}_1 - \underline{V}_1)$ and $(\overline{V}_2 - \underline{V}_2)$. It is not a universal characteristic. Three quantities can be positive, negative, or even positive over part of the composition range and negative over the rest, depending on the different system.

The partial molar enthalpy of a substance in a binary mixture (Bangqing et al., 2010) can be obtained by the same method, using the enthalpy change on mixing (or heat of mixing) value. Measurements are usually made using a steady-state flow calorimeter as shown in Fig. 4-4. Two streams, one of pure fluid 1 and the second of pure fluid 2, both at a temperature T and a pressure p, enter this steady-state mixing device, and a single mixed stream, also at T and p, leaves.

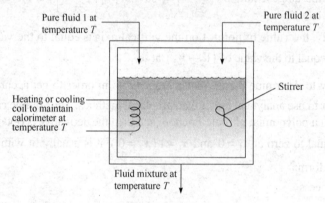

Fig.4-4 Isothermal flow calorimeter

When the component of the heat of mixing is known, $\overline{H}_1$ and $\overline{H}_2$ can be calculated in a manner similar to partial molar volumes. So,

$$\Delta \underline{H} - x_1 \left(\frac{\partial \Delta \underline{H}}{\partial x_1} \right)_{T,p} = \overline{H}_2 - \underline{H}_2 \qquad (4\text{-}97)$$

and

$$\Delta \underline{H} - x_2 \left(\frac{\partial \Delta \underline{H}}{\partial x_2} \right)_{T,p} = \Delta \underline{H} + x_2 \left(\frac{\partial \Delta \underline{H}}{\partial x_1} \right)_{T,p} = \overline{H}_1 - \underline{H}_1 \qquad (4\text{-}98)$$

So that either the computational or graphical method may also be used to calculate the partial molar enthalpy (Bejan et al., 2006).

Table 4-5 and Fig. 4-5 contain the mixed thermodynamic data of the water-methanol system. The data are used to calculate the partial molar enthalpies (Arenosa et al., 1979). Note that one feature of the heat of mixing data (Christensen et al.,1984) of Fig. 4-6 is that it is skewed, with the largest absolute value at $x_1 = 0.73$ (and not $x_1 = 0.5$).

Table 4-5 Heat of mixing value for the water (1)+methanol (2) system at $T = 19.69°C$

x_1	$Q^+(CH_3OH)/(kJ/mol)$	$Q = \Delta_{mix}\underline{H}/(kJ/mol)$
0.05	−0.134	−0.127
0.10	−0.272	−0.245
0.15	−0.419	−0.356
0.20	−0.569	−0.455
0.25	−0.716	−0.537
0.30	−0.862	−0.603
0.35	−1.017	−0.661
0.40	−1.197	−0.718
0.45	−1.398	−0.769
0.50	−1.632	−0.816
0.55	−1.896	−0.853
0.60	−2.218	−0.887
0.65	−2.591	−0.907
0.70	−3.055	−0.917
0.75	−3.666	−0.917
0.80	−4.357	−0.871
0.85	−5.114	−0.767
0.90	−5.989	−0.599
0.95	−6.838	−0.342

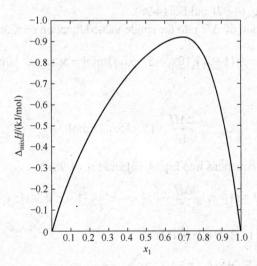

Fig.4-5 Heat of mixing data for the water (1) + methanol (2) system at $T=19.69°C$

The partial molar properties of any extension function equations have the following conclusions.

$$\Delta\underline{M} - x_2\left(\frac{\partial\Delta\underline{M}}{\partial x_2}\right)_{T,p} = \overline{M}_1(T,p,x_1) - \underline{M}_1(T,p) \qquad (4\text{-}99)$$

and

$$\Delta \underline{M} - x_1 \left(\frac{\partial \Delta \underline{M}}{\partial x_1} \right)_{T,p} = \overline{M}_2(T,p,x_1) - \underline{M}_2(T,p) \qquad (4\text{-}100)$$

If M is any molar property of the mixture:

$$\overline{M}_1(T,p,x_1) = \underline{M}(T,p,x_1) - x_2 \left[\frac{\partial \underline{M}(T,p,x_1)}{\partial x_2} \right]_{T,p} \qquad (4\text{-}101)$$

and

$$\overline{M}_2(T,p,x_1) = \underline{M}(T,p,x_1) - x_1 \left[\frac{\partial \underline{M}(T,p,x_1)}{\partial x_1} \right]_{T,p} \qquad (4\text{-}102)$$

EXAMPLE 4-3

At 298 K and 1×10^5 Pa, the relationship between mixing heat of component 1 and component 2 and the compositions of the mixture are as follows: $\Delta H = x_1 x_2 (10 x_1 + 5 x_2)$ J/mol. At the same temperature and pressure, the molar enthalpy of pure liquid is $H_1 = 418$ J/mol, $H_2 = 628$ J/mol, respectively. Calculate infinite dilution partial molar enthalpy $\overline{H}_1^\infty$ and $\overline{H}_2^\infty$ at 298 K and 1×10^5 Pa.

SOLUTION

The relationship between molar property and concentration of solution is known, which can be solved according to Eq. (4-97) and Eq. (4-98).

Arrange the expression of ΔH into the single-valued function relation of x_1,

$$\Delta H = x_1(1-x_1)(10x_1 + 5 - 5x_1) \text{ J/mol} = 5(x_1 - x_1^3) \text{ J/mol}$$

Take the derivative of x_1,

$$\frac{\mathrm{d}\Delta H}{\mathrm{d}x_1} = (5 - 15 x_1^2) \text{ J/mol}$$

Substitute the above two equations into Eq. (4-97) and Eq. (4-98):

$$\Delta \overline{H}_1 = \Delta H + (1-x_1) \frac{\mathrm{d}\Delta H}{\mathrm{d}x_1} = \left[5(x_1 - x_1^3) + (1 - x_1)(5 - 15 x_1^2) \right] \text{J/mol}$$

$$= \left[5(1 - 3x_1^2 + 2x_1^3) \right] \text{J/mol}$$

$$\Delta \overline{H}_2 = \Delta H + x_1 \frac{\mathrm{d}\Delta H}{\mathrm{d}x_1} = \left[5(x_1 - x_1^3) - x_1(5 - 15 x_1^2) \right] \text{J/mol} = 10 x_1^3 \text{ J/mol}$$

Partial molar enthalpy of components 1 and 2,

$$\Delta \overline{H}_1 = H_1 + 5(1 - 3x_1^2 + 2x_1^3) \text{ J/mol} = \left[418 + 5(1 - 3x_1^2 + 2x_1^3) \right] \text{J/mol}$$

$$\Delta \overline{H}_2 = H_2 + 10 x_1^3 \text{ J/mol} = 628 \text{ J/mol} + 10 x_1^3 \text{ J/mol}$$

The partial molar enthalpy for infinite dilution of component 1 and 2,

$$\bar{H}_1^\infty = \lim_{x_1 \to 0} \bar{H}_1 = 423 \text{ J/mol}$$

$$\bar{H}_2^\infty = \lim_{x_1 \to 1} \bar{H}_2 = 638 \text{ J/mol}$$

4.6 Gibbs Free Energy and Fugacity of a Component in a Mixture

The most important part in the thermodynamic analysis of mixtures is the partial molar properties of each substance in the mixture. Partial molar Gibbs free energy is an important theoretical basis for the study of phase equilibrium and chemical equilibrium, so the study of partial molar Gibbs free energy has special significance (Smith et al., 2005).

4.6.1 Ideal Gas Mixture

The density of ideal gas mixture is very low, and the interaction between molecules is not obvious. An ideal gas is an imaginary gas in which the molecules themselves have no volume and no interaction between molecules. The partial pressure of each component in an ideal gas is equal to the product of the mole fraction of the component and the total pressure.

$$pV^{\text{igm}} = (n_1 + n_2 + \cdots)RT = \sum_{j=1}^C n_j RT = nRT \quad (4\text{-}103)$$

$$U^{\text{igm}}(T, n_1, n_2, \cdots) = \sum_{j=1}^C n_j \underline{U}_j^{\text{ig}}(T)$$

$$\bar{U}_i^{\text{igm}}(T, x_1, x_2, \cdots) = \left[\frac{\partial U^{\text{igm}}(T, n_1, n_2, \cdots)}{\partial n_i} \right]_{T,p,n_{j \neq i}} = \left[\frac{\partial \sum_{j=1}^C n_j \underline{U}_j^{\text{ig}}(T)}{\partial n_i} \right]_{T,p,n_{j \neq i}} = \underline{U}_i^{\text{ig}}(T) \quad (4\text{-}104)$$

The partial molar internal energy of component i in the ideal gas mixture at a certain temperature is equal to the pure component molar internal energy of component i in the ideal gas at the same temperature.

$$\bar{V}_i^{\text{igm}}(T, p, x_1, x_2, \cdots) = \left[\frac{\partial V^{\text{igm}}(T, p, n_1, n_2, \cdots)}{\partial n_i} \right]_{T,p,n_{j \neq i}} = \left(\frac{\partial \sum_{j=1}^C n_j \frac{RT}{p}}{\partial n_i} \right)_{T,p,n_{j \neq i}} \quad (4\text{-}105)$$

$$= \frac{RT}{p} = V_i^{\text{ig}}(T, p)$$

The partial molar volume of component i in an ideal gas mixture at a certain temperature and pressure is equal to the molar volume of the pure component as an ideal gas at same temperature and pressure.

Considering the process of forming an ideal gas mixture at certain T and p from a collection of pure ideal gases.

$$\Delta V_{\text{mix}}^{\text{igm}} = V^{\text{igm}}(T,p,n_1,n_2,\cdots) - \sum_{i=1}^{C} n_i \underline{V}_i(T,p)$$

$$= \sum_{i=1}^{C} n_i \overline{V}_i(T,p,x_1,\cdots,x_{C-1}) - \sum_{i=1}^{C} n_i \underline{V}_i(T,p) \qquad (4\text{-}106)$$

$$= \sum_{i=1}^{C} n_i \underline{V}_i(T,p) - \sum_{i=1}^{C} n_i \underline{V}_i(T,p) = 0$$

$$\Delta U_{\text{mix}}^{\text{igm}} = 0 \qquad (4\text{-}107)$$

Similarly,

$$\Delta H_{\text{mix}}^{\text{igm}} = \Delta U_{\text{mix}}^{\text{igm}} + p\Delta V_{\text{mix}}^{\text{igm}} = 0 \qquad (4\text{-}108)$$

The partial pressure of component i in gas mixture,

$$p_i = x_i p \qquad (4\text{-}109)$$

For the ideal gas mixture,

$$p_i^{\text{igm}}(n,V,T,x_1,x_2,\cdots) = \frac{n_i}{\sum_{j=1}^{C} n_j} p = \frac{n_i}{\sum_{j=1}^{C} n_j}\left(\sum_{j=1}^{C} n_j \frac{RT}{V}\right) = \frac{n_i RT}{V}$$

$$= p^{\text{ig}}(n_i,V,T) \qquad (4\text{-}110)$$

Therefore, the partial pressure of component i in the ideal gas mixture is equal to the pressure that the mole number n_i of component i is separately contained in the same V at the certain T and p. Since there is no energy of interaction in an ideal gas mixture, the effect on each component of forming an ideal gas mixture at certain temperature and total pressure is equal to reducing the pressure from p to p_i.

$$d\underline{S} = \frac{C_p}{T}dT - \left(\frac{\partial \underline{V}}{\partial T}\right)_p dp = -\left(\frac{\partial \underline{V}}{\partial T}\right)_p dp$$

$$\overline{S}_i^{\text{igm}}(T,p,x_1,x_2,\cdots) - \underline{S}_i^{\text{ig}}(T,p) = \int_p^{p_i} -\left(\frac{\partial \underline{V}}{\partial T}\right)_p dp = \int_p^{p_i}\left(-\frac{R}{p}\right)dp$$

$$= -R\ln\frac{p_i}{p} = -R\ln x_i \qquad (4\text{-}111)$$

or,

$$\overline{S}_i^{\text{igm}}(T,V,x_1,x_2,\cdots) - \underline{S}_i^{\text{ig}}(T,V_i)$$

$$= R\ln\frac{V}{V_i} = -R\ln x_i \qquad (4\text{-}112)$$

Consequently,

$$\Delta S_{mix}^{igm} = \sum_{i=1}^{C} n_i \left[\overline{S}_i^{igm}(T,p,x_1,x_2,\cdots) - \underline{S}_i^{ig}(T,p) \right] = \sum_{i=1}^{C} n_i \left[-R\ln x_i \right]$$

$$= -R\sum_{i=1}^{C} n_i \ln x_i \quad \Delta \underline{S}^{igm} = \frac{\Delta S^{igm}}{n} = -R\sum_{i=1}^{C} x_i \ln x_i \quad (4\text{-}113)$$

Ideal gas mixture is completely mixed or random mixture. According to the changes of energy, volume and entropy during mixing, other thermodynamic properties (Kimura et al., 1983) of ideal gas mixture can be easily calculated. Table 4-6 is the properties of ideal gas mixture.

$$\overline{G}_i^{igm}(T,p,x_1,x_2,\cdots) = \overline{H}_i^{igm}(T,p,x_1,x_2,\cdots) - T\overline{S}_i^{igm}(T,p,x_1,x_2,\cdots)$$
$$= \underline{H}_i^{ig}(T,p) - T\left[\underline{S}_i^{ig}(T,p) - R\ln x_i \right] = \underline{G}_i^{ig}(T,p) + RT\ln x_i \quad (4\text{-}114)$$

$$\Delta G_{mix}^{igm} = \sum_{i=1}^{C} \left\{ x_i \left[\overline{G}_i^{igm}(T,p,x_1,x_2,\cdots) - \underline{G}_i^{ig}(T,p) \right] \right\} = RT \sum_{i=1}^{C} x_i \ln x_i \quad (4\text{-}115)$$

Table 4-6 Properties of ideal gas mixture (mixing at certain T and p)

Internal Energy	$\overline{U}_i^{igm}(T,x_1,x_2,\cdots) = \underline{U}_i^{ig}(T)$	$\Delta \underline{U}^{igm} = 0$
Enthalpy	$\overline{H}_i^{igm}(T,x_1,x_2,\cdots) = \underline{H}_i^{ig}(T)$	$\Delta \underline{H}^{igm} = 0$
Volume	$\overline{V}_i^{igm}(T,x_1,x_2,\cdots) = \underline{V}_i^{ig}(T)$	$\Delta \underline{V}^{igm} = 0$
Entropy	$\overline{S}_i^{igm}(T,p,x_1,x_2,\cdots) = \underline{S}_i^{ig}(T,p) - R\ln x_i$	$\Delta \underline{S}^{igm} = -R\sum_{i=1}^{C} x_i \ln x_i$
Gibbs free energy	$\overline{G}_i^{igm}(T,p,x_1,x_2,\cdots) = \underline{G}_i^{ig}(T,p) + RT\ln x_i$	$\Delta \underline{G}^{igm} = RT\sum_{i=1}^{C} x_i \ln x_i$
Helmholtz energy	$\overline{A}_i^{igm}(T,p,x_1,x_2,\cdots) = \underline{G}_i^{ig}(T,p) + RT\ln x_i$	$\Delta \underline{A}^{igm} = RT\sum_{i=1}^{C} x_i \ln x_i$
Internal Energy	$\underline{U}^{igm}(T,x_1,x_2,\cdots) = \sum x_i \underline{U}_i^{ig}(T)$	
Enthalpy	$\underline{H}^{igm}(T,p,x_1,x_2,\cdots) = \sum x_i \underline{H}_i^{ig}(T,p)$	
Volume	$\underline{V}^{igm}(T,p,x_1,x_2,\cdots) = \sum x_i \underline{V}_i^{ig}(T,p)$	
Entropy	$\underline{S}^{igm}(T,p,x_1,x_2,\cdots) = \sum x_i \underline{S}_i^{ig}(T,p) - R\sum x_i \ln x_i$	
Gibbs free energy	$\underline{G}^{igm}(T,p,x_1,x_2,\cdots) = \sum x_i \underline{G}_i^{ig}(T,p) + RT\sum x_i \ln x_i$	
Helmholtz energy	$\underline{A}^{igm}(T,p,x_1,x_2,\cdots) = \sum x_i \underline{A}_i^{ig}(T,p) + RT\sum x_i \ln x_i$	

4.6.2 Ideal Mixture and Excess Mixture Properties

We choose a state for each system in which the thermodynamic properties are reasonably known and then try to estimate how the departure of the real system from the ideal system state affects the system properties.

The properties of real fluids can be calculated as the sum of an ideal gas plus the departure from ideal gas behavior. The ideal mixture (which can be a gas or liquid mixture) is defined as:

$$\bar{H}_i^{im}(T,p,\underline{x}) = \underline{H}_i(T,p) \tag{4-116}$$

And,

$$\bar{V}_i^{im}(T,p,x_1,\underline{x}) = \underline{V}_i(T,p) \tag{4-117}$$

This is available to all temperatures, pressures and compositions. Unlike the ideal gas mixture, here neither $\underline{H}_i$ nor $\underline{V}_i$ is an ideal gas property (Nagata et al., 1977—1978). There are no volume or enthalpy changes on the formation of an ideal mixture from its pure components at the certain temperature and pressure.

$$\Delta V_{mix}^{im}(T,p,x_1,\cdots) = \sum_{i=1}^{C} n_i \left[\bar{V}_i^{im}(T,p,x_1,\cdots) - \underline{V}_i(T,p) \right] = 0 \tag{4-118}$$

$$\Delta H_{mix}^{im}(T,p,x_1,\cdots) = \sum_{i=1}^{C} n_i \left[\bar{H}_i^{im}(T,p,x_1,\cdots) - \underline{H}_i(T,p) \right] = 0 \tag{4-119}$$

Therefore, there is no change in volume or enthalpy when an ideal mixture is formed from pure components at certain temperature and pressure. Since $\bar{V}_i^{im} = \underline{V}_i$ at all temperatures, pressures and compositions,

$$\bar{f}_i^{im}(T,p,\underline{x}) = x_i f_i(T,p)$$

The following equation can be easily obtained:

$$\bar{U}_i^{im}(T,p,x_1,\cdots) = \underline{U}_i(T,p)$$

Since $\bar{V}_i^{im}(T,p,x_1,\cdots) = \underline{V}_i(T,p)$ are suitable to all temperatures, pressures, and compositions,

$$RT\ln\left[\frac{\hat{f}_i^{im}(T,p,x_1,\cdots)}{x_i f_i(T,p)}\right] = \int_0^p \left(\bar{V}_i^{im} - \underline{V}_i\right)\mathrm{d}p = 0$$

$$\hat{f}_i^{im}(T,p,x_1,\cdots) = x_i f_i(T,p)$$

And so,

$$\varphi_i^{im} = \frac{\hat{f}_i^{im}}{x_i p} = \frac{x_i f_i(T,p)}{x_i p} = \frac{f_i(T,p)}{p} = \varphi_i(T,p)$$

We know that,

$$\bar{G}_i(T,p,x_1,\cdots) - \underline{G}_i(T,p) = RT\ln\left[\frac{\hat{f}_i(T,p,x_1,\cdots)}{f_i(T,p)}\right]$$

$$\bar{G}_i^{im}(T,p,x_1,\cdots) - \underline{G}_i(T,p) = RT\ln\left[\frac{\hat{f}_i^{im}(T,p,x_1,\cdots)}{f_i(T,p)}\right] = RT\ln\left[\frac{x_i f_i(T,p)}{f_i(T,p)}\right] \tag{4-120}$$

$$= RT\ln x_i$$

Also,

Chapter 4　The Thermodynamics of Multicomponent Mixtures

$$\overline{G}_i^{\text{im}}(T,p,x_1,\cdots) = \underline{G}_i(T,p) + RT\ln x_i$$

$$\overline{A}_i^{\text{im}}(T,p,x_1,\cdots) = \underline{A}_i(T,p) + RT\ln x_i$$

$$\overline{S}_i^{\text{im}}(T,p,x_1,\cdots) = \underline{S}_i(T,p) - R\ln x_i \tag{4-121}$$

And,

$$\underline{U}^{\text{im}}(T,p,x_1,\cdots) = \sum x_i \underline{U}_i(T,p)$$

$$\underline{V}^{\text{im}}(T,p,x_1,\cdots) = \sum x_i \underline{V}_i(T,p)$$

$$\underline{H}^{\text{im}}(T,p,x_1,\cdots) = \sum x_i \underline{H}_i(T,p) \tag{4-122}$$

We get,

$$\underline{S}^{\text{im}}(T,p,x_1,\cdots) = \sum x_i \underline{S}_i(T,p) - R\sum x_i \ln x_i$$

$$\underline{G}^{\text{im}}(T,p,x_1,\cdots) = \sum x_i \underline{G}_i(T,p) + RT\sum x_i \ln x_i$$

$$\underline{A}^{\text{im}}(T,p,x_1,\cdots) = \sum x_i \underline{A}_i(T,p) + RT\sum x_i \ln x_i \tag{4-123}$$

An ideal mixture identically satisfies the Gibbs-Duhem equation.

$$\underline{S}^{\text{im}}dT - \underline{V}^{\text{im}}dp + \sum_{i=1}^{C} x_i d\overline{G}_i^{\text{im}}$$

$$= \sum_{i=1}^{C} x_i \overline{S}_i^{\text{im}} dT - \sum_{i=1}^{C} x_i \overline{V}_i^{\text{im}} dp + \sum_{i=1}^{C} x_i d\left[\underline{G}_i(T,p) + RT\ln x_i\right]$$

$$= \sum_{i=1}^{C} x_i \underline{S}_i dT - R\sum_{i=1}^{C} x_i \ln x_i dT - \sum_{i=1}^{C} x_i \underline{V}_i dp \tag{4-124}$$

$$+ \sum_{i=1}^{C} x_i d\underline{G}_i + R\sum_{i=1}^{C} x_i \ln x_i dT + RT\sum_{i=1}^{C} x_i d\ln x_i$$

$$= \sum_{i=1}^{C} x_i \left(\underline{S}_i dT - \underline{V}_i dp + d\underline{G}_i\right) + RT\sum_{i=1}^{C} dx_i$$

$$\underline{S}_i dT - \underline{V}_i dp + \underline{G}_i = 0$$

$$\sum_{i=1}^{C} dx_i = d\sum_{i=1}^{C} x_i = d(1) = 0 \tag{4-125}$$

So,

$$\underline{S}^{\text{im}}dT - \underline{V}^{\text{im}}dp + \sum_{i=1}^{C} x_i d\overline{G}_i^{\text{im}} = \sum_{i=1}^{C} x_i \left(\underline{S}_i dT - \underline{V}_i dp + d\underline{G}_i\right) + RT\sum_{i=1}^{C} dx_i = 0$$

Any change in the properties of mixture $\Delta \underline{M}_{\text{mix}}$ can be expressed by the properties of the ideal mixture plus the following additional terms:

$$\Delta \underline{M}_{\text{mix}}(T,p,x_1,\cdots)$$
$$= \Delta \underline{M}_{\text{mix}}^{\text{im}}(T,p,x_1,\cdots) + \left[\Delta \underline{M}_{\text{mix}}(T,p,x_1,\cdots) - \Delta \underline{M}_{\text{mix}}^{\text{im}}(T,p,x_1,\cdots)\right] \tag{4-126}$$
$$= \Delta \underline{M}_{\text{mix}}^{\text{im}}(T,p,x_1,\cdots) + \Delta \underline{M}_{\text{mix}}^{\text{E}}$$

$$\Delta \underline{M}_{\text{mix}}^{\text{E}} = \Delta \underline{M}(T,p,x_1,\cdots) - \Delta \underline{M}_{\text{mix}}^{\text{im}}(T,p,x_1,\cdots) = \sum_{i=1}^{C} x_i \bar{M}_i - \sum_{i=1}^{C} x_i \bar{M}_i^{\text{im}} = \underline{M}^{\text{E}}$$
$$= \sum_{i=1}^{C} x_i \left(\bar{M}_i - \bar{M}_i^{\text{im}} \right) \tag{4-127}$$

$\Delta \underline{M}_{\text{mix}}^{\text{E}}$ is the excess mixing property, the change in $\underline{M}$ that occurs on mixing in addition to that would occur if an ideal mixture were formed at certain temperature and pressure.

Excess properties $\Delta \underline{M}^{\text{E}}$ are generally complicated, nonlinear functions of the composition, temperature, and pressure are usually must be obtained from experiment. However, the excess properties are to be small (particularly when $\underline{M}$ is the Gibbs free energy or entropy) compared with $\Delta \underline{M}^{\text{im}}$ so that even an approximate theory for $\Delta \underline{M}^{\text{E}}$ may be sufficient to calculated $\Delta \underline{M}$ with reasonable accuracy.

Partial molar excess quantity

$$\bar{M}_i^{\text{E}} = \left[\frac{\partial (n \Delta \underline{M}^{\text{E}})}{\partial n_i} \right]_{T,p,n_{j \neq i}} = \left\{ \frac{\partial \sum_{k=1}^{C} \left[n_k \left(\bar{M}_k - \bar{M}_k^{\text{im}} \right) \right]}{\partial n_i} \right\}_{T,p,n_{j \neq i}} \tag{4-128}$$

$$\bar{M}_i^{\text{E}} = \bar{M}_i - \bar{M}_i^{\text{im}} \tag{4-129}$$

$\bar{M}_i$ is equal to the partial molar Gibbs free energy,

$$\bar{G}_i^{\text{E}} = \left(\bar{G}_i - \bar{G}_i^{\text{im}} \right) = \left(\bar{G}_i - \bar{G}_i^{\text{igm}} \right) + \left(\bar{G}_i^{\text{igm}} - \bar{G}_i^{\text{im}} \right)$$
$$= \left(\bar{G}_i - \bar{G}_i^{\text{igm}} \right) + \left(\underline{G}_i^{\text{ig}} + RT\ln x_i - \underline{G}_i - RT\ln x_i \right)$$
$$= \left(\bar{G}_i - \bar{G}_i^{\text{igm}} \right) + \left(\underline{G}_i^{\text{ig}} - \underline{G}_i \right) = RT\ln \left(\frac{\hat{f}_i}{x_i f_i} \right) \tag{4-130}$$
$$= RT\ln \left(\frac{\hat{\varphi}_i}{\varphi_i} \right) = \int_0^p \left(\bar{V}_i - \underline{V}_i \right) dp$$

If equations of state (or certain volume data) are available for both the pure fluid and the mixture, the integral can be evaluated. However, for liquid mixtures that cannot be described by the equation of state, common thermodynamic method is to define an activity coefficient $\gamma_i(T,p,x_1,\cdots)$, which is a function of temperature, pressure, and composition.

$$\hat{f}_i^{\text{L}}(T,p,x_1,\cdots) = x_i \gamma_i(T,p,x_1,\cdots) f_i^{\text{L}}(T,p) \tag{4-131}$$

where the superscript L represents the liquid phase.

$$\bar{G}_i^{\text{E}} = RT\ln \left(\frac{\hat{f}_i}{x_i f_i} \right) = RT\ln \gamma_i(T,p,x_1,\cdots) = \left(\frac{\partial n \underline{G}^{\text{E}}}{\partial n_i} \right)_{T,p,n_{j \neq i}} \tag{4-132}$$

The calculation of the activity coefficient for each component in mixture is an important step in phase equilibrium calculations.

EXAMPLE 4-4

The expression of excess Gibbs function of liquid argon (1) - methane (2) system is $\dfrac{G^E}{RT} = x_1 x_2 \left[A + B(1-2x_1) \right]$. The coefficients A and B are as follows:

T/K	A	B
109.0	0.3036	−0.0169
112.0	0.2944	0.0118
115.74	0.2804	0.0546

Calculate (a) Mixing heat of equimolar mixture; (b) Excess line.

SOLUTION

(a)

$$\frac{H^E}{R} = -\left[\frac{\partial (G^E/RT)}{\partial T} \right]_{P,\{x\}} = -x_1 x_2 \frac{\mathrm{d}A}{\mathrm{d}T} + (1-2x_1)\frac{\mathrm{d}B}{\mathrm{d}T}$$

Among them,

$$\frac{\mathrm{d}A}{\mathrm{d}T} \approx \frac{A_3 - A_1}{T_3 - T_1} = \frac{0.2804 - 0.3036}{115.74 - 109.0} = -0.00344$$

$$\frac{\mathrm{d}B}{\mathrm{d}T} \approx \frac{B_3 - B_1}{T_3 - T_1} = \frac{0.0546 + 0.0169}{115.74 - 109.0} = 0.0106$$

We obtain

$$\frac{H^E}{R} \approx 0.00344 x_1 x_2 + 0.0106(1-2x_1)$$

(b)

$$H^E = G^E + TS^E$$

$$\frac{H^E}{RT} = \frac{G^E}{RT} + \frac{S^E}{R} \Rightarrow \frac{S^E}{R} = -x_1 x_2 \left[A + B(1-2x_1) \right] + \left[-x_1 x_2 \frac{\mathrm{d}A}{\mathrm{d}T} + (1-2x_1)\frac{\mathrm{d}B}{\mathrm{d}T} \right]$$

$$= 0.0236 x_1^2 x_2 - 0.310 x_1 x_2 - 0.0212 x_1 + 0.0106$$

4.6.3 Partial Molar Gibbs Free Energy and Fugacity

In the previous chapter, the equation of state is defined as the mathematical relationship between p, V and T. If any two are specified, the third property can be calculated. The vapor-liquid equilibrium in pure fluid is determined by the equation of state with Gibbs free energy. We start from,

$$\underline{G}^{\mathrm{L}}(T,p) = \underline{G}^{\mathrm{V}}(T,p)$$

$$\mathrm{d}\underline{G} = -\underline{S}\mathrm{d}T + \underline{V}\mathrm{d}p$$

$$\left(\frac{\partial \underline{G}}{\partial T}\right)_p = -\underline{S}$$

$$\left(\frac{\partial \underline{G}}{\partial p}\right)_T = \underline{V}$$

Integration between any two pressures p_1 and p_2 (at certain temperature) yields

$$\underline{G}(T_1, p_2) - \underline{G}(T_1, p_1) = \int_{p_1}^{p_2} \underline{V}\mathrm{d}p \tag{4-133}$$

If the fluid under consideration can be regards as an ideal gas

$$\underline{G}^{\mathrm{ig}}(T_1, p_2) - \underline{G}^{\mathrm{ig}}(T_1, p_1) = \int_{p_1}^{p_2} \frac{RT}{p}\mathrm{d}p \tag{4-134}$$

We obtain

$$\left[\underline{G}(T_1, p_2) - \underline{G}^{\mathrm{ig}}(T_1, p_2)\right] - \left[\underline{G}(T_1, p_1) - \underline{G}^{\mathrm{ig}}(T_1, p_1)\right] = \int_{p_1}^{p_2}\left(\underline{V} - \frac{RT}{p}\right)\mathrm{d}p \tag{4-135}$$

We set p_1 to zero, all fluids are ideal gases when $p = 0$. The fugacity coefficient can be obtained from the following equation:

$$\varphi = \frac{f}{p} = \mathrm{e}^{\frac{\underline{G}(T,p) - \underline{G}^{\mathrm{ig}}(T,p)}{RT}} = \mathrm{e}^{\frac{\int_0^p \left(\underline{V} - \frac{RT}{p}\right)\mathrm{d}p}{RT}} \tag{4-136}$$

And when $f \to p$ as $f \to 0$. It can be seen that fugacity is the dimension of pressure. The fugacity is equal to the pressure at pressures low enough that the fluid approaches the ideal gas state. Gibbs function is one of the most important thermodynamic functions (Wark et al., 1995). In order to calculate the Gibbs function of components conveniently, fugacity is introduced.

It can be proved that

$$RT\left(\frac{\partial \ln f}{\partial p}\right)_T = \underline{V} = \left(\frac{\partial \underline{G}}{\partial p}\right)_T$$

$$\left(\frac{\partial \ln \varphi}{\partial T}\right)_p = -\frac{\underline{H}(T,p) - \left[\underline{H}^{\mathrm{ig}}(T,p)\right]}{RT^2}$$

For mixture,

$$\mathrm{d}G = -S\mathrm{d}T + V\mathrm{d}p + \sum_{i=1}^{C} \overline{G}_i \mathrm{d}n_i \tag{4-137}$$

The exchange property of the second derivative of Gibbs free energy is written as:

$$\left.\frac{\partial}{\partial n_i}\right|_{T,p,n_{j\neq i}}\left(\frac{\partial G}{\partial T}\right)_{p,n_j} = \left.\frac{\partial}{\partial T}\right|_{p,n_j}\left(\frac{\partial G}{\partial n_i}\right)_{T,p,n_{j\neq i}} \tag{4-138}$$

Chapter 4 The Thermodynamics of Multicomponent Mixtures

$$\left.\frac{\partial}{\partial n_i}\right|_{T,p,n_{j\neq i}}(-S)=\left.\frac{\partial}{\partial T}\right|_{p,n_j}(\bar{G}_i)$$

$$\bar{S}_i=-\left(\frac{\partial \bar{G}_i}{\partial T}\right)_{p,n_j}$$

Similarly

$$\left.\frac{\partial}{\partial n_i}\right|_{T,p,n_{j\neq i}}\left(\frac{\partial G}{\partial p}\right)_{T,n_j}=\left.\frac{\partial}{\partial p}\right|_{T,n_j}\left(\frac{\partial G}{\partial n_i}\right)_{T,p,n_{j\neq i}} \tag{4-139}$$

Similarly

$$\left.\frac{\partial}{\partial n_i}\right|_{T,p,n_{j\neq i}}(V)=\left.\frac{\partial}{\partial p}\right|_{T,n_j}(\bar{G}_i)$$

$$\bar{V}_i=\left(\frac{\partial \bar{G}_i}{\partial p}\right)_{T,n_j} \tag{4-140}$$

$$\bar{G}_i(T_1,p_2,x)-\bar{G}_i(T_1,p_1,x)=\int_{p_1}^{p_2}\bar{V}_i\,dp \tag{4-141}$$

The fugacity of component i in a mixture is defined as follows:

$$\begin{aligned}\hat{f}_i(T,p,x_1,\cdots)&=x_ip\mathrm{e}^{\frac{\bar{G}_i(T,p,x_1,\cdots)-\bar{G}_i^{\mathrm{igm}}(T,p,x_1,\cdots)}{RT}}\\ &=x_ip\mathrm{e}^{\frac{\int_0^p(\bar{V}_i-\bar{V}_i^{\mathrm{ig}})dp}{RT}} \\ &=p\mathrm{e}^{\frac{\bar{G}_i(T,p,x_1,\cdots)-\underline{G}_i^{\mathrm{ig}}(T,p)}{RT}}\end{aligned} \tag{4-142}$$

The fugacity of component i in a mixture, $\hat{f}_i$ is not a partial molar fugacity. It is easily shown that $\hat{f}_i(T,p,x_1,\cdots)\to x_ip=p_i$ ($p\to 0$).

Also, the fugacity coefficient for a component in a mixture is defined as

$$\begin{aligned}\hat{\varphi}_i&=\frac{\hat{f}_i}{x_ip}\\ &=\mathrm{e}^{\frac{\bar{G}_i(T,p,x_1,\cdots)-\bar{G}_i^{\mathrm{igm}}(T,p,x_1,\cdots)}{RT}}\\ &=\mathrm{e}^{\frac{\int_0^p(\bar{V}_i-\bar{V}_i^{\mathrm{ig}})dp}{RT}}\end{aligned} \tag{4-143}$$

Similarly, the multi-component of $RT\left(\dfrac{\partial \ln f}{\partial p}\right)_T=\underline{V}=\left(\dfrac{\partial \underline{G}}{\partial p}\right)_T$ obtained by differentiating $\ln \hat{f}_i$ about pressure at a certain temperature and composition, is

$$RT\left(\frac{\partial \ln \hat{f}_i}{\partial p}\right)_{T,x_1,\cdots}=\bar{V}_i=\left(\frac{\partial \bar{G}_i}{\partial p}\right)_{T,x_1,\cdots} \tag{4-144}$$

To relate the fugacity of pure component i to the fugacity of component i in a mixture,

$$RT\left(\frac{\partial \ln f_i}{\partial p}\right)_T = \underline{V}_i \tag{4-145}$$

We can demonstrate that

$$RT\left(\frac{\partial \ln \hat{f}_i}{\partial p}\right)_{T,x_1,\cdots} = \overline{V}_i \tag{4-146}$$

So,

$$RT\left(\frac{\partial \ln \hat{f}_i}{\partial p}\right)_{T,x_1,\cdots} - RT\left(\frac{\partial \ln f_i}{\partial p}\right)_T = \overline{V}_i - \underline{V}_i \tag{4-147}$$

Integrate between $p = 0$ and random p, so:

$$RT\ln\left[\frac{\hat{f}_i(T,p,x_1,\cdots)}{\hat{f}_i(T,p\to 0,x_1,\cdots)}\right] - RT\ln\left[\frac{f_i(T,p)}{f_i(T,p\to 0)}\right] = \int_0^p (\overline{V}_i - \underline{V}_i)\mathrm{d}p \tag{4-148}$$

$p \to 0$, $\hat{f}_i(T,p,x_1,\cdots) \to x_i p$, $f \to p$

$$RT\ln\left[\frac{\hat{f}_i(T,p,x_1,\cdots)}{x_i p}\right] - RT\ln\left[\frac{f_i(T,p)}{p}\right] = \int_0^p (\overline{V}_i - \underline{V}_i)\mathrm{d}p \tag{4-149}$$

$$RT\ln\left[\frac{\hat{f}_i(T,p,x_1,\cdots)}{x_i f_i(T,p)}\right] = \int_0^p (\overline{V}_i - \underline{V}_i)\mathrm{d}p \tag{4-150}$$

$\hat{f}_i$ and f_i are related through the integral pressures of the difference between the component partial molar and pure component molar volumes.

The relationship between fugacity $\hat{f}_i$ and temperature is shown in the following equation:

$$\hat{f}_i(T,p,x_1,\cdots) = x_i p e^{\frac{\overline{G}_i(T,p,x_1,\cdots) - \overline{G}_i^{\mathrm{igm}}(T,p,x_1,\cdots)}{RT}} \tag{4-151}$$

$$\ln\frac{\hat{f}_i(T,p,x_1,\cdots)}{x_i p} = \frac{\overline{G}_i(T,p,x_1,\cdots) - \overline{G}_i^{\mathrm{igm}}(T,p,x_1,\cdots)}{RT} \tag{4-152}$$

$$\left[\frac{\partial \ln(\hat{f}_i/x_i p)}{\partial T}\right]_{p,\underline{x}} = -\frac{\overline{G}_i - \overline{G}_i^{\mathrm{igm}}}{RT^2} + \frac{1}{RT}\left[\frac{\partial(\overline{G}_i - \overline{G}_i^{\mathrm{igm}})}{\partial T}\right]_{p,\underline{x}} \tag{4-153}$$

$$\left[\frac{\partial \ln \hat{\varphi}_i}{\partial T}\right]_{p,\underline{x}} = -\frac{\overline{G}_i - \overline{G}_i^{\mathrm{igm}}}{RT^2} + \frac{1}{RT}\left[\frac{\partial(\overline{G}_i - \overline{G}_i^{\mathrm{igm}})}{\partial T}\right]_{p,\underline{x}} \tag{4-154}$$

$$\left[\frac{\partial \ln \hat{\varphi}_i}{\partial T}\right]_{p,\underline{x}} = -\frac{(\overline{G}_i - \overline{G}_i^{\mathrm{igm}}) - T\left[\frac{\partial(\overline{G}_i - \overline{G}_i^{\mathrm{igm}})}{\partial T}\right]_{p,\underline{x}}}{RT^2} \tag{4-155}$$

$$\overline{S}_i = -\left(\frac{\partial \overline{G}_i}{\partial T}\right)_{p,\underline{x}} \tag{4-156}$$

$$\overline{S}_i^{igm} = -\left(\frac{\partial \overline{G}_i^{igm}}{\partial T}\right)_{p,\underline{x}} \tag{4-157}$$

$$\begin{aligned}\left[\frac{\partial \ln \hat{\varphi}_i}{\partial T}\right]_{p,\underline{x}} &= -\frac{\left(\overline{G}_i - \overline{G}_i^{igm}\right) + T\left(\overline{S}_i - \overline{S}_i^{igm}\right)}{RT^2} \\ &= -\frac{\left(\overline{G}_i + T\overline{S}_i\right) - \left(\overline{G}_i^{igm} + T\overline{S}_i^{igm}\right)}{RT^2} \\ &= -\frac{\left(\overline{H}_i - \overline{H}_i^{igm}\right)}{RT^2}\end{aligned} \tag{4-158}$$

The change of partial molar Gibbs free energy of components can be expressed between the two states with the certain temperature and pressure, but not

$$\Delta \overline{G}_i = \overline{G}_i\left(T, p, x_1^{II}, \cdots\right) - \overline{G}_i\left(T, p, x_1^{I}, \cdots\right) \tag{4-159}$$

where the superscripts I and II denote phases of composition.

$$\hat{f}_i(T, p, x_1, \cdots) = x_i p e^{\frac{\overline{G}_i(T,p,x_1,\cdots) - \overline{G}_i^{igm}(T,p,x_1,\cdots)}{RT}} \tag{4-160}$$

$$\begin{aligned}\overline{G}_i(T, p, x_1, \cdots) &= \overline{G}_i^{igm}(T, p, x_1, \cdots) + RT\ln\left[\frac{\hat{f}_i(T, p, x_1, \cdots)}{x_i p}\right] \\ &= \overline{G}_i^{igm}(T, p, x_1, \cdots) + RT\ln\hat{\varphi}_i(T, p, x_1, \cdots)\end{aligned} \tag{4-161}$$

$$\overline{G}_i^{igm}(T, p, x_1, \cdots) = \underline{G}_i^{ig}(T, p) + RT\ln x_i \tag{4-162}$$

$$\begin{aligned}\Delta \overline{G}_i &= \overline{G}_i\left(T, p, x_1^{II}, \cdots\right) - \overline{G}_i\left(T, p, x_1^{I}, \cdots\right) \\ &= \overline{G}_i^{igm}\left(T, p, x_1^{II}, \cdots\right) + RT\ln\left[\frac{\hat{f}_i\left(T, p, x_1^{II}, \cdots\right)}{x_i^{II} p}\right] \\ &\quad - \overline{G}_i^{igm}\left(T, p, x_1^{I}, \cdots\right) - RT\ln\left[\frac{\hat{f}_i\left(T, p, x_1^{I}, \cdots\right)}{x_i^{I} p}\right] \\ &= \underline{G}_i^{ig}(T, p) + RT\ln x_i^{II} + RT\ln\left[\frac{\hat{f}_i\left(T, p, x_1^{II}, \cdots\right)}{x_i^{II} p}\right] \\ &\quad - \underline{G}_i^{ig}(T, p) - RT\ln x_i^{I} - RT\ln\left[\frac{\hat{f}_i\left(T, p, x_1^{I}, \cdots\right)}{x_i^{I} p}\right]\end{aligned} \tag{4-163}$$

$$\Delta \bar{G}_i = \bar{G}_i\left(T,p,x_1^{II},\cdots\right) - \bar{G}_i\left(T,p,x_1^{I},\cdots\right)$$
$$= RT\ln\left[\frac{\hat{f}_i\left(T,p,x_1^{II},\cdots\right)}{\hat{f}_i\left(T,p,x_1^{I},\cdots\right)}\right] = RT\ln\left[\frac{x_i^{II}\hat{\varphi}_i\left(T,p,x_1^{II},\cdots\right)}{x_i^{I}\hat{\varphi}_i\left(T,p,x_1^{I},\cdots\right)}\right] \quad (4\text{-}164)$$

A special case of the equation is when I is a pure component:

$$\bar{G}_i(T,p,x_1,\cdots) - \underline{G}_i(T,p) = RT\ln\left[\frac{\hat{f}_i(T,p,x_1,\cdots)}{f_i(T,p)}\right]$$
$$= RT\ln\left[\frac{x_i\hat{\varphi}_i(T,p,x_1,\cdots)}{\varphi_i(T,p)}\right] \quad (4\text{-}165)$$

The total fugacity of the mixture f_m is defined by the following equation

$$f_m = pe^{\frac{G(T,p,x_1,\cdots) - G^{igm}(T,p,x_1,\cdots)}{RT}} \quad (4\text{-}166)$$

It can be calculated that

$$\ln\hat{\varphi}_i(T,p,x_1,\cdots) = \ln\left[\frac{\hat{f}_i(T,p,x_1,\cdots)}{x_i p}\right]$$
$$= \frac{1}{RT}\int_0^p \left(\bar{V}_i - \underline{V}^{ig}\right)dp$$
$$= \frac{1}{RT}\int_{\underline{V}=\infty}^{\underline{V}=ZRT/p}\left[\frac{RT}{\underline{V}} - n\left(\frac{\partial p}{\partial n_i}\right)_{T,V,n_{j\neq i}}\right]d\underline{V} - \ln Z \quad (4\text{-}167)$$

For a pure fluid,

$$\ln\varphi = \ln\left[\frac{f(T,p)}{p}\right] = \frac{1}{RT}\int_{\underline{V}=\infty}^{\underline{V}=\frac{ZRT}{p}}\left(\frac{RT}{\underline{V}} - p\right)d\underline{V} - \ln Z + (Z-1) \quad (4\text{-}168)$$

The fugacity function is introduced here because its relationship with Gibbs free energy is convenient to deal with phase equilibrium.

4.7 Application of Aspen Plus to Thermodynamic Properties of multicomponent Mixtures

The laboratory needs to prepare antifreeze containing 30% (mol%) methanol (1) + 70% H_2O (2). Determine the mixed volume of methanol and water at 25℃. Partial molar volumes of methanol and water at 25℃, 30% (mol%) methanol are known: $\bar{V}_1 = 38.632 \text{ cm}^3/\text{mol}$, $\bar{V}_2 = 17.765 \text{ cm}^3/\text{mol}$.

SOLUTION

$$V = \sum x_i \overline{V}_i = x_1 \overline{V}_1 + x_2 \overline{V}_2 = (0.3 \times 38.632 + 0.7 \times 17.765) \text{cm}^3/\text{mol} = 24.025 \text{ cm}^3/\text{mol}$$

SIMULATION

Input water and methanol into Aspen respectively.

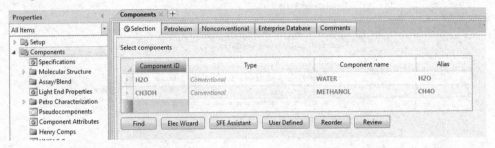

Select NRTL physical parameters.

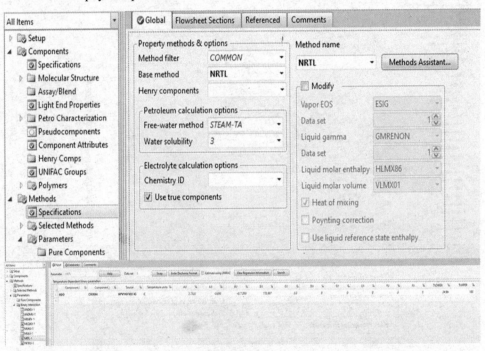

Make 7 kmol/h water and 3 kmol/h methanol enter the mixer at the same time.

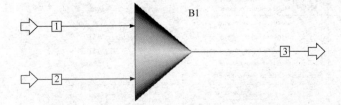

Results.
Mix volume.

```
<< All Blocks Reinitialized >>

<< Run reinitialized 17:44:13 Wed Dec 8, 2021>>

->Processing input specifications ...

  Flowsheet Analysis :

  COMPUTATION ORDER FOR THE FLOWSHEET:
  B1

->Calculations begin ...

    Block: B1         Model: MIXER
->Simulation calculations completed ...


    *** No Warnings were issued during Input Translation ***

    *** No Errors or Warnings were issued during Simulation ***

->Generating results ...
```

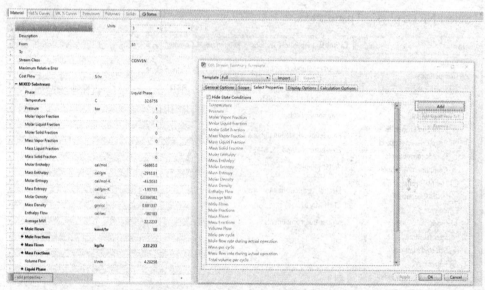

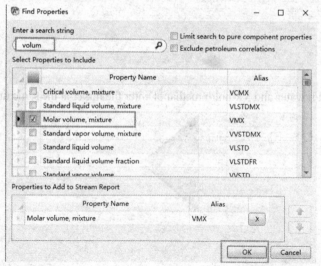

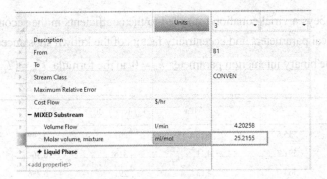

CONCLUSION

This chapter mainly discusses the content of multi-component mixture thermodynamics and summarizes the development of the theory of multi-component thermodynamics. The main purpose is to apply the basic theory of thermodynamics to solve problems in the chemical industry rationally, which involves many interesting phenomena. Partial molar Gibbs free energy is a thermodynamic function in equilibrium calculation. The methods are used to estimate partial molar Gibbs free energy in gas, liquid and solid mixtures. Thermodynamics has many applications in explaining and predicting the great diversity of physical, chemical, and biochemical equilibria that occur in mixture.

EXERCISES

1. A laboratory needs to prepare 3×10^{-3} m³ aqueous solution containing 20% (mass fraction) methanol as antifreeze. Try to figure out what volume of methanol at 20℃ is needed to mix with water. The partial molar volume of 20% (mass fraction) methanol solution at 20℃, $\overline{V_1} = 37.8$ cm³/mol, $\overline{V_2} = 18$ cm³/mol ; At 20℃, the volume of pure methanol $V_1 = 40.46$ cm³/mol, and that of pure water $V_2 = 18.04$ cm³/mol.

2. The enthalpy of a two-component liquid mixture at 298 K and 1.0133×10^5 Pa can be expressed by the following formula:

$$H = 100x_1 + 150x_2 + x_1x_2(10x_1 + 5x_2)$$

In the formula, the unit of H is J/mol. Try to determine under this temperature and pressure state

(a) $\overline{H_1}$ and $\overline{H_2}$ represented by x_1;

(b) The value of pure component enthalpy H_1 and H_2;

(c) $\overline{H_1}^\infty$ and $\overline{H_2}^\infty$ for infinite dilution of liquid.

3. The evaporation coefficients of methyl ethyl ketone (1) and toluene (2) in the isomolar mixture of methyl ethyl ketone (1) and toluene (2) were calculated at 323 K and 25 kPa. Suppose

the gas mixture obeys a virial equation truncated to the coefficients in the second dimension. The values of the critical parameters and eccentricity factors of the known substances are shown in the table below, let the binary interaction parameter $k_{ij}=0$ in the formula $T_{cij}=\left(T_{ci}T_{cj}\right)^{1/2}\left(1-k_{ij}\right)$.

ij	T_{cij} /K	p_{cij} /MPa	V_{cij} /(cm³/mol)	Z_{cij}	ω_{ij}
11	535.6	4.15	267	0.249	0.293
22	591.7	4.11	316	0.364	0.257
12	563.0	4.13	291	0.256	0.293

4. The fugacity of component 1 in binary solution at 39℃ and 2 MPa is

$$\hat{f}_1 = 6x_1 - 9x_1^2 + 4x_1^3$$

Where x_1 below is the mole fraction of component 1; The unit $\hat{f}_1$ is MPa. Try it at the above temperature and pressure:

(1) Fugacity and fugacity coefficient of pure component 1;
(2) Henry's coefficient k_1 for component 1;
(3) The relation between activity coefficient γ_1 and x_1 (the standard state of component 1 is based on lewis-Randall's rule).

5. At $303\text{ K}, 1.013\times10^5\text{ Pa}$, the volume data for the liquid mixtures of benzene (1) and cyclohexane (2) can be expressed by quadratic equation $V = \left(109.4 - 16.8x_1 - 2.64x_1^2\right)\times 10^{-3}\text{ cm}^3/\text{mol}$.

Where x_1 is the mole fraction of benzene, and the unit of V is m³/kmol. Try to find the expressions of $\bar{V}_1$, $\bar{V}_2$ and ΔV under $303\text{ K}, 1.013\times 10^6\text{ Pa}$ (the standard state is based on the Lewis-Randall rule).

6. The definition equation of heat capacity at constant pressure is known as $C_p = (\partial H/\partial T)_p$, Try to prove that $\bar{C}_{p,i} = \left(\dfrac{\partial \bar{H}_i}{\partial T}\right)_{p,x}$.

7. At 25℃ and 0.1 MPa, the measured partial molar volume of component 2 of the methanol (1)+water (2) binary system is approximately

$$\bar{V}_2 = 18.1 - 3.2x_1^2 \text{ cm}^3/\text{mol}$$

Knowing that the molar volume of pure methanol is $V_1 = 40.7\text{ cm}^3/\text{mol}$, try to find the partial molar volume of methanol and the molar volume of the mixture under this condition.

8. Knowing the average molar volume $V_{C_4H_{10}} = 0.119\times 10^{-3}\text{ m}^3/\text{mol}$ and saturated vapor pressure $p^S_{C_4H_{10}} = 1.574\times 10^6\text{ Pa}$ of 360.96 K liquid isobutane, Calculate the fugacity of liquid isobutane at 361.96 K, $1.02\times 10^7\text{ Pa}$. Critical parameters and eccentricity factor of isobutene: $T_c = 408.1\, K$, $p_c = 408.1\, K$, $\omega = 0.176$.

9. Try to deduce the relationship between $\dfrac{\Delta G^{id}}{RT}$ of the ideal solution and the composition x_i

from the Lewis-Randall rule $\hat{f}_i^{id} = f_i x_i$.

10. It is known that the activity coefficients of infinite dilution of the acetone(1)+benzene(2) binary system at 45℃ are $\gamma_1^\infty = 1.65$ and $\gamma_2^\infty = 1.52$, respectively. Assuming that the solution obeys the Margules equation, try to find its Margules equation and activity coefficient.

REFERENCES

[1] Abello L. Excess heats of binary systems containing benzene hydrocarbons and chloroform or methylchloroform. I. Experimental results[J]. J. Chim. Phys. Phys.-Chim. Biol, 1973, 70: 1355.

[2] Albuquerque L, Ventura C, Goncalves R. Refractive indices, densities, and excess properties for binary mixtures containing methanol, ethanol, 1,2-ethanediol, and 2-methoxyethanol[J]. Journal of Chemical & Engineering Data, 1996, 41(4): 685-688.

[3] Arce A, Blanco A, Soto A, et al. Densities, refractive indices, and excess molar volumes of the ternary systems water + methanol + 1-octanol and water + ethanol + 1-octanol and their binary mixtures at 298.15 K[J]. Journal of Chemical & Engineering Data, 1993, 38(2): 336-340.

[4] Arenosa R L, Menduina C, Tardajos G, et al. Excess enthalpies at 298.15 K of binary mixtures of cyclohexane with n-alkanes[J]. The Journal of Chemical Thermodynamics, 1979, 11(2): 159-166.

[5] Bangqing, Ni, Liyan, et al. Thermodynamic properties of the binary mixtures of 1,2-dichloroethane with chlorobenzene and bromobenzene from (298.15 to 313.15)K[J]. Journal of Chemical & Engineering Data, 2010, 55(10): 4541-4545.

[6] Bejan A. Advanced engineering thermodynamics[M]. 3rd ed. New York: Wiley Interscience, 2006.

[7] Christensen C, Gmehling J, Holderbaum T. Heats of mixing data collection[M]. DECHEMA, Deutsche Gesellschaft für Chemisches Apparatewesen, Distributed exclusively by Scholium International, 1984.

[8] Fumio, Kimura, George C, Benson. Excess enthalpies of binary mixtures of 2-methyl-1-pentanol with hexane isomers[J]. Fluid Phase Equilibria, 1982, 8(1): 107-112.

[9] Nagata I, Asano H, Fujiwara K. Excess enthalphies for systems of 2-propanol-benzene-methylcyclohexane[J]. Fluid Phase Equilibria, 1977, 1(3): 211-217.

[10] Smith J M, Ness V, Abbott M M. Introduction to chemical engineering thermodynamics. McGraw-Hill, 2005.

[11] Smith V C, Robinson R L. Vapor-liquid equilibriums at 25.deg. in the binary mixtures formed by hexane, benzene, and ethanol[J]. Journal of Chemical & Engineering Data, 2002, 15(3): 391-395.

[12] Treszczanowicz A J, Pawlowski T S, Treszczanowicz T, et al. Excess volume of the 1propanol + 1alkene systems in terms of an equation of state with association[J]. Journal of Chemical & Engineering Data, 2010, 55(12): 5478-5482.

[13] Wark, K, Jr. Advanced thermodynamics for engineers[M]. New York: McGraw-Hill, 1995.

[14] Williamson A G, Scott R L. Heats of mixing of non-electrolyte solutions. I. Ethanol + benzene and methanol + benzene[J]. The Journal of Physical Chemistry, 1960, 64(4): 440-442.

Chapter 5
Phase Equilibrium

Learning Objectives

- Calculate the fugacity of a pure compound at any T and p using an equation of state.
- Use fugacity to estimate vapor pressure.
- Use the Poynting correction to estimate the fugacity of liquids and solids at elevated pressures.
- It should be noted that the liquid and vapor phases of a mixture in equilibrium almost never have the same composition (when they have the same composition, it is called azeotrope).
- Read the mixed data sheet and phase plot to understand the difference between the vapor-liquid equilibrium and the T-xy and p-xy plots.
- Know when the simplest model of vapor-liquid equilibrium of mixtures, namely Raoul law, will work and when it will not work.
- It is easier to determine when to separate the mixture of two substances by distillation method and when it is difficult to separate.

5.1 Phase Equilibrium for a Single-Component System

Thermodynamic properties of fluids are divided into two categories: directly measurable and not directly measurable. Thermodynamic properties that can be directly measured are: temperature, volume, pressure, heat capacity, enthalpy, etc. Thermodynamic properties that cannot be directly measured are: internal energy, Helmholtz free energy A, entropy, Gibbs free energy G, etc. The unmeasurable thermodynamic properties can be calculated by establishing a relationship with the measurable thermodynamic properties, which is the greatness and cleverness of thermodynamics. It is very important to establish the functional relationship between these two kinds of thermodynamic properties.

5.1.1 Mathematical Models of Phase Equilibrium

In this chapter, some useful methods for predicting or modeling the phase transition conditions of pure compounds are investigated.

Mathematical Expression of the Equilibrium Criterion

In this section, the mathematical model of phase equilibrium is established. In practice, fugacity is an important physical quantity to solve specific problems, whether it is pure compounds in this chapter or subsequent mixtures. Gibbs free energy lays the foundation for the equilibrium model. Fugacity is derived from Gibbs free energy according to the definition introduced in next section.

As shown in Fig.5-1, the liquid and vapor phase of water are in equilibrium at $p=1$ atm and $T=100\,°C$. The addition of the heat causes a small amount of liquid to boil.

- The increase in heat makes a small amount of liquid boil. So, the quality of vapor-liquid mixture is improved.
- Since the molar volume of saturated steam is larger than that of saturated liquid, the vapor-liquid mixture expands slightly. The volume of vapor-liquid mixture changes while the temperature and pressure remain unchanged.
- S, U and H in the liquid and vapor phases remain unchanged in this process, which are intensive properties and are not affected by the change of the amount of matter in each phase. Gibbs phase rule shows that there are two degrees of freedom ($F=2$) in a single phase. If the two strong properties (T and p) remain unchanged, the other intensive properties will not change.
- Because the S, U and H of vapor phase are higher than liquid phase, the total entropy, internal energy and enthalpy of the system are increased by the transformation from liquid phase to vapor phase.

The conversion of liquid to vapor increases the total entropy, internal energy, and enthalpy of the system. How about Gibbs free energy?

By the definition of the Gibbs free energy $\underline{G} = \underline{H} - T\underline{S}$.

In the boiling process, the total enthalpy and entropy of the system are increased, and the Gibbs free energy cannot be determined.

Because G is a natural function of pressure and temperature, it is uniquely certain. When the saturated liquid transforms into saturated vapor, V, U, H and S increase, but the pressure and temperature remain unchanged ($dp = 0$, $dT = 0$), so the molar Gibbs free energy of the whole system remains unchanged.

$$d\underline{G} = -\underline{S}dT + \underline{V}dp \qquad (5\text{-}1)$$

$$d\underline{G}^{sys} = 0 \qquad (5\text{-}2)$$

Fig.5-1 Piston-cylinder device holding a steam-water VLE mixture at atmospheric pressure

Saturated liquid and saturated vapor phases must have identical values of molar Gibbs free energy. Because if they are different, the molar Gibbs free energy of the system changes as the liquid boils.

$$\underline{G}^V = \underline{G}^L \tag{5-3}$$

This observation generalizes to any phase equilibrium.

The criterion for phase equilibrium in a pure compound can be expressed as follows.

In vapor-liquid equilibrium, $\underline{G}^V = \underline{G}^L$

In liquid-solid equilibrium, $\underline{G}^L = \underline{G}^S$

In vapor-solid equilibrium, $\underline{G}^V = \underline{G}^S$

The triple point of a compound is the specific temperature and pressure at which $\underline{G}^V = \underline{G}^L = \underline{G}^S$.

What if $\underline{G}^V$, $\underline{G}^L$, and $\underline{G}^S$ are all different from each other at a specific temperature and pressure? Solve this problem by considering a closed system, which keeps constant and uniform temperature and pressure.

$$\frac{dG}{t} = -TS_{gen} \tag{5-4}$$

$\frac{dG}{t}$ is the rate of change of Gibbs free energy in the system. Eq. (5-4) reveals that any process occurring in a closed system at fixed T, p will decrease the Gibbs free energy of the system. Equilibrium will occur when $\underline{G}$ reaches a minimum. The most stable phase is the one that corresponds to the minimum value of $\underline{G}$ at equilibrium.

Chemical Potential

This section introduces a new property that is useful in the analysis of equilibrium. It is known from the previous section that Gibbs free energy is a basic property of the equilibrium model, so it's necessary to quantify the Gibbs free energy. The chemical potential (μ_i) of a compound i is defined as the partial derivative of Gibbs free energy with respect to the number of moles of compound i:

$$\mu_i = \left(\frac{\partial G}{\partial n_i}\right)_{T, p, n_{j \neq i}} = \overline{G}_i \tag{5-5}$$

Chemical potential indicates the change in Gibbs free energy of a system when a small amount of compound i is added, if everything else (temperature, pressure, and the amount of all compounds other than i) is constant.

For a pure compound, the value of $\underline{G}$ has nothing to do with the number of moles

$$\mu_i = \underline{G} \tag{5-6}$$

For mixtures, chemical potential is a function of composition as well as temperature and pressure. The μ_i of mixture is necessary to establish a phase equilibrium model, which is not necessarily equal to $\underline{G}_i$.

The Clapeyron Equation

When a pure compound exists in two phases at equilibrium, the molar Gibbs free energy of each of the phases are identical. Example 5-1 uses this fact to estimate the unknown vapor pressure.

EXAMPLE 5-1

Estimating the vapor pressure of water
For liquid water at $T = 60°C$ the vapor pressure is $p^{sat} = 7.6$ kPa , and $\Delta \underline{H}^{vap} = 46.94$ kJ/mol .
Using only the given information, estimate the vapor pressures of water at $T = 70°C$.

SOLUTION

Step 1 Apply the equilibrium criterion
The vapor pressure corresponds to the boiling point one to one, and the vapor pressure increases with the increase of temperature. Fig. 5-2 is only a rough graph. Given a point on the graph, the position of other points can be estimated based on the known information. Water reaches a vapor-liquid equilibrium at $T = 60°C$ and $p^{sat} = 7.6$ kPa , so $\underline{G}^V = \underline{G}^L$ at this temperature and pressure. Liquid and vapor phase of water in vapor-liquid equilibrium have different values of $\underline{G}^V$ and $\underline{G}^L$ at different temperatures and pressures, but $\underline{G}^V = \underline{G}^L$. And again, for any point on the vapor pressure curve, $\underline{G}^V = \underline{G}^L$.

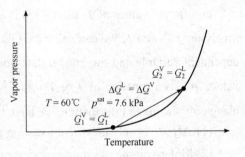

Fig.5-2 Sketch of vapor pressure vs. temperature

As we move from one point to another along the vapor pressure curve, the change in $\underline{G}^V$ must be the same as the change in $\underline{G}^L$.

$$d\underline{G}^L = d\underline{G}^V \qquad (5\text{-}7)$$

To determine the relationship between vapor pressure and temperature, we must relate G to them.

Step 2 Relate Gibbs free energy to pressure and temperature
dG is related to temperature and pressure across fundamental physical relations

$$\underline{V}^L dp^{sat} - \underline{S}^L dT = \underline{V}^V dp^{sat} - \underline{S}^V dT \qquad (5\text{-}8)$$

Rearrange the formula

$$\left(\underline{V}^L - \underline{V}^V\right) dp^{sat} = \left(\underline{S}^L - \underline{S}^V\right) dT \qquad (5\text{-}9)$$

$$\frac{dp^{sat}}{dT} = \frac{\left(\underline{S}^L - \underline{S}^V\right)}{\left(\underline{V}^L - \underline{V}^V\right)} = \frac{\left(\underline{S}^V - \underline{S}^L\right)}{\left(\underline{V}^V - \underline{V}^L\right)} \qquad (5\text{-}10)$$

The slope of the curve ($\frac{dp^{sat}}{dT}$) is obtained from Eq. (5-10), where $\Delta \underline{S}^{vap}$ is unknown while $\Delta \underline{H}^{vap}$ is known.

We can relate them by vapor pressure curves $\underline{G}^V = \underline{G}^L$ and the definitions of $\underline{G} = \underline{H} - T\underline{S}$

$$\underline{G}^L = \underline{H}^L - T\underline{S}^L = \underline{H}^V - T\underline{S}^V = \underline{G}^V \tag{5-11}$$

Rearrange equation Eq. (5-11) in terms of Eq. (5-10)

$$T\left(\underline{S}^V - \underline{S}^L\right) = \underline{H}^V - \underline{H}^L \tag{5-12}$$

Replace molar entropy in Eq. (5-10) with molar enthalpy

$$\frac{dp^{sat}}{dT} = \frac{\left(\underline{S}^V - \underline{S}^L\right)}{\left(\underline{V}^V - \underline{V}^L\right)} = \frac{\underline{H}^V - \underline{H}^L}{T\left(\underline{V}^V - \underline{V}^L\right)} = \frac{\Delta \underline{H}^{vap}}{T\left(\underline{V}^V - \underline{V}^L\right)} \tag{5-13}$$

Step 3 Make simplifying assumptions

Eq. (5-12) is known as the Clapeyron equation. In this case, our goal is to determine the p^{sat} at different values of T. Eq. (5-13) is a first-order ordinary differential equation (ODE) involving p^{sat} and T. We can solve the ODE with T as the independent variable, p^{sat} as the dependent variable and the known data point ($p^{sat} = 7.6$ kPa when $T = 60°C$) as the initial value. p^{sat} is a function of T or $T = f(T)$. However, if we attempt to integrate Eq. (5-12) using only the given information, we have two obstacles:

(1) $\Delta \underline{H}^{vap}$ is only known at a certain point on the vapor pressure curve.

(2) Molar volume of liquid and vapor is unknown.

For the first problem, it is possible to assume that $\Delta \underline{H}^{vap}$ is a constant. The vapor pressure decreases with increasing boiling point. Vapor pressure approaches zero at the critical point where vapor and liquid are indistinguishable. However, in order to solve differential Eq. (5-1), it is possible to express $\Delta \underline{H}^{vap}$ as a mathematical function of either T or p, so that variables can be separated and integrated.

Near the non-critical point, the molar volume of the liquid is smaller than the molar volume of the vapor, so the quantity $\underline{V}^V - \underline{V}^L$ can be considered the same as $\underline{V}^V$, unless five-digit accuracy is required. $\underline{V}^L$ is negligible compared to $\underline{V}^V$. Therefore

$$\frac{dp^{sat}}{dT} = \frac{\Delta \underline{H}^{vap}}{T\left(\underline{V}^V - \underline{V}^L\right)} \approx \frac{\Delta \underline{H}^{vap}}{T\underline{V}^V} \tag{5-14}$$

To solve Eq. (5-14), the vapor can be assumed as an ideal gas, and V, T and p can be associated with the equation of state, and then dT and dp^{sat} can be integrated.

$$\frac{dp^{sat}}{dT} = \frac{\Delta \underline{H}^{vap}}{T\left(\dfrac{RT}{p^{sat}}\right)} \tag{5-15}$$

$$\frac{dp^{sat}}{p^{sat}} = \frac{\Delta \underline{H}^{vap}}{RT^2} dT \tag{5-16}$$

Step 4 Integrate

The differential equation is solved directly by integrating Eq. (5-16) from known data points ($T = 60°C$, $p = 7.6$ kPa) to various unknown points (T_2, p_2) on the vapor pressure curve. R and $\Delta \underline{H}^{vap}$ are treated as constant.

$$\int_{p_1}^{p_2} \frac{dp^{sat}}{p^{sat}} = \int_{T_1}^{T_2} \frac{\Delta \underline{H}^{vap}}{RT^2} dT \tag{5-17}$$

$$\ln \frac{p_2^{sat}}{p_1^{sat}} = -\frac{\Delta \underline{H}^{vap}}{R} \left(\frac{1}{T_2} - \frac{1}{T_1} \right) \tag{5-18}$$

Step 5 Plug in known values

If $T_2 = 70°C$, the only unknown in Eq. (5-18) is p_2^{sat}. Thus,

$$\ln \left(\frac{p_2^{sat}}{7.6 \text{ kPa}} \right) = \left[\frac{-46.49 \text{ J/mol}}{8.314 \text{ J/(mol·K)}} \right] \left[\frac{1}{(65+273.15)\text{K}} - \frac{1}{(60+273.15)\text{K}} \right]$$

$$p_2^{sat} = 7.6 \text{ kPa}$$

The vapor pressure at some other unknown temperature can be calculated in a similar way.

The Clausius Clapeyron equation is very useful, but it can only be used to simulate the phase equilibrium at low pressure. The next example examines the relationship between vapor pressure and temperature at high pressure, also taking water as an example.

EXAMPLE 5-1 illustrates a special case commonly used, the Clausius-Clapeyron equation:

$$\ln \frac{p_2^{sat}}{p_1^{sat}} = -\frac{\Delta \underline{H}^{vap}}{R} \left(\frac{1}{T_2} - \frac{1}{T_1} \right) \tag{5-19}$$

When using the Clausius-Clapeyron equation, it is necessary to consider the assumptions made:

(1) $\Delta \underline{H}^{vap}$ is constant for temperature and pressure.

(2) $\underline{V}^L$ is negligible in comparison with $\underline{V}^V$.

(3) $\underline{V}^V$ can be simulated by the ideal gas law.

The second and third assumptions are reasonable at low pressure.

The Clausius-Clapeyron equation is very useful, but only for phase equilibria at low pressure.

If T_1 and T_2 are close together and neither of them are near the critical point, the first hypothesis is reasonable.

The Shortcut Equation

EXAMPLE 5-2

Estimating the high-temperature vapor pressure of water

At $T = 179.88°C$, the vapor pressure of water is $p = 1000$ kPa. At these conditions, $\Delta \hat{H}^{vap} = 2014.59$ kJ/kg, $\hat{V}^V = 0.1944$ m^3/kg, and $\hat{V}^L = 0.001127$ m^3/kg.

A. Using only this information, estimate the vapor pressure of water at 230°C, 250°C and 300°C.

B. Compare the value obtained in part A to the data in the steam tables.

A different method is used to estimate several unknown vapor pressures from a known data point.

SOLUTION

Solution A:

Step 1 Compare physical situation to assumptions of Clausius-Clapeyron equation

The Clausius-Clapeyron equation simulates the vapor phase using the ideal gas law, which is valid at low pressure. Here, the data point we know is the vapor pressure at 1000 kPa, and the vapor pressure is going to be higher, because it's going to happen at a higher temperature. Therefore, the Clausius-Clapeyron equation cannot be expected to be valid under these pressures. We need a different approach to model that does not use the ideal gas law.

The derived Clapeyron equation in Example 5-1 does not depend on any assumptions and can be used as a starting point.

$$\frac{dp^{sat}}{dT} = \frac{\Delta \underline{H}^{vap}}{T\left(\underline{V}^V - \underline{V}^L\right)} \tag{5-20}$$

Step 2 Relate molar volume to temperature and pressure

We can relate $\underline{V}^V$ and $\underline{V}^L$ to pressure and temperature through the compressibility. Factor $Z = \dfrac{p\underline{V}}{RT}$:

$$\frac{dp^{sat}}{dT} = \frac{\Delta \underline{H}^{vap}}{T\left(\dfrac{Z^V RT}{p^{sat}} - \dfrac{Z^L RT}{p^{sat}}\right)} \tag{5-21}$$

$$\frac{dp^{sat}}{dT} = \frac{p^{sat} \Delta \underline{H}^{vap}}{RT^2 \Delta Z^{vap}} \tag{5-22}$$

Step 3 Make a simplifying assumption

As the pressure increase approaches the critical point, $\Delta \underline{H}^{vap}$ becomes very small. Similarly, as the temperature and pressure approach the critical point, the compression factors (Z) of the saturated liquid and saturated vapor approach as well, since they both converge at Z_c. Thus, while $\Delta \underline{H}^{vap}$ and ΔZ^{vap} alone are not constant, it is logical to surmise that this ratio remains constant, since both quantities will approach at the same point.

$$\frac{\Delta \underline{H}^{vap}}{\Delta \underline{Z}^{vap}} \approx \text{constant as } p \text{ approaches } p_c \tag{5-23}$$

Step 4 Integrate the Clapeyron equation

The Clapeyron equation can be solved by using simplified assumptions

$$\int_{p_1}^{p_2} \frac{dp^{sat}}{p^{sat}} = \frac{\Delta \underline{H}^{vap}}{R\Delta \underline{Z}^{vap}} \int_{T_1}^{T_2} \frac{dT}{T^2} \tag{5-24}$$

$$\ln\left(\frac{p_2^{sat}}{p_1^{sat}}\right) = \frac{-\Delta \underline{H}^{vap}}{R\Delta \underline{Z}^{vap}} \times \left(\frac{1}{T_2} - \frac{1}{T_1}\right) \tag{5-25}$$

Step 5 Calculate ΔZ^{vap}

Applying the definition of Z gives

$$Z^V = \frac{pV}{RT} = \frac{1000 \text{ kPa} \times 0.1944 \text{ m}^3/\text{kg} \times 18.015 \text{ g/mol} \times \frac{1 \text{ kg}}{1000 \text{ g}}}{\left[8314 \text{ kPa} \cdot \text{cm}^3/(\text{mol} \cdot \text{K})\right] \times (179.88 + 273.15) \text{K} \times \left(\frac{1 \text{ m}}{100 \text{ cm}}\right)^3} = 0.9298$$

$$Z^L = \frac{pV}{RT} = \frac{1000 \text{ kPa} \times 0.001127 \text{ m}^3/\text{kg} \times 18.015 \text{ g/mol} \times \frac{1 \text{ kg}}{1000 \text{ g}}}{\left[8314 \text{ kPa} \cdot \text{cm}^3/(\text{mol} \cdot \text{K})\right] \times (179.88 + 273.15) \text{K} \times \left(\frac{1 \text{ m}}{100 \text{ cm}}\right)^3}$$

$$\Delta Z^{vap} = 0.9298 - 0.539 = 92.44$$

Step 6 Apply known values

We use the known data point at T_1, p_1 to find the vapor pressure at $T_2 = 230^\circ\text{C}$:

$$\ln\left(\frac{p_2^{sat}}{1000 \text{ kPa}}\right) = \frac{-2014.59 \text{ kJ/kg} \times 18.015 \text{ g/mol} \times \frac{1 \text{ kg}}{1000 \text{ g}}}{8.314 \text{ J/(mol} \cdot \text{K)} \times 0.9244 \times \frac{1 \text{ kJ}}{1000 \text{ J}}} \times \left[\frac{1}{(230+273.15)\text{K}} - \frac{1}{(179.88+273.15)\text{K}}\right]$$

$$p_2^{sat} = 2695 \text{ kPa}$$

Vapor pressures for other temperatures are computed in an analogous manner and the results are summarized in Table 5-1.

Table 5-1 Comparison to data of vapor pressures for water as predicted using Eq. (5-25)

Temperature/°C	Estimated p^{sat}/kPa	p^{sat} from Steam Table/kPa
230	2695	2795
250	4044	3976
300	8887	8588
350	17210	16530

Solution B:

The model results are compared with the data in the steam table. As can be seen from Table 5-1, with the increase of temperature difference of T_2 and T_1, the accuracy of prediction decreases somewhat, but even when $T_2 = 350°C$, the error is only 4%. Considering that this is a good result, we base our calculations on a single data point and extrapolate over a large temperature range of 160°C. By comparison, using the same Clausius-Clapeyron equation and the known data point yield estimate of $p^{sat} = 13880$ kPa at $T_2 = 350°C$, and the error is 16%.

EXAMPLE 5-2 illustrates a strategy for establishing vapor-liquid equilibrium under high pressure. Assuming $\Delta \underline{H}^{vap} / \Delta Z^{vap}$ is a constant that is proved valid for H₂O over the temperature and pressure ranges tested in the example. This assumption comes down to Eq. (5-25), as follows

$$\ln\left(\frac{p_2^{sat}}{p_1^{sat}}\right) = \frac{-\Delta \underline{H}^{vap}}{R\Delta Z^{vap}}\left(\frac{1}{T_2} - \frac{1}{T_1}\right) \tag{5-26}$$

The critical point and acentric factor have been tabulated for many compounds. Start by specifying "State 1" (p_1 and T_1) as equal to the critical point:

$$\ln\left(\frac{p_2^{sat}}{p_c}\right) = \frac{-\Delta \underline{H}^{vap}}{R\Delta Z^{vap}}\left(\frac{1}{T_2} - \frac{1}{T_c}\right) \tag{5-27}$$

And express "State 2" again in terms of reduced properties. By definition, $p = p_c p_r$ and $T = T_c T_r$

$$\ln\left(\frac{p_r^{sat} p_c}{p_c}\right) = \frac{-\Delta \underline{H}^{vap}}{R\Delta Z^{vap}}\left(\frac{1}{T_r T_c} - \frac{1}{T_c}\right) \tag{5-28}$$

$$\ln\left(p_r^{sat}\right) = \frac{-\Delta \underline{H}^{vap}}{RT_c \Delta Z^{vap}}\left(\frac{1}{T_r} - 1\right) \tag{5-29}$$

Converting from natural logarithms into common (base 10) logarithms:

$$\frac{\log_{10}\left(p_r^{sat}\right)}{\log_{10} e} = \frac{-\Delta \underline{H}^{vap}}{RT_c \Delta Z^{vap}}\left(\frac{1}{T_r} - 1\right) \tag{5-30}$$

$$\log_{10}\left(p_r^{sat}\right) = \frac{-(\log_{10} e)\Delta \underline{H}^{vap}}{RT_c \Delta Z^{vap}}\left(\frac{1}{T_r} - 1\right) \tag{5-31}$$

The acentric factor is a known value of p_r^{sat}, since it is defined from the vapor pressure at a reduced temperature $T_r = 0.7$ as

$$\omega = -1 - \log_{10}\left(\frac{p^{sat}}{p_c}\right)\bigg|_{T_r=0.7} \tag{5-32}$$

Introducing the acentric factor:

$$-(\omega+1) = \log_{10}\left(\frac{p^{sat}}{p_c}\right)\bigg|_{T_r=0.7} = \frac{-(\log_{10} e)\Delta \underline{H}^{vap}}{RT_c \Delta Z^{vap}} \times \left(\frac{1}{0.7} - 1\right) \tag{5-33}$$

Which can be rearranged as

$$\frac{1+\omega}{\frac{1}{0.7}-1} = \frac{(\log_{10}e)\Delta \underline{H}^{vap}}{RT_c\Delta Z^{vap}} \qquad (5\text{-}34)$$

This modeling approach is based on the premise that the dimensionless quantity on the right-hand side is approximately constant, since $\Delta \underline{H}^{vap}$ and ΔZ^{vap} both tend to be zero at the critical point. Eq. (5-34) uses the acentric factor to say that this constant is approximately equal to Eq. (5-35):

$$\frac{(\log_{10}e)\Delta \underline{H}^{vap}}{RT_c\Delta Z^{vap}} = \frac{7}{3}(\omega+1) \qquad (5\text{-}35)$$

By introducing this approximation into Eq. (5-31), the shortcut vapor pressure equation is obtained.

$$\log_{10}\left(p_r^{sat}\right) = \frac{7}{3}(\omega+1)\times\left(1-\frac{1}{T_r}\right) \qquad (5\text{-}36)$$

The known data point in Example 5-2 ($p^{sat}=1000$ kPa $T=179.88°C$) corresponds to $T_r=0.7$ ($T_r=T/T_c=453.03$ K$/647.1$ K$=0.700$). Thus, the acentric factor for water is:

$$\omega = -1-\log_{10}\left(\frac{1000 \text{ kPa}}{22060 \text{ kPa}}\right)\bigg|_{T_r=0.7} = 0.344$$

The shortcut vapor pressure equation is valid at high pressures, and is therefore a useful complement to the Clausius-Clapeyron equation, which is valid at ideal gas conditions.

Elliott and Lira recommend that the shortcut equation can be applied in the range $0.5 < T_r < 1.0$.

The Antoine Equation

Antoine equation has been shown to be useful for fitting vapor pressure vs. temperature data for a wide variety of compounds.

$$\log_{10}\left(p^{sat}\right) = A - \frac{B}{T+C} \qquad (5\text{-}37)$$

The form of Antoine Equation is somewhat similar to that of the Clapeyron equation, but it does not have the strict theoretical basis of the Clapeyron equation(Fig.5-3).

The Antoine equation constants are best regarded as empirical. Since they are derived from real data, they will produce excellent vapor pressure estimates over the recommended temperature range. However, one should not extrapolate beyond the temperature range of the validated constants, as there is no theoretical reason to assume that the vapor pressure follows Antoine Equation at all temperatures.

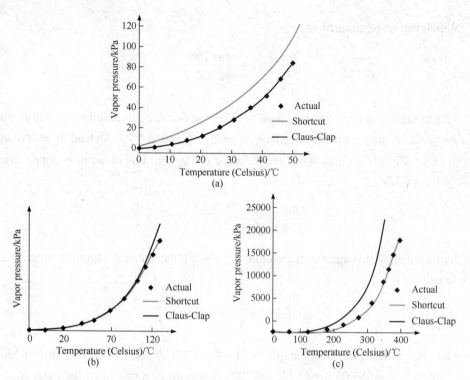

Fig.5-3 Vapor pressure of water vs. temperature: Comparisons of the predictions from Clausius-Clapeyron equation and the shortcut equation

EXAMPLE 5-3

Calculating vapor pressure with the Antoine equation

A process of using ethylbutyl butyrate as solvent is designed. Ensure that ethylbutyl butyrate does not boil under process conditions as part of the safety analysis. Use the Antoine equation to find:

(1) The vapor pressure at $T = 100°C$.
(2) The normal boiling point.

SOLUTION

Step 1 Look up data

For ethylbutyl butyrate, the Antoine constants are

$$A = 6.96, \ B = 1424.26, \text{ and } C = 213.21$$

Step 2 Solve Antoine equation

The data relate temperature expressed in degrees Celsius to pressure in mmHg.

$$\log_{10}\left(p^{\text{sat}}\right) = 6.96 - \frac{1424.26}{100°C + 213.21}$$

$$p^{\text{sat}} = 257.04 \text{ mmHg}$$

Step 3 Back-solve Antoine equation

Here we wish to know the boiling temperature that corresponds to a pressure of 1atm, or 760 mmHg

$$\log_{10}(760 \text{ mmHg}) = 6.96 - \frac{1424.26}{T+213.21}$$

$$T = 135.87°C$$

5.1.2 Fugacity and Its Use in Modeling Phase Equilibrium

Some strategies for estimating pressure and temperature when phase equilibria (primarily vapor-liquid equilibria) occur have been demonstrated and molar enthalpy (H) and molar entropy (S) changes have been modeled using equations of state. In fact, the entire vapor-liquid equilibrium curve can be predicted without any information other than the equation of state describing the vapor-liquid phase.

Calculating Changes in Gibbs Free Energy

The phase equilibrium criterion for pure compounds states that for the phase in equilibrium, the mole Gibbs free energy of each phase must be the same. For vapor-liquid equilibrium, $\underline{G}^V = \underline{G}^L$.

If G can be calculated for the liquid or vapor phase at any temperature and pressure from the equation of state, then the temperature and pressure can be determined when $\underline{G}^V = \underline{G}^L$. Because the mathematical relationship already had been obtained between G, T and p. The basic relationship of properties as follow:

$$d\underline{G} = \underline{V}dp - \underline{S}dT$$

However, a practical difficulty arises when working with $\underline{G}$ (at least directly). This will be illustrated by considering an isothermal change in $\underline{G}$.

EXAMPLE 5-4

Calculating Gibbs free energy on an isothermal path
In specifying the molar Gibbs free energy of a compound, the liquid phase at $T = 300$ K and $p = 100$ kPa is used as the reference state. Therefore $\underline{G}^L = 0$ at $T = 300$ K and $p = 100$ kPa. It has been estimated that for pure vapor at $T = 300$ K and $p = 100$ kPa, $\underline{G}^V = 5.0 \times 10^5$ kPa·cm^3/mol. The molar volume of the liquid at the reference state is $\underline{V}^L = 125$ cm^3/mol.

A. Estimate the liquid phase $\underline{G}^L$ at $T = 300$ K and $p = 1$ kPa, relative to this reference state.
B. Estimate the vapor phase $\underline{G}^V$ at $T = 300$ K and $p = 1$ kPa, relative to this reference state.
C. Estimate the vapor pressure at $T = 300$ K.

SOLUTION

Step 1 Mathematically relate G to p

Writing an exact derivative firstly that links the intensive property, the change in Gibbs free energy (dG), to the two most intensive properties (in this case T and p).

$$d\underline{G} = \left(\frac{\partial G}{\partial p}\right)_T dp + \left(\frac{\partial G}{\partial T}\right)_p dT \tag{5-38}$$

Then, $\left(\frac{\partial G}{\partial p}\right)_T$ and $\left(\frac{\partial G}{\partial T}\right)_p$ are re-expressed in terms of temperature and pressure. In this case, there is no math to do, just recognize the basic properties of G as

$$d\underline{G} = \underline{V}dp - \underline{S}dT \tag{5-39}$$

All three parts of this problem involve an isothermal path, so the equation for $d\underline{G}$ simplifies to

$$d\underline{G} = \underline{V}dp \tag{5-40}$$

Solution A:

Step 2 Integrate for liquid phase

Assume $\underline{V}$ is a constant for the liquid phase. The $\underline{G}_1^L$ (The molar Gibbs free energy of the liquid at $T = 300$ K and $p = 100$ kPa) is the reference state and it is 0 by definition. $\underline{G}_2^L$ is the Gibbs free energy of the liquid at $T = 300$ K and $p = 1$ kPa and that is what we are trying to determine.

$$\int_{\underline{G}_1^L=0}^{\underline{G}_2^L=\underline{G}_2^L} d\underline{G} = \int_{p_1=100\,\text{kPa}}^{p_2=1\,\text{kPa}} \underline{V}^L dp \tag{5-41}$$

$$\underline{G}_2^L - \underline{G}_1^L = \underline{V}^L (p_2 - p_1) \tag{5-42}$$

$\underline{G}_2^L = 125 \text{ cm}^3/\text{mol} \times (1 \text{ kPa} - 100 \text{ kPa}) = -12375 \text{ kPa} \cdot \text{cm}^3/\text{mol}$

Solution B:

Step 3 Integrate for vapor phase

Since we are examining a vapor at pressures below atmospheric, we will use the ideal gas law.

$$\int_{\underline{G}_1^V=500000\,\text{kPa}\cdot\text{cm}^3/\text{mol}}^{\underline{G}_2^V=\underline{G}_2^V} d\underline{G} = \int_{p_1=100\,\text{kPa}}^{p_2=1\,\text{kPa}} \underline{V}^V dp$$

$$\int_{\underline{G}_1^V=500000\,\text{kPa}\cdot\text{cm}^3/\text{mol}}^{\underline{G}_2^V=\underline{G}_2^V} d\underline{G} = \int_{p_1=100\,\text{kPa}}^{p_2=1\,\text{kPa}} \frac{RT}{p} dp$$

$$\underline{G}_2^V - \underline{G}_1^V = RT \ln\left(\frac{p_2}{p_1}\right)$$

$\underline{G}_2^V = 500000 \text{ kPa} \cdot \text{cm}^3/\text{mol} + 8314 \text{ kPa} \cdot \text{cm}^3/(\text{K} \cdot \text{mol}) \times 300 \text{ K} \times \ln\left(\frac{1 \text{ kPa}}{100 \text{ kPa}}\right)$

$= -10986200 \text{ kPa} \cdot \text{cm}^3/\text{mol}$

Solution C:

Step 4 Find pressure at which $\underline{G}^V = \underline{G}^L$

Notice that for $T = 300$ K, $\underline{G}^L < \underline{G}^V$ at $p = 100$ kPa and $\underline{G}^V < \underline{G}^L$ at $p = 1$ kPa; this means the compound is a vapor at $p = 1$ kPa and a liquid at $p = 100$ kPa. The vapor pressure is the unique pressure at which $\underline{G}^V = \underline{G}^L$.

$$\underline{G}_2^L = \underline{G}_2^V$$

$$125 \text{ cm}^3/\text{mol} \times (p_2 - 100 \text{ kPa}) = 500000 \text{ kPa} \cdot \text{cm}^3/\text{mol} + 8314 \text{ kPa} \cdot \text{cm}^3/\text{K} \times 300 \text{ K} \times \ln\left(\frac{p_2}{100 \text{ kPa}}\right)$$

$$p_2 = 81.76 \text{ kPa}$$

Therefore, if it is the correct formula for $\underline{G}_2^V$ and it is the correct formula for $\underline{G}_2^L$, then the vapor pressure of the compound at $T = 300$ K is $p^{\text{sat}} = 81.76$ kPa. However, Fig. 5-4 reveals some of the practical challenges of computing using mole Gibbs free energy. In the vapor phase, $\underline{G}$ is very sensitive to pressure, and in fact, $\underline{G}$ goes to 2 as p goes to 0. In contrast, the liquid phase G is relatively insensitive to pressure. At $p^{\text{sat}} = 81.76$ kPa, four significant values of vapor pressure were reported. It might more reasonably say that $p^{\text{sat}} = 82$ kPa $p^{\text{sat}} = 81.8$ kPa. However, $\underline{G}^V$ is very sensitive to pressure, and if substitute $p^{\text{sat}} = 82$ kPa $p^{\text{sat}} = 81.8$ kPa into equation, these will not be necessarily recognized as correct answers; the molar Gibbs free energy of the two phases is significantly different, as shown in Table 5-2.

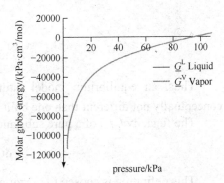

Fig.5-4 Molar Gibbs free energy vs. pressure at $T = 300$ K for the compound examined in Example 5-4

Table 5-2 Values of $\underline{G}^L$ and $\underline{G}^V$ obtained from Step 2 and Step 3, for some values of pressure that are identical to two significant figures

Pressure/kPa	$\underline{G}^L/(\text{kPa} \cdot \text{cm}^3/\text{mol})$	$\underline{G}^V/(\text{kPa} \cdot \text{cm}^3/\text{mol})$
81.75	−2281	−2592
81.76	−2280	−2287
81.80	−2275	−1067
82.00	−2250	5024

Example 5-4 illustrates that the vapor pressure can be estimated from a simple model using the equilibrium criterion $\underline{G}^V = \underline{G}^L$: It can be assumed that the vapor is an ideal gas and the liquid is an incompressible fluid. It can be easily estimated the vapor pressure using more complex models. In Step 1 and Step 2, any equation of state can be used to relate V and p. However, Table 5-2 shows how a seemingly insignificant change in p can lead to a significant change in $\underline{G}$ for the vapor phase. And furthermore, we proved mathematically that as pressure goes to zero, $\underline{G}^V$ goes to minus infinity. This aspect of $\underline{G}$ will become a more important issue when we begin to model mixtures. Previous section illustrates how we use the remaining molar Gibbs free energy, rather than the molar Gibbs free energy itself, as the basis for phase equilibrium calculations.

Mathematical Definition of Fugacity

As pressure goes to 0, the value of $\underline{G}^V$ goes to minus infinity. However, consider the residual mole of Gibbs free energy: $\underline{G}^R = \underline{G} - \underline{G}^{ig}$. As the pressure goes down, the real stuff behaves more and more like an ideal gas. Thus, as pressure decreases, $\underline{G}$ and $\underline{G}^{ig}$ respectively tend to negative infinity, and as pressure goes to 0, the difference between them ($\underline{G} - \underline{G}^{ig}$) tends to 0 (as do all residual properties). So

$$\lim_{p \to 0} \underline{G} = -\infty \tag{5-43}$$

$$\lim_{p \to 0} \left(\underline{G} - \underline{G}^{ig} \right) = 0 \tag{5-44}$$

Thus, an equilibrium model built upon the residual molar Gibbs free energy is conceptually not different than one built on the molar Gibbs free energy itself.

The fugacity (f) of a pure substance is defined through

$$d\underline{G} = RTd(\ln f) \tag{5-45}$$

This definition is chosen to mirror equation

$$d\underline{G}^{ig} = RTd(\ln p) \tag{5-46}$$

Which is derived for an ideal gas undergoing an isothermal process.

The residual molar Gibbs free energy

$$d\left(\underline{G} - \underline{G}^{ig}\right) = RT\left[d(\ln f) - d(\ln p)\right] \tag{5-47}$$

When integrated and rearranged, this gives rise to the fugacity equation, which is most useful for calculating the fugacity of pure compounds.

$$\ln\left(\frac{f}{p}\right) = \frac{\underline{G} - \underline{G}^{ig}}{RT} = \frac{\underline{G}^R}{RT} \tag{5-48}$$

Equation can be applied to a substance in any phase: vapor, liquid, or solid.

The fugacity is defined here in terms of pressure.

Under ideal gas conditions, $f^V = p$ (where f^V represents the fugacity of a pure vapor).

Fugacity is a convenient property for use in the modeling of mixtures.

The equilibrium criterion also can be expressed in terms of fugacity because fugacity is related directly to the residual molar Gibbs free energy.

$f^V = f^L$ at liquid-vapor equilibrium.

$f^V = f^S$ at solid-vapor equilibrium.

$f^L = f^S$ at solid-liquid equilibrium.

Therefore, we can combine the expressions of the enthalpy and entropy residual functions with Eq. (5-48), and associate fugacity with measurable properties.

$$\ln\left(\frac{f}{p}\right) = \frac{G - G^{ig}}{RT} = \frac{\underline{H} - \underline{H}^{ig}}{RT} - \frac{T(\underline{S} - \underline{S}^{ig})}{RT} \tag{5-49}$$

$$\underline{H} - \underline{H}^{ig} = \int_{T=T, p=0}^{T=T, p=p} \left[\underline{V} - T\left(\frac{\partial \underline{V}}{\partial T}\right)_p\right] dp \tag{5-50}$$

$$\underline{S} - \underline{S}^{ig} = \int_{T=T, p=0}^{T=T, p=p} \left[\left(\frac{\partial \underline{V}}{\partial T}\right)_p - \frac{R}{p}\right] dp \tag{5-51}$$

When we merge integrals, $\left(\frac{\partial \underline{V}}{\partial T}\right)_p$ terms cancel each other out and leave with

$$\ln\left(\frac{f}{p}\right) = \frac{1}{RT}\int_{T=T, p=0}^{T=T, p=p}\left[\underline{V} - T\left(\frac{\partial \underline{V}}{\partial T}\right)_p\right]dp - \frac{T}{RT}\int_{T=T, p=0}^{T=T, p=p}\left[\left(\frac{\partial \underline{V}}{\partial T}\right)_p - \frac{R}{p}\right]dp$$

$$= \frac{1}{RT}\int_{T=T, p=0}^{T=T, p=p}\left[\underline{V} - \frac{RT}{p}\right]dp \tag{5-52}$$

The dimensionless ratio f/p is called the fugacity coefficient and is given the symbol φ:

$$\varphi = \frac{f}{p} \tag{5-53}$$

The criterion for vapor-liquid equilibrium can be expressed as either $f^V = f^L$ or $\varphi^V = \varphi^L$.

$$\ln\left(\frac{f}{p}\right) = \ln\varphi = \int_{T=T, p=0}^{T=T, p=p}\left[\frac{\underline{V}}{RT} - \frac{1}{p}\right]dp = \int_{T=T, p=0}^{T=T, p=p}\left[\frac{Z-1}{p}\right]dp \tag{5-54}$$

Start with the alternative ($d\underline{V}$) expressions for the residual molar enthalpy and entropy

$$\ln\left(\frac{f}{p}\right) = \ln\varphi = (Z-1) - \ln Z + \int_{T=T, \underline{V}=\infty}^{T=T, \underline{V}=\underline{V}}\left[\frac{1-Z}{\underline{V}}\right]d\underline{V} \tag{5-55}$$

5.2 Vapor-Liquid Equilibrium

5.2.1 Motivational Example

Boiling a pot of water at 1 atm, the boiling point is 100°C. Suppose there is an equimolar mixture of water and another compound, such as ethanol, in the pot at a pressure of 1 atm. What is the boiling temperature of the mixture? Also, will the composition of the steam leaving the pot be equimolar?

The experiment shows that a liquid mixture of equimolar water and ethanol will boil at

80℃ and 1atm, while the vapor phase composition will not be equal mole (like liquid), but 65% mole ethanol. These results are contrary to our intuitive approach, but also offer a huge possibility. The different vapor phase components can be obtained by boiling a liquid mixture of one component. Thus, our steam component will be enriched in ethanol in the above example, separating some ethanol from water. This is one of the most important and obvious unit operations in chemical engineering: distillation column. Obviously, it is very important to understand why our intuitive approach does not work and develop a method to predict and model the results of any interested steam liquid mixture in the chemical processing industry.

In order to understand the experimental results described in the previous paragraph, we must go back to the phenomena of evaporation and boiling to consider the nature of this process at the molecular level. Suppose there is a pure compressed liquid in a fixed volume container. The container is in a controlled temperature and constant hot bath. The liquid fills all the spaces in the container. If not, will the liquid evaporate? There's no room for steam to fill because it is a closed system. However, assuming that the container is connected to the second container through a partition and the contents of the second container are discharged (Fig. 5-5), the pressure of the second container is 0 atm. What happens when the separator is removed? The liquid will begin to expand to the evacuation space and some of the liquid will evaporate, as any vapor bubble will not attempt to stop from the surface of the liquid (there will be no pressure, no force to stop them) above the liquid. The expansion and evaporation of the liquid will continue until some vapor begins to condense. A dynamic equilibrium is established when the rate of evaporation equals the rate of condensation. The pressure exerted by the vapor on the liquid surface is now in equilibrium with the pressure exerted by the liquid phase. This pressure is called vapor pressure. It is the only pressure that liquid and vapor can exist in equilibrium. It is a function of system temperature.

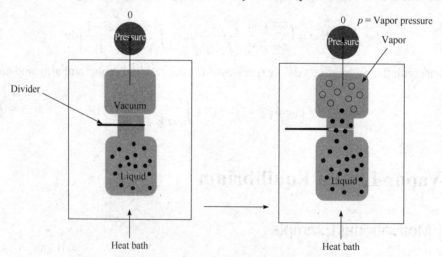

Fig.5-5 A liquid expanding and vaporizing into the evacuated space

The first container is a mixture of two liquids rather than a pure compound. The first container is again connected to the second container (its contents are emptied) through a divider. When the separator is removed, there is expansion and evaporation, and the dynamic balance is finally established. But what is the final pressure of the system? We already know that when we want to calculate the properties of a mixture, we need to use the mole fraction to calculate the percentage of each substance in the mixture. Eq. (5-56) introduces the molar enthalpy of the mixture of two components (a and b) as

$$\underline{H} = x_a \underline{H}_a + x_b \underline{H}_b \qquad (5\text{-}56)$$

Is the pressure of the liquid mixture the sum of the vapor pressures of the pure components (weighted by their mole fractions) as follows?

$$p = x_a p_a^{sat} + x_b p_b^{sat} \qquad (5\text{-}57)$$

The interaction of the liquid phase in the mixture affects the combined vapor pressure of the mixture. Due to the diversity of this behavior and the importance of vapor-liquid equilibrium in the separation of chemicals, it is essential for chemical engineers to easily extract appropriate information from data in tabular and graphical formats for these systems. We will describe this in the next section.

5.2.2 *Raoult's Law* and the Presentation of Data

Gibbs phase rule was introduced before, which provides a strong constraint on the number of variables for the system by understanding the number of components and equilibrium phases.

$$F = C - \pi + 2 \qquad (5\text{-}58)$$

We asked a question before: what is the boiling point of water? Through Gibbs phase rule analysis, we find that F (degree of freedom) is equal to 1 atm boiling point. Therefore, there is no unique answer to the question about the boiling point of water, and we need to specify another strengthening variable because $F=1$. What is the boiling point of water at 1 atm? The answer is 100 ℃. If we specify a different pressure, the answer is a different temperature. For convenience, we can prepare a table listing the boiling point of water as a function of pressure. Pressure is an independent variable and temperature is a dependent variable.

For mixtures, like ethanol and water, what's their boiling point? We use Gibbs phase rule to get $F = 2$ ($C = 2$; $\pi = 2$). Therefore, we need to fix two intensive variables. We can ask what is the boiling point of the mixture of ethanol and water at $p=1$ atm and 50% ethanol? The answer is about 80 ℃. If we change the values of one or two of the strong variables (the system pressure and the mole fraction of ethanol in the liquid phase), the boiling point will change. We can represent the data in graphical form instead of tabular form. In the rest of this section, we will present VLE data by using tables and graphs.

Take a mixture of R-134a (1) (1,1,1,2-tetrafluoroethane (CH_2FCF_3)) and R-245fa (2) (1,1,1,3,3-pentafluoropropane ($C_3H_3F_5$)) as an example. In order to effectively represent the vapor-liquid equilibrium of this mixture, it is necessary to keep one of the variables unchanged to display the data in tabular form because there are two degrees of freedom. Typically, this type of data is obtained by keeping the temperature or pressure constant and showing changes in other variables as all liquid components are mixed within the range. For this reason, Table 5-3 shows how the system pressure will change when the composition of the mixture is changed at a constant temperature (293.15 K). These endpoints usually provide pure components for the two species in the mixture. When $x_1 = 0$ (the first row of Table 5-3), pure ($x_2 = 1$) is obtained. This row shows the vapor pressure of at 293.15 K. Similarly, the fifth line of Table 5-3 is pure R-134a, showing the vapor pressure at 293.15 K. For completeness, we provide values for x_1 and y_2 in Table 5-3.

Table 5-3 Vapor-liquid equilibrium data for the mixture R-134a (1) + R-245fa (2) at 293.15 K

Temperature/K	Pressure/kPa	x_1	x_2	y_1	y_2
293.15	123	0	1	0	1
293.15	202	0.18	0.82	0.48	0.52
293.15	315	0.45	0.55	0.76	0.24
293.15	439	0.72	0.28	0.91	0.09
293.15	571	1	0	1	0

Besides fixing temperature and changing liquid mole fraction, pressure and liquid mole fraction can also be fixed. The vapor-liquid equilibria of paraxylene (1) (C_8H_{10}) and butyl butyrate (2) ($C_8H_{16}O_2$) at 101.3 kPa are shown in Table 5-4.

Table 5-4 Vapor-liquid equilibrium data for the mixture paraxylene (1) + butyl butyrate (2) at 101.3 kPa

Pressure/kPa	Temperature/℃	x_1	x_2	y_1	y_2
101.3	138.26	1	0	1	0
101.3	139.57	0.85	0.15	0.92	0.08
101.3	140.32	0.82	0.18	0.89	0.11
101.3	142.43	0.70	0.30	0.80	0.20
101.3	143.79	0.63	0.37	0.76	0.24
101.3	147.53	0.48	0.52	0.67	0.33
101.3	150.72	0.37	0.63	0.57	0.43
101.3	160.32	0.12	0.88	0.23	0.77
101.3	161.47	0.09	0.91	0.19	0.81
101.3	164.98	0	1	0	1

For Table 5-4, the first and last lines indicate the vapor pressures for pure components of paraxylene (first line) and butyl butyrate (last line). The inner rows represent the composition of the liquid in equilibrium with the vapor at that temperature and pressure.

In addition to tabular form, VLE data are often displayed in graphical form. For example, in Table 5-3, we give the data of R-134a (1) + R-245fa (2) mixture. We change the composition of liquid phase and keep the temperature constant ($T = 293.15$ K). The resulting pressure and vapor phase composition are shown in Table 5-3. In order to present the same data graphically, we usually represent fixed things in the graph or graph title (here temperature), and plot the system pressure as a function of liquid and vapor mole fractions. This is often called a "p-xy" plot, as shown in Fig. 5-6.

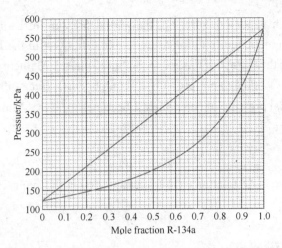

Fig.5-6 $p\text{-}xy$ plot for the R-134a (1) R-245fa (2) system at 293.15K

Fig. 5-6 depicts a system with a temperature equal to 293.15 K. The curve itself contains the information in Table 5-3, that is the vapor and liquid reach equilibrium at this temperature and pressure. The area above the upper curve is the area of the single liquid phase without vapor phase, while the area below the lower curve is the area of the single vapor phase without liquid phase. The next step is to describe some of the key features of Fig. 5-6 in more detail, including what it means between curves.

The longitudinal axis on the left ($x_1 = 0$) describes the pure components. $p = 123\,\text{kPa}$ is the vapor pressure at 293.15 K. If the pressure is higher than the vapor pressure of pure, like 200 kPa, then there is only one equilibrium phase. At this point, if a small amount of R-134a (isothermal) is added to the system, it will move horizontally to the right side of Fig. 5-6, and there is still liquid at equilibrium phase. However, as more R-134a continues to be added to the system, it will continue to move horizontally to the right side of Fig. 5-6, and eventually a curve will appear at $x_1 = 0.17$. This is called the bubble point curve, which describes the composition of the first vapor bubble at equilibrium with the liquid.

In a mixture, each component does not exist on the surface alone. Therefore, each one can't play its full vapor pressure. The vapor pressure exerted by each component will be a function of its surface coverage (composition) and the interaction between different components. Assuming it allows us to simply use the weighted average of the vapor pressures of the pure components, we can use the ideal solution here. Since our ideal solution model provides a good estimate of the bubble point pressure of the mixture as $x_1 = 0.17$, can it also help us to estimate the vapor composition in equilibrium with the liquid mixture? The pressure of the system in which the liquid interacts as an ideal solution is 200 kPa. This is a medium low pressure at which many (though not all) systems can reasonably be modeled as ideal gas. We will examine the vapor phase by assuming the behavior of the ideal gas, and then test the accuracy of our final answer. The sum of the partial pressures of the ideal gas mixture is equal to the total pressure of the system, we can determine the components because we know from the model that the total pressure of the system is 1.99 kPa. Since liquid is well modeled as an ideal solution, the contribution of each component to the bubble point pressure of the mixture is based on the proportion of its own molecules on the surface of the mixture multiplied by its own vapor pressure. So, the partial pressure of a substance is the vapor pressure of its pure component multiplied by its mole fraction in the liquid (we use liquid composition because the surface is formed in the liquid phase). Mathematically, this becomes $p_1 = x_1 p_1^{sat}$, where p_1^{sat} is the vapor pressure of component 1. Because we think of vapor as an ideal gas, the partial pressure of component 1 becomes $p_1 = py_1$. Make the partial pressure of component 1 equal.

$$py_1 = x_1 p_1^{sat} \tag{5-59}$$

Find vapor phase mole fraction (y_1) from the following formula

$$py_1 = x_1 p_1^{sat} \tag{5-60}$$

Put in the numbers to get:

$$y_1 = \frac{0.17 \times (571 \text{ kPa})}{199 \text{ kPa}} = 0.49$$

If we assume that the mixture of R-134a + R-245fa behaves as an ideal gas in the vapor phase, the mole fraction of the vapor and liquid equilibrium, and the composition of the liquid is $x_1 = 0$. The temperature is 293 K, and $y_1 = 0.49$. Let's go back to the experimental data in Fig. 5-6. To find the vapor composition in equilibrium with the liquid phase, just move horizontally from the bubble point curve (at 200 kPa and $x_1 = 0.17$) to the point where the next curve intersects.

This point, about $y_1 = 0.49$, is called dew point, and the curve itself is dew point curve. Dew point is where the vapor and the first drop of liquid (or dew) appear. When we compare the results of the ideal gas model $y_1 = 0.49$ (calculated above) with the actual values of $y_1 = 0.49$, this is a good estimate. The ideal solution assumption of liquid, combined with the ideal gas assumption of vapor, is a very fast and popular method to estimate the vapor-liquid equilibrium of mixtures.

The assumption of an ideal solution for the liquid, combined with the assumption of an ideal gas for the vapor, is called *Raoult's Law* and can be written for each component individually as

$$y_1 = \frac{x_1 p_1^{sat}}{p} \tag{5-61}$$

$$y_2 = \frac{x_2 p_2^{sat}}{p}$$

If one adds these equations together, the sum becomes

$$y_1 + y_2 = 1 = \frac{x_1 p_1^{sat}}{p} + \frac{x_2 p_2^{sat}}{p} \tag{5-62}$$

Solving for the system pressure yields

$$p = x_1 p_1^{sat} + x_2 p_2^{sat} \tag{5-63}$$

$$p = x_1 p_1^{sat} + (1-x_1) p_2^{sat} \tag{5-64}$$

$$p = x_1 \left(p_1^{sat} - p_2^{sat} \right) + p_2^{sat} \tag{5-65}$$

Eq. (5-65) is a straight line, if someone plots p as a function of x_1, where the slope is $p_1^{sat} + p_2^{sat}$, and the intercept of the p-axis is p_2^{sat}. This is an important result because *Raoult's law p* and the x_1 curve (which by definition is the bubble point curve) must be a straight line. The R-134a + R-245fa system can well approximate *Raoult's law*, as shown in Fig. 5-6. It has a bubble point curve, which looks like a line. We just need to start with the mixture (30% mole R-134a) and move 200 kPa horizontally (in both directions) until it intersects the bubble point curve and dew point curve. Where the bubble point curve intersects ($x_1 = 0.17$) is the composition of liquid phase, and where the dew point curve intersects ($y_1 = 0.49$) is the composition of vapor phase. These two points are at the same pressure and temperature, which is required for phase equilibrium.

Just like constant temperature data has graphic representation, constant pressure data also has graphic representation. In Fig. 5-7, we graphically show the data in Table 5-4, which plots the temperature as a function of equilibrium liquid and vapor composition, called the *T-xy* plot. Like *p-xy* plot, *T-xy* plot has pure component information at $x_1 = 0$ (pure butyl butyrate) and $x_1 = 1$ (pure paraxylene). The boiling point of paraxylene at 101.3 kPa is slightly higher than 130 ℃, as shown in Fig. 5-7 (at $x_1 = 1$).

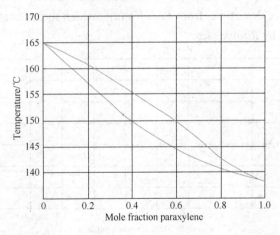

Fig.5-7 *T-xy* plot for the paraxylene (1) + butyl butyrate (2) system at 101.3 kPa

Now that we have learned how to read *p-xy* and *T-xy* plots, we will introduce a variety of different systems to show the behavioral diversity that can occur when the system does not behave as ideal as R-134a + R-245fa or butyl butyrate + paraxylene. In Fig. 5-8, we show the system of n-pentane (1) (C_5H_{12}) and methanol (2) (CH_3OH) at 422.6 K. This is a hydrocarbon and an alcohol. They have not only different sizes, but also many different types of self-interaction. Hydrocarbons are non-polar and only cause dipole interaction. On the other hand, methanol is a polar compound which can form hydrogen bond. Obviously, this is a system that can't be simulated well with ideal solution hypothesis. Therefore, we expect *Raoult's law* to be a bad approximation to simulate the behavior of the system (let alone that the system is at high pressure, which makes the ideal gas part of *Raoult's law* unattractive). The black line in Fig. 5-8 is the model prediction of bubble and dew point curve of the system, and obviously, the *Raoult's law* is not applicable to the system. The actual phase behavior is represented by grey curve, and the reading method is the same as that in Fig. 5-6. The curve above is the bubble point curve, and the curve below is the dew point curve. However, unlike Fig. 5-6, this curve has a point where the liquid and vapor compositions are the same, namely at $x_1 = y_1 = 0.53$. This is called azeotrope, and the system with azeotrope is called azeotrope system. Knowledge of azeotrope location is important when designing distillation columns that separate two components by boiling point. Typical chemical engineering courses include at least one course on separation. In this course, you will learn that azeotropic composition limits the purity of separation, because the composition of liquid phase and vapor phase is the same. Therefore, chemical engineers have developed different methods to overcome this limitation, such as adding a third chemical to the system, changing the system temperature, and even trying different separation methods (such as based on freezing point rather than boiling point).

One of the main failures of *Raoult's law* of the system is that the azeotrope is not produced in the model, while the azeotrope is observed in the experiment. In addition, the system pressure is about 2500 kPa for the blends, while the pressure predicted by *Raoult's law* is slightly lower than 1500 kPa. If the pressure of the system is higher than that predicted by *Raoult's law*, the system will deviate from *Raoult's law*, which means that the pressure predicted by *Raoult's law* is greater (or positive). Note that as we will show later, some systems show negative deviations from *Raoult's law*, which means that the system pressure is less than predicted by *Raoult's law*.

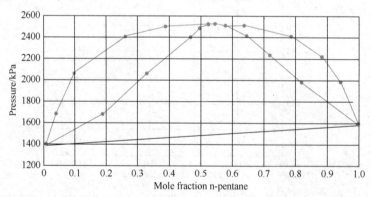

Fig.5-8 *p-xy* plot for the n-pentane (1) 1 methanol (2) system at 422.6K [*Raoult's Law* is the black curve (both bubble- point and dew-point curves are shown) in the figure]

Mechanically, why does a mixture that does not conform to the ideal solution assumption produce a pressure higher than the predicted pressure of the ideal solution? In order to explore this problem in more detail, we need to go back to the concept of vapor pressure. As we know from the previous discussion on vapor pressure, a substance with high vapor pressure is volatile, which means that it is easy to evaporate from the liquid phase to the vapor phase. Of course, this is related to the interaction at the molecular level in the liquid phase (especially at the surface). When the mixture behaves as an ideal solution, the similar size and interaction of the molecules involved means that the snapshot (interaction and size) of the mixture will be similar to that of any pure component. A good example of an ideal solution, as we discussed earlier, is a mixture of butyl butyrate and paraxylene. However, when you have an unsatisfactory solution, the situation changes. The overall interaction between molecules in a mixture is usually less favorable than that of any pure component, because the different interactions between two different molecules are less energetically favorable than those between similar substances. This reduction in attractive interactions (relative to pure components) results in a more volatile system. The n-pentane + methanol system in Fig. 5-8 is a good example of non-ideal solution. Therefore, for a given component, the system pressure is higher than that of the ideal solution, so there is a positive deviation from *Raoult's law*.

On the other hand, there are some systems that can form hydrogen bonds between different components in the mixture, but not separate pure components. This additional increase in attractive interaction (relative to the pure component alone) results in less volatility of the system. Therefore, the system will show a negative deviation from *Raoult's law*. Acetone $(CH_3)_2CO$ + chloroform $(CHCl_3)$ is such a system, as shown in Fig. 5-9. Here, the pure components cannot form hydrogen bonds, but they will participate in the interaction of hydrogen bonds. Since this is an unusual set of conditions, the negative deviation from *Raoult's law* is far less than the positive deviation.

Systems that exhibit a pressure above that predicted by *Raoult's Law* are said to have "positive deviations" from *Raoult's Law*, meaning the pressure is greater (or positive) relative to what *Raoult's Law* would predict.

Some systems will show negative deviations from *Raoult's Law*, meaning that the system pressure is less than that predicted by *Raoult's Law*.

In Fig. 5-8 and Fig. 5-9, the azeotropes of the two systems appear near the middle of the composition region of *p-xy* plot. However, this is not necessarily the case. For example, the studied water (1) + ethanol (2) system shows a positive deviation from *Raoult's law*, and almost all its azeotropic components are ethanol, as shown in Fig. 5-10. Indeed, as the temperature drops, the system becomes no azeotrope.

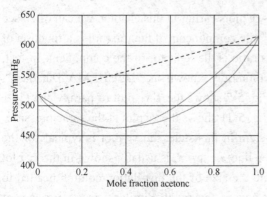
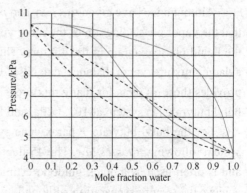

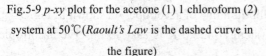

Fig.5-9 *p-xy* plot for the acetone (1) 1 chloroform (2) system at 50℃ (*Raoult's Law* is the dashed curve in the figure)

Fig.5-10 *p-xy* plot for the water (1) 1 ethanol (2) system at 30℃

Distribution Coefficients, Relative Volatility, and *xy* plots

It is common to use the terms "partition coefficient," "distribution coefficient," or "K-factor" to describe the ratio of the mole fraction of a component in one phase relative to another phase.

$$K_i = \frac{y_i}{x_i} \tag{5-66}$$

Since paraxylene is more volatile (relative to butyl butyrate), it will have a *K*-factor larger than 1 (it appears preferentially in the vapor), while butyl butyrate will have a *K*-factor less than 1.

The relative volatility is

$$\alpha_{ij} = \frac{K_i}{K_j} \tag{5-67}$$

The relative volatility is one measure to describe the ease of separation of two components through distillation. Very clearly, the *K*-factors and the relative volatility are composition dependent quantities. Relative volatility, usually expressed by the *K*-factor of the more volatile component in the molecule, is a measure of the ease of separating two components by distillation. In short, the higher the relative volatility between the two components, the higher the efficiency of distillation separation. In general, more efficient separation requires smaller distillation columns, less energy consumption, etc. For example, for the ethanol and water system shown in Fig. 5-10, we find that the relative volatility at 30℃ and 7 kPa is equal to $a_{e,w} = 7.36$, which means that ethanol and water system is easier to separate than butyl butyrate-paraxylene system. However, this is a state dependent observation because the ethanol and water system (as shown in Fig. 5-10) has an azeotrope, which makes it very difficult to separate when they are close to the azeotrope (because the relative volatility of the azeotrope is 1). *K*-factor and relative volatility are components related.

Although we have given the *T-xy* and *p-xy* plots earlier in this chapter, we can often see that the vapor phase diagram of the more volatile components of the mixture is a function of the liquid phase diagram in the mixture. This ratio is the *K*-factor of the component. In fact, a large mixture vapor balance database, commonly known as "DECHEMA Database", provides these *xy* plots for many mixtures. Fig. 5-7 shows the *T-xy* plot of paraxylene (1) + butyl butyrate (2) system at 101.3 kPa, and Fig. 5-11 shows the *xy* plot at the same pressure.

For the butyl butyrate + paraxylene system in this state, the *xy* plot is typical for the system following *Raoult's law*. The 45-degree line, i.e, $y = x$, is usually shown in the *xy* plot and is often used as a quick guide to estimate the size of the distillation column needed to achieve the required separation, although it is beyond this range. Similarly, in the presence of azeotrope, the 45-degree line bisects the equilibrium line of azeotrope. As shown in Fig. 5-12, an example of the n-pentane + methanol system at 422.6 K is easy to observe azeotropic components at about 55% n-pentane.

 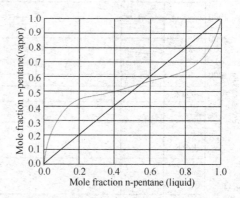

Fig.5-11 *xy* plot for the paraxylene (1) + butyl butyrate (2) system at 101.3 kPa

Fig.5-12 *xy* plot for the n-pentane (1) + methanol (2) system at 422.6K(Note the azeotrope at about 55% n-pentane)

5.2.3 Mixture Critical Points

It is usually convenient to show the VLE data of a system in multiple states on the same graph. This allows a quick check of the effect of specific state changes (temperature or pressure) on the existing phase equilibria. For example, Fig. 5-13 shows the changes of trifluoromethane (1) (CHF_3) and trifluorochloromethane (2) ($CClF_3$) systems at four temperatures, ranging from about 75 K. On the vertical axis (left and right), you can see how the vapor pressure of the pure component changes with temperature. Because of this great change in vapor pressure, the data is usually best represented by a semi logarithmic graph, where the pressure axis is logarithmic. It is noted that the system exhibits azeotropes at all four temperatures, and the azeotrope composition shifts from 55% mole of trifluoromethane at 199 K to 65% mole of trifluoromethane at 273 K. Azeotropic components are represented by filled circles.

The second component in the mixture is trifluorochloromethane, which has one chlorine and three fluorine atoms connected to one carbon atom. If we replace this chlorine atom with fluorine atom to produce tetrafluoromethane, also known as carbon tetrafluoride, what effect will it have on the mixture? The trifluoromethane (1) + tetrafluoromethane (2) system is shown in Fig. 5-14. The simple exchange of these two halogens (chlorine for fluorine) affects the phase equilibrium in an important way. We can't calculate the vapor pressure of pure components by looking at two longitudinal axes at the same time. We can see that the left axis (corresponding to pure tetrafluoromethane) is between 225 K and 255 K. This is because the critical temperature of CF_4 is 227 K, so this will affect the critical point of the mixture. It is worth noting that the critical temperature of trifluoromethane is 302 K, while that of trifluoromethane is 299 K. Therefore, for the state shown in Fig. 5-13, no problem of the critical point of the mixture is observed.

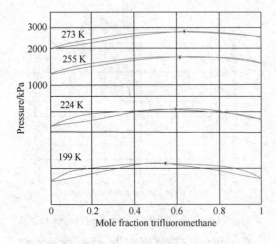 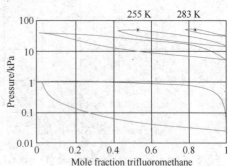

Fig.5-13 *p-xy* plot for the trifluoromethane (1) + chlorotrifluoromethane (2) system at four different temperatures. The filled circles are the azeotropic compositions

Fig.5-14 *p-xy* plot for the trifluoromethane (1) + tetrafluoromethane (2) system at four different temperatures

5.2.4 Lever Rule and the Flash Problem

Think about the discussion that defines bubble points and dew points earlier. For the bubble point, we have a balanced mixture, except for the first bubble, all of which are liquids. So, the phase point is on the bubble point curve. Similarly, the dew point describes a system that is a balanced mixture of all vapor except the first drop of liquid (or dew). Between bubbles and dew points, each equilibrium phase of different quantities (mass or mole) can be achieved. At the bubble point, the system is about 100% liquid and single bubble vapor. At dew point, it is about 100% of the water vapor and a drop of liquid. This will tell us the amount of each phase, depending on the position of the equilibrium point on the connection (called the tie line) connecting the two phases. This can be solved in two ways: (1) lever rules and (2) substance balance.

Lever rule is a simple geometric argument, which utilizes the linear relationship between composition and mass (or mole). If the composition on the *p-xy* or *T-xy* plot is expressed in mole fraction, the resulting lever rule is for mole. Or, if the component is a mass fraction, the lever rule that is generated is mass. Since lever rules work in the same way as *p-xy* or *T-xy* plots, we will describe the *p-xy* plots in the following example.

EXAMPLE 5-5

Lever Rule

If mixing 4 moles of methanol and 6 moles of N-butyl alcohol together at 303.15 K and 10.0 kPa, what are the composition and amounts of the phase(s) present at equilibrium? Use the lever rule.

p-xy plot for the methanol (1) + N-butyl alcohol (2) system at 303.15 K.

SOLUTION

Step 1 Where are the points on the mixed phase diagram? Liquid, vapor, or both?

Fortunately, we have a phase diagram of the system, which contains the states (T and p) associated with the problem. When we mix the pure compounds together, the composition of the mixture is 40% moles of methanol. In order to avoid confusion with our nomenclature for the mole fractions of liquid (x) and vapor (y), we can call this total component w. Therefore, the total component is $w_1 = 0.4$. When we look at Fig. 5-15, this is a

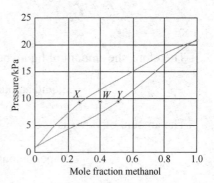

Fig.5-15 p-xy plot for the methanol (1) + N-butyl alcohol (2) system at 303.15 K

fixed temperature (303.15 K), we can find the system in the figure at 10.0 kPa and $w_1 = 0.4$. We see that it's between the bubble and the dew point curve, so we have two phases (liquid and vapor). If we describe the bubble point curve of the composition and temperature of the system at 11.0 kPa: we have a liquid composition of 10 moles, $x_1 = 0.4$. Similarly, if we are below the bubble point curve, at this composition and temperature at 9.0 kPa: we have 10 moles of vapor, and the composition $y_1 = 0.4$. However, neither case is here, we need to determine the composition and quantity of the liquid, as well as the vapor pressure.

Step 2 Application of the lever rule

Under the given temperature, pressure and composition, there is no single equilibrium phase in the mixture, but the system is divided into two phases. Specifically, the mixture splits into a liquid phase at $x_1 = 0.28$. and a vapor phase at $y_1 = 0.52$. The lever rule will help us find the amount of each phase (here is the number of moles). Let x denote the point where the line intersects the leftmost curve. In our problem, this is $p = 10$ at $x_1 = 0.28$ tie lines. Let y denote the point where the tie line intersects the rightmost curve. In our problem, this is $y_1 = 0.52$, $p = 10$. Let w represent the composition of the total mixture at $p = 10$ and $w_1 = 0.4$. The quantity of the leftmost phase (liquid phase is p-xy plot; for the vapor of a T-xy plot), we give the ratio of the length of line WY to the length of line xy, multiplied by the quantity (n) of the whole system

$$\text{Liquid amount} = n\left(\overline{WY} / \overline{XY}\right)$$

Likewise, the vapor amount (for a p-xy plot) is given as

$$\text{Vapor amount} = n\left(\overline{XW} / \overline{XY}\right)$$

Alternatively, the amount of vapor can be determined by subtracting the amount of liquid from the whole system by a simple mass balance.

Step3 Using the lever rule to obtain the amounts of each equilibrium phase

Implementing equations for the distances relevant in Fig. 5-15, we arrive at

$$\overline{XW} = 0.5 \text{ cm}.$$

$$\overline{XY} = 1.0 \text{ cm}$$

$$\overline{WY} = 0.5 \text{ cm}$$

Therefore, the amount of liquid becomes

$$\text{Liquid amount} = 10 \text{ mol} \times \frac{0.5 \text{ cm}}{1 \text{ cm}} = 5 \text{ mol}$$

The amount of vapor is

$$\text{Vapor amount} = 10 \text{ mol} \times \frac{0.5 \text{ cm}}{1 \text{ cm}} = 5 \text{ mol}$$

5.3 Theory and Model of Vapor Liquid Equilibrium of Mixtures: Modified *Raoult's law* Method

In Chapter 4, we lay the foundation for the establishment of vapor-liquid equilibrium model of mixture, that is, the assumption of ideal gas and ideal solution-*Raoult's law*. However, as we can see in Chapter 4, the assumption of *Raoult's law* is very limited, so this model can't be used in practice, only for a small part of the mixture. In Chapter 4, the technology we focus on is not the ideal solution, but the real solution. This modeling method is generally called modified *Raoult's law*.

5.3.1 Examples of Incentives

Bioethanol is ethanol created from a biomass source. In this process, cellulose is extracted from biomass and converted to monosaccharides by enzymes (or other processes), the sugar is then fermented into ethanol by yeast, water is used as a detergent. It is very important to remove water from bioethanol. Therefore, it is necessary to understand the phase equilibrium of ethanol + water mixture.

Referring to Fig. 5-16, the vapor-liquid equilibrium of water + ethanol system at 30℃ is given. The solid line is the experimental data, and the dotted line is the model prediction (*Raoult's law*). Obviously, this model does not describe the system accurately. Finally, when we size the separation equipment in the factory (for example, in the case of bioethanol separation), we need experimental data. So, if we need experimental data, what is the focus of modeling?

The answer to this question comes from combinatorics. It is estimated that biological systems can create about 10^{60} compounds, which is an elusive number (Dobson, 2004). How can we think of the number of binary, ternary and other mixtures we can create? Of course, we can't evaluate all the properties of all the mixtures under all the conditions of interest (temperature, pressure and composition) through experiments. However, if we have a way to estimate some of the properties of some mixtures under various conditions, we can make smarter process related decisions ahead of time rather than having to perform costly (time and money) experiments - at least at the beginning. Therefore, we are motivated to make models that are not subject to specific conditions (for example, ideal solutions), which is the focus of this chapter. In order to do this, we need to go back and build a foundation for the basic principles of mixture phase equilibrium, so that we can build our modeling method to describe the experimental results more accurately. In short, we need to address the "cause" of phase equilibria in mixtures, as we did when we discussed the pure component phase equilibrium.

5.3.2 Phase Equilibrium of Mixture

Let's go back to Chapter one, where we identified three driving forces for system change: mechanical, thermal, and chemical. In the previous work, we established that a pure compound existed as two phases in equilibrium, and all the driving forced must be 0. Therefore, the two phases must have the same temperature (no thermal driving force), the same pressure (no mechanical driving force) and the same molar Gibbs free energy (no chemical driving force). Do these constraints hold for mixtures? Let's consider a thought experiment.

In Fig.5-16, we have a closed system that surrounds two open systems (α and β). Each open system contains two components (dark and dark). Recall that an open system means that mass and energy can flow through the boundary of the system (represented by a dotted box). If we say that the two open systems are in phase equilibrium, it means that the temperatures and pressure of the two phases (α and β) are the same. We know that for pure components in phase equilibrium, the Gibbs free energy of phases is the same.

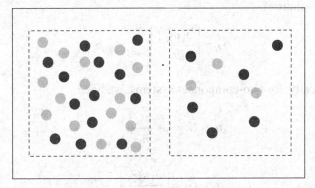

Fig.5-16 Two open systems (α and β) that are encased in a closed system

The Gibbs free energy of a two-component mixture can be expressed as

$$n\underline{G} = f(T, p, n_1, n_2) \tag{5-68}$$

$$d(n\underline{G}) = \left[\frac{\partial(n\underline{G})}{\partial T}\right]_{p,n_1,n_2} dT + \left[\frac{\partial(n\underline{G})}{\partial p}\right]_{T,n_1,n_2} dp + \left[\frac{\partial(n\underline{G})}{\partial n_1}\right]_{T,p,n_2} dn_1 + \left[\frac{\partial(n\underline{G})}{\partial n_2}\right]_{T,p,n_1} dn_2$$

$$\tag{5-69}$$

According to

$$n\underline{S} = -\left[\frac{\partial(n\underline{G})}{\partial T}\right]_p$$

$$nV = \left[\frac{\partial(nG)}{\partial p}\right]_T$$

However, since the first two terms of Eq. (5-69) remain unchanged, they are mathematically equivalent to the equation. Therefore, we can simplify Eq. (5-69) to

$$d(nG) = (-nS)dT + (nV)dp + \left[\frac{\partial(nG)}{\partial n_1}\right]_{T,p,n_2} dn_1 + \left[\frac{\partial(nG)}{\partial n_2}\right]_{T,p,n_1} dn_2 \qquad (5\text{-}70)$$

This expression is more compact, but let's look at the last two terms. The first thing to notice is that it looks like a partial mole. Recall that we defined the partial molar M of component i.

$$\bar{M}_i = \left[\frac{\partial(nM)}{\partial n_i}\right]_{T,p,n_{j\neq i}} \qquad (5\text{-}71)$$

Therefore, the last two terms in Eq. (5-70) are partial molar Gibbs free energy. However, this partial derivative has an additional name called chemical potential (μ_i). The chemical potential of pure substances has been introduced in the previous section. For pure components, we know that the chemical potential is equal to the molar Gibbs free energy. However, this is not the case for mixtures where the chemical potential of a component in the mixture is equal to the partial molar Gibbs free energy of the component (not the molar Gibbs free energy of the solution).

$$\bar{G}_i = \left[\frac{\partial(nG)}{\partial n_i}\right]_{T,p,n_{j\neq i}} = \mu_i \qquad (5\text{-}72)$$

More specifically, for two-component systems, we have

$$\mu_1 = \bar{G}_1 = \left[\frac{\partial(nG)}{\partial n_1}\right]_{T,p,n_2} \qquad (5\text{-}73)$$

$$\mu_2 = \bar{G}_2 = \left[\frac{\partial(nG)}{\partial n_2}\right]_{T,p,n_1} \qquad (5\text{-}74)$$

Now we can replace it with Eq. (5-74).

$$d(nG) = (-nS)dT + (nV)dp + \mu_1 dn_1 + \mu_2 dn_2 \qquad (5\text{-}75)$$

Going back to Fig. 5-16, if we say that our whole system is in equilibrium, we can now write Eq. (5-75) for each of our three systems (population and two open systems - α and β). Here, we already know that the temperature and pressure of the two open systems (α and β) are the same, so we don't need to mark T and p.

Overall (closed):

$$d(nG) = (-nS)dT + (nV)dp + \mu_1 dn_1 + \mu_2 dn_2 \qquad (5\text{-}76)$$

α (open):

$$\mathrm{d}(n\underline{G})^\alpha = (-n\underline{S})^\alpha \, \mathrm{d}T + (n\underline{V})^\alpha \, \mathrm{d}p + \mu_1^\alpha \, \mathrm{d}n_1^\alpha + \mu_2^\alpha \, \mathrm{d}n_2^\alpha \tag{5-77}$$

β (open):

$$\mathrm{d}(n\underline{G})^\beta = (-n\underline{S})^\beta \, \mathrm{d}T + (n\underline{V})^\beta \, \mathrm{d}p + \mu_1^\beta \, \mathrm{d}n_1^\beta + \mu_2^\beta \, \mathrm{d}n_2^\beta \tag{5-78}$$

So, n_1 and n_2 are constants in this closed system, both $\mathrm{d}n_1$ and $\mathrm{d}n_2$ are 0. For a closed system, no mass can leave the boundary of the system. This simplifies the closed expression to:

Overall (closed):

$$\mathrm{d}(n\underline{G}) = (-n\underline{S})\mathrm{d}T + (n\underline{V})\mathrm{d}p \tag{5-79}$$

Any change of Gibbs free energy of open system α or β is a part of the change of the whole system. Therefore,

$$\mathrm{d}(n\underline{G}) = \mathrm{d}(n\underline{G})^\alpha + \mathrm{d}(n\underline{G})^\beta \tag{5-80}$$

$$(-n\underline{S})\mathrm{d}T + (n\underline{V})\mathrm{d}p = (-n\underline{S})^\alpha \, \mathrm{d}T + (n\underline{V})^\alpha \, \mathrm{d}p + \mu_1^\alpha \mathrm{d}n_1^\alpha + \mu_2^\alpha \mathrm{d}n_2^\alpha + (-n\underline{S})^\beta \, \mathrm{d}T$$
$$+ (n\underline{V})^\beta \, \mathrm{d}p + \mu_1^\beta \mathrm{d}n_1^\beta + \mu_2^\beta \mathrm{d}n_2^\beta$$

The results can be obtained by inserting Eq. (5-77), Eq. (5-78) and Eq. (5-79) into Eq. (5-80). For molar entropy and molar volume, the same additive relationship can be obtained.

The above expression can be simplified to:

$$0 = \mu_1^\alpha \mathrm{d}n_1^\alpha + \mu_2^\alpha \mathrm{d}n_2^\alpha + \mu_1^\beta \mathrm{d}n_1^\beta + \mu_2^\beta \mathrm{d}n_2^\beta \tag{5-81}$$

By paying attention to the following relation, we can further simplify Eq. (5-82). For example, if a mole of component 1 leaves phase β and enter the phase α, the change of the mole number of components in phase β (which will be negative) is equivalent to changing the phase of component 1 in phase α (positive). From a mathematical point of view, this is expressed as

$$\mathrm{d}n_1^\alpha = -\mathrm{d}n_1^\beta \tag{5-82}$$

$$\mathrm{d}n_2^\alpha = -\mathrm{d}n_2^\beta \tag{5-83}$$

In the same way, Eq. (5-81) is simplified as

$$0 = \mu_1^\alpha \mathrm{d}n_1^\alpha + \mu_2^\alpha \mathrm{d}n_2^\alpha - \mu_1^\beta \mathrm{d}n_1^\alpha - \mu_2^\beta \mathrm{d}n_2^\alpha \tag{5-84}$$

Thus,

$$0 = \left(\mu_1^\alpha - \mu_1^\beta\right)\mathrm{d}n_1^\alpha + \left(\mu_2^\alpha - \mu_2^\beta\right)\mathrm{d}n_2^\alpha \tag{5-85}$$

From the most general point of view, there are only two ways to satisfy this equation. First, we can make sum equal to 0. This is certainly a mathematical solution to Eq. (5-85), but it will exclude the crossing connection phase α and β. The possibility of mass transfer at the system boundary. From the point of view of experimental observation, and from the point of view of phase equilibrium that we introduced in Chapter 5, we know that this solution does not happen. Another general result of Eq. (5-85) is to force every item in brackets to be 0 (while allowing $\mathrm{d}n_1^\alpha$ and $\mathrm{d}n_2^\alpha$ to be general).

$$\left(\mu_1^\alpha - \mu_1^\beta\right) = 0 \tag{5-86}$$

$$\left(\mu_2^\alpha - \mu_2^\beta\right) = 0 \tag{5-87}$$

Thus,

$$\mu_1^\alpha = \mu_1^\beta \tag{5-88}$$

$$\mu_2^\alpha = \mu_2^\beta \tag{5-89}$$

Eq. (5-88) and Eq. (5-89) constitute a powerful result and show that in a given system, the chemical potential of a substance has the same value in any phase in equilibrium. When a system has a chemical potential difference in one or more species, it is not in equilibrium, which creates the mass transfer driving force as the system equilibrium (in the same way as the temperature thermal equilibrium driving force and pressure are in dynamic mechanical equilibrium). This equality of individual chemical potentials is called chemical equilibrium condition.

The chemical potential of a species has the same value in any phases that are in equilibrium in each system.

$$\mu_i^\alpha = \mu_i^\beta \tag{5-90}$$

This equality of chemical potentials for the individual species is called the chemical equilibrium condition.

5.3.3 Fugacity of Mixture

It is easy for us to get lost in the maze of column crossing, underline, subscript, superscript and so on. Therefore, in order to improve the clarity, the introduction of mixture fugacity in this section will be carried out in a systematic way.

In Chapter 5, pure component fugacity is introduced as a substitute for Gibbs free energy.

$$d\underline{G} = RT d\ln f \tag{5-91}$$

Similarly, we can define the fugacity of the components in the mixture as

$$d\mu_i = RT d\ln \hat{f}_i \tag{5-92}$$

Note that the mixed fugacity sign of component i has a carat (^) to distinguish it from pure component fugacity. Similarly, we do not use the kPa here, because the fugacity of the mixture of component i is a mixture property, but not a partial molar property.

In this section, two things will be related to the fugacity of component i. First, we'll set it up for the modeling phase, which we'll see through the components for the rest of this chapter. Second, we will rewrite the phase equilibrium requirements of mixtures in terms of unstable components, just like our pure component instability in Chapter 5. First, let's go back to the definition of the fugacity of component i shown in Eq. (5-90). We can integrate this expression from the pure component ($x_i = 1$) to the mixture component at the same temperature and pressure.

$$\int_{\mu_i(T,p)}^{\mu_i(T,p,\underline{x})} d\mu_i = \int_{\ln f_i(T,p)}^{\ln \hat{f}_i(T,p,\underline{x})} RT d\ln \hat{f}_i \qquad (5\text{-}93)$$

Perform integration generation:

$$\mu_i(T,p,\underline{x}) - \mu_i(T,p) = RT \ln \hat{f}_i(T,p,\underline{x}) - RT \ln f_i(T,p) \qquad (5\text{-}94)$$

Because the sign of chemical potential is the same, whether it is pure component or mixture, we can use the excess molar Gibbs free energy of pure component to replace the chemical potential of pure component to avoid some confusion. Therefore,

$$\mu_i(T,p,\underline{x}) = \underline{G}_i + RT \ln \left[\frac{\hat{f}_i(T,p,\underline{x})}{f_i(T,p)} \right] = \overline{G}_i \qquad (5\text{-}95)$$

Note that for Eq. (5-71), the chemical potential of component i in the mixture is also equal to the partial molar Gibbs free energy of that component, which is expressed in Eq. (5-95).

With Eq. (5-95), we can explore two important models, our vapor-liquid equilibrium model, that is, the ideal solution of liquid and the ideal gas of vapor.

If we want to know the chemical potential of substance i in a mixture modeled as an ideal solution, we have

$$\mu_i^{im}(T,p,\underline{x}) = \underline{G}_i + RT \ln \left[\frac{\hat{f}_i^{im}(T,p,\underline{x})}{f_i(T,p)} \right] \qquad (5\text{-}96)$$

In order to simplify Eq. (5-96), we will reconsider the expression of molar Gibbs free energy of ideal solution.

In combination with Eq. (5-95), we can see that

$$\overline{G}_i^{im} = \underline{G}_i + RT \ln x_i \qquad (5\text{-}97)$$

Or in general, Eq. (5-96) and Eq. (5-97) are the same, because the chemical potential of the ideal solution of component i is equal to the partial molar Gibbs free energy of the ideal solution of component i. By comparing the terms between Eq. (5-96) and Eq. (5-97), we can see that

$$\frac{\hat{f}_i^{im}(T,p,\underline{x})}{f_i(T,p)} = x_i \qquad (5\text{-}98)$$

This relationship is usually written as

$$\hat{f}_i^{im}(T,p,\underline{x}) = x_i f_i(T,p) \qquad (5\text{-}99)$$

This is known as the *Lewis Randall rule*. It says that the fugacity of the component i modeled as an ideal solution is equal to the fugacity of its pure component at mixing temperature and pressure multiplied by its mole fraction.

Replace this relationship back to the Eq. (5-96) yield equation.

$$\mu_i^{ID}(T,p,\underline{x}) = \underline{G}_i + RT \ln x_i \qquad (5\text{-}100)$$

Let's find the partial molar Gibbs free energy of the ideal gas component more directly. This was done earlier in Chapter 5.

$$\mu_i^{ig}(T,p,\underline{y}) = \underline{G}_i^{ig} + RT\ln\left[\frac{\hat{f}_i^{ig}(T,p,\underline{y})}{f_i^{ig}(T,p)}\right] \qquad (5\text{-}101)$$

Since the chemical potential of the ideal gas of component *i* is equal to the partial molar Gibbs free energy of the ideal gas of component *i*, we can equate Eq. (5-100) and Eq. (5-101).

$$\frac{\hat{f}_i^{ig}(T,p,\underline{y})}{f_i^{ig}(T,p)} = y_i \qquad (5\text{-}102)$$

From Chapter 5, we know that the fugacity of the pure component of an ideal gas is exactly the system pressure; therefore, we can model the fugacity of component *i* of the mixture as the ideal gas.

$$\hat{f}_i^{ig}(T,p,\underline{y}) = f_i^{ig}(T,p)y_i = py_i \qquad (5\text{-}103)$$

Replace Eq. (5-103) with the yield of Eq. (5-105).

$$\mu_i^{ig}(T,p,\underline{y}) = \underline{G}_i^{ig} + RT\ln\left(\frac{py_i}{p}\right) \qquad (5\text{-}104)$$

$$\mu_i^{igm}(T,p,\underline{y}) = \underline{G}_i^{ig} + RT\ln y_i \qquad (5\text{-}105)$$

In retrospect, so far, we have determined how to write the chemical potential of component *i* in the mixture under the ideal solution and ideal gas model. In addition, we also determined how to write the fugacity of the mixture of component *i* in the state of ideal solution Eq. (5-97) and ideal gas Eq. (5-101). Our next step includes determining the relationship, which will help us model when the system deviates from the behavior of ideal solution and ideal gas.

In order to capture the deviation from the ideal solution, we use the excess property. Therefore, we can use Eq. (5-95) and Eq. (5-104) to calculate the excess chemical potential of component *i* in the mixture. Therefore,

$$\mu_i^E(T,p,x) = \mu_i(T,p,x) - \mu_i^{im}(T,p,x)$$
$$= \underline{G}_i + RT\ln\left[\frac{\hat{f}_i(T,p,x)}{f_i(T,p)}\right] - \left(\underline{G}_i + RT\ln x_i\right) \qquad (5\text{-}106)$$

$$\mu_i^E(T,p,x) = RT\ln\left[\frac{\hat{f}_i(T,p,x)}{x_i f_i(T,p)}\right] = RT\ln\gamma_i \qquad (5\text{-}107)$$

Within logarithm, we define the activity coefficient of component *i*, γ_i as the ratio of the fugacity of component *i* to its ideal solution value (i.e., *Lewis Randall rule*). The activity coefficient represents the deviation from the ideal solution.

Let me put it another way, now we move to the vapor phase and look for deviations from the ideal gas. To do this, we use the remaining attributes. Therefore, we can use Eq. (5-95) and Eq. (5-104) to calculate the residual chemical potential of component *i* in the mixture. Therefore,

Chapter 5　Phase Equilibrium

$$\mu_i^R(T,p,\underline{y}) = \mu_i(T,p,\underline{y}) - \mu_i^{ig}(T,p,\underline{y})$$
$$= \underline{G}_i + RT\ln\left[\frac{\hat{f}_i(T,p,\underline{y})}{f_i(T,p)}\right] - \left(\underline{G}_i^{ig} + RT\ln y_i\right) \quad (5\text{-}108)$$

$$\mu_i^R(T,p,\underline{y}) = \left(\underline{G}_i - \underline{G}_i^{ig}\right) + RT\ln\left[\frac{\hat{f}_i(T,p,\underline{y})}{y_i f_i(T,p)}\right] \quad (5\text{-}109)$$

Unlike the excess chemical potential, the molar Gibbs free energy of the pure component can be eliminated in Eq. (5-107). We must consider the pure component deviation from the ideal gas state. However, as we have done in Chapter 5, the equation we need is Eq. (5-110).

$$\left(\underline{G}_i - \underline{G}_i^{ig}\right) = RT\ln\left[\frac{f_i(T,p)}{p}\right] \quad (5\text{-}110)$$

Therefore

$$\mu_i^R(T,p,y) = RT\ln\left[\frac{f_i(T,p)}{p}\right] + RT\ln\left[\frac{\hat{f}_i(T,p,y)}{y_i f_i(T,p)}\right] \quad (5\text{-}111)$$

Rate of return combined with logarithm.

$$\mu_i^R(T,p,y) = RT\ln\left[\frac{\hat{f}_i(T,p,y)}{y_i p}\right] = RT\ln\hat{\varphi}_i \quad (5\text{-}112)$$

Within the logarithm, we define the mixture fugacity coefficient of component i, $\hat{\varphi}_i$ as the ratio of the fugacity of the mixture of component i to its ideal gas value, with the sign of carat above, to distinguish it from the fugacity coefficient of pure component introduced in Chapter 5.

$$\hat{f}_i(T,p,y) = \hat{f}_i^{ig}\hat{\varphi}_i = y_i p \hat{\varphi}_i \quad (5\text{-}113)$$

As we can see, the fugacity coefficient of the mixture captures the deviation from the behavior of the ideal gas. So, before we leave this part and continue to use the activity coefficient and fugacity coefficient to build the VLE model, we need to rewrite the chemical equilibrium requirement about fugacity.

The chemical equilibrium condition of phase equilibrium in the mixture is

$$\mu_i^\alpha = \mu_i^\beta \quad (5\text{-}114)$$

So, we can use Eq. (5-93) to do a direct substitution.

$$\underline{G}_i^\alpha + RT\ln\left[\frac{\hat{f}_i(T,p,\underline{x})^\alpha}{f_i(T,p)^\alpha}\right] = \underline{G}_i^\beta + RT\ln\left[\frac{\hat{f}_i(T,p,\underline{x})^\beta}{f_i(T,p)^\beta}\right] \quad (5\text{-}115)$$

Here, the temperature and pressure of the two phases (α and β) are the same, because the equality of temperature and pressure is also the requirement of phase equilibrium. In addition, because the pure components ($\underline{G}_i$ and f_i) are calculated at the same temperature and pressure, they will be the same.

$$RT\ln \hat{f}_i(T,p,\underline{x})^\alpha = RT\ln \hat{f}_i(T,p,\underline{x})^\beta \tag{5-116}$$

Then,

$$\hat{f}_i(T,p,\underline{x})^\alpha = \hat{f}_i(T,p,\underline{x})^\beta \tag{5-117}$$

This tells us that the chemical equilibrium condition of the mixture is equivalent to the fugacity condition of the mixture, like that of the pure component. Please note that this result is not new information "in addition to" chemical equilibrium conditions. The result is the same, but it's written differently.

In Chapter 5, we introduce the fugacity coefficient as well as the pure component fugacity. We find that the equivalence of fugacity coefficient is another way to describe the chemical equilibrium constraints of pure components. However, this is not the case for mixtures, which is defined by the fugacity coefficient in the mixture. Specifically, by substituting Eq. (5-114) into Eq. (5-110), we get

$$\hat{f}_i(T,p,y)^\alpha = \hat{f}_i(T,p,y)^\beta \tag{5-118}$$

Since the pressures of the two phases are equal at equilibrium, we obtain equation. Therefore, when the fugacity coefficient of mixture is used, the chemical equilibrium conditions include mole fraction. Moreover, as we see in Chapter 5, the mole fractions of different phases are almost always unequal at equilibrium (except in azeotropes).

Now that we have used fugacity to describe the phase equilibrium conditions of mixtures and the framework for describing the behavior of deviating from ideal solutions and gases, we can begin to model carefully.

5.3.4 Gamma-Phi Modeling

In establishing the phase equilibrium model of the mixture, we always start with the formal phase equilibrium condition, that is, one component of the mixture has the same mixture fugacity in any phase it appears Eq. (5-118). For specific situations in which liquid and steam mixtures are balanced, we have:

$$\hat{f}_i^L = \hat{f}_i^V \tag{5-119}$$

In order to describe the fugacity of component i in the liquid phase, we use an ideal solution as a reference. Therefore, by Eq. (5-120), we have:

$$\hat{f}_i^L(T,p,x) = x_i f_i(T,p)\gamma_i \tag{5-120}$$

Where is the behavior of deviating from the ideal solution captured by the activity coefficient γ_i?

Similarly, in order to describe the fugacity of component i in the vapor phase, we use an ideal gas as a reference.

$$\hat{f}_i^V(T,p,y) = y_i p \hat{\varphi}_i \tag{5-121}$$

The deviation from the ideal gas behavior is captured by the mixture fugacity coefficient of component i, φ_i (phi).

Using the left side of Eq. (5-121), Eq. (5-120) and the right of Eq. (5-121), we have:

$$\hat{f}_i^V(T,p,y) = y_i p \hat{\varphi}_i \tag{5-122}$$

This is the "gamma-phi" equation, which is a common method to simulate the gas-liquid equilibrium behavior of the mixture. The next step in this chapter is to establish a model for how to best (or most easily) describe the liquid and vapor phases using Eq. (5-121).

5.3.5 *Raoult's law* Revisited

The simplest model for vapor-liquid equilibrium of mixtures (which we have described before) is *Raoult's law*. Using this method, we simulate liquid as ideal solution and vapor as ideal gas. We got this fugacity in the previous section, and we provide them here again. For ideal solutions, we have:

$$\hat{f}_i^L = f_i^{im} = x_i f_i \tag{5-123}$$

This means that the activity coefficient of element i is equal to 1 when it is simulated as an ideal solution.

Similarly, the fugacity of a component in the vapor mixture simulated as an ideal gas is also changed to

$$\hat{f}_i^V = f_i^{ig} = y_i p \tag{5-124}$$

This means that the fugacity coefficient of component i is equal to 1.

Replace these results with Eq. (5-119).

$$x_i f_i(T,p) = y_i p \tag{5-125}$$

We can calculate the fugacity of pure component i in the liquid phase at the mixture temperature and the vapor pressure of pure component at that temperature, and then adjust the mixture pressure by Poynting correction.

$$x_i f_i^{sat}(T,p_i^{sat}) e^{\frac{V_i(p-p_i^{sat})}{RT}} = y_i p \tag{5-126}$$

Because the fugacity of the pure component is equal to the pressure multiply the fugacity coefficient.

$$x_i p_i^{sat} \varphi_i^{sat}(T,p_i^{sat}) e^{\frac{V_i(p-p_i^{sat})}{RT}} = y_i p \tag{5-127}$$

However, as we can see in Eq. (5-127), the fugacity value of liquid at high pressure is obviously different from the vapor pressure at that temperature. Because we have assumed that vapor is an ideal gas at low pressure. Therefore, the fugacity coefficient will be very close to 1, and so will *Poynting* correction. Their product can be assumed to be:

$$\varphi_i^{sat} e^{\frac{V_i(p-p_i^{sat})}{RT}} \sim 1 \tag{5-128}$$

Eq. (5-127) is reduced to *Raoult's law*, which we have described and used in Chapter 5. Therefore,

$$x_i p_i^{sat} = y_i p \qquad (5\text{-}129)$$

Recall that *Raoult's law* works only for systems with similar sizes and intermolecular interactions. We can see that the system of benzene + toluene accords with *Raoult's law*, but the system of ethanol + water does not. However, there is a specific condition that most systems have a component that behaves according to *Raoult's law* at low to medium pressure regardless of whether these molecules have similar size or similar intermolecular interactions, and are close to the end points of pure components.

5.3.6 *Henry's law*

Consider Fig. 5-17, which is a mixture of two kinds of molecules. It was observed that the system was concentrated in light molecules and diluted in dark molecules. Many light molecules do not have a nearest neighbor, which is the dark molecule. In fact, many light molecules are not affected by dark molecules at all, because there are very few dark molecules. In this case, the light molecule is basically interacting with other light molecule "of similar size, similar intermolecular interaction". The existence of this system does not affected the molecule by the presence of dark molecules (by a good approximation) and is linearly proportional to the amount of its existence (pure component fugacity with a proportional constant).

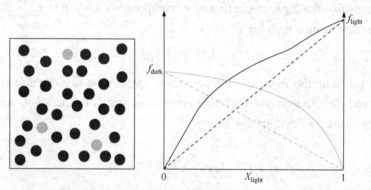

Fig.5-17 A cartoon representation showing a mixture of light and dark molecules (concentrated in light) as well as the mixture fugacity of the light and dark molecules as a function of composition

The dashed lines indicate the *Lewis- Randall law*.

In other words, the light molecule well simulates the ideal solution in this state. Fig. 5-17 shows a cartoon of the variation of the fugacity of light molecules in solution with the composition. Therefore, when the concentration of light molecules is very high, *Lewis Randall law* holds for light molecules.

As we mentioned earlier, there are so many dark molecules that most light molecules are not surrounded by any dark molecules in fact, they are not significantly affected by any other dark molecules. Similarly, when the concentration of dark molecules is low, one dark molecule will hardly be surrounded by another dark molecule, and will not be affected by any dark molecule. It's like every dark molecule is an ocean of light molecules. Therefore, the influence of each dark molecule is not (by a good approximation) affected by the rest of the dark molecules, and its fugacity is linearly proportional to its concentration. However, in this case, the scale constant is not the pure compositional fugacity of the dark molecule, but some terms that include how a single dark molecule interacts with the ocean of light molecules.

This proportional constant, known as Henry constant (H), is unique to the type of molecules in the mixture and the temperature of the system. Then we write the fugacity of the dark molecule according to *Henry's law*.

$$\hat{f}_{\text{dark}} = x_{\text{dark}} H_{\text{dark,light}} \tag{5-130}$$

Although we have spent some time describing and calculating the results of *Raoult's law* in the previous chapter, this is our first introduction to *Henry's law*. So, while *Henry's law* of course applies to the dilution region, for mixtures that we have encountered, this method is usually reserved for describing the small amount of gas dissolved in a liquid. *Henry's law* tells us that when a gas (a gas is above its critical temperature, but at a sufficiently low pressure) meets the liquid and is in equilibrium, a small amount of gas will dissolve in the liquid. It is better to think of it as a "gas dissolved in a liquid" rather than condensing into a liquid phase, because gases (by definition) cannot condense. Our daily life is surrounded by gases rather than vapors (oxygen, nitrogen, carbon dioxide, etc.). So, *Henry's law* provides a good model for how much of these gases are in liquids, such as water. In fact, when you boil water in a pan on the stove and heat it (but far below the boiling point), the small bubbles that leave the liquid are the dissolved gases in the water, and they leave the liquid phase. This common observation indicates that at higher temperatures, the oxygen concentration in water is lower than at lower temperatures.

EXAMPLE 5-6

In an intermediate calculation to size an absorber, you need to determine the partial pressure of benzene in the vapor phase, in equilibrium with a liquid composed of 99.9 mol% water and 0.1 mol% benzene. The infinite dilution activity coefficient for benzene in water at this temperature is 2500.

$$p_b^{\text{sat}} = 0.1 \text{ atm}$$

Modified *Raoult's Law*

$$\hat{f}_i^{\text{L}} = \hat{f}_i^{\text{V}}$$

For the liquid,

$$\hat{f}_i^{\text{L}}(T, p, \underline{x}) = x_i f_i(T, p) \gamma_i$$

The activity coefficient of component i is the factor that accounts for the deviations to ideal solution behavior.

Replace the pure component fugacity and assume an ideal gas for the vapor phase.

$$x_i \gamma_i f_i^{\text{sat}}(T, p_i^{\text{sat}}) e^{\frac{V_i(p-p_i^{\text{sat}})}{RT}} = y_i p$$

In the previous section, we explored both *Raoult's Law*, which describes the vapor-liquid equilibrium for an ideal solution in equilibrium with an ideal gas, and *Henry's Law* for dissolved gases in liquids. The next step on our journey in phase equilibrium modeling of mixtures is to examine the much more common scenario: vapor-liquid equilibrium involving non-ideal solutions and (eventually) non-ideal gases.

Since the deviations from ideal solution have a much bigger impact on modeling the phase behavior than the deviations from ideal gas, we will focus on strategies to account for non-idealities in the liquid phase.

Write the mixture fugacity of component i in terms of the pure component fugacity, composition, and the activity coefficient.

Since we have assumed ideal gas for the vapor, the pressure is going to be low.

$$\varphi_i^{\text{sat}}(T, p_i^{\text{sat}}) e^{\frac{V_i(p-p_i^{\text{sat}})}{RT}} \sim 1$$

$$x_i \gamma_i p_i^{\text{sat}} \varphi_i^{\text{sat}} e^{\frac{V_i(p-p_i^{\text{sat}})}{RT}} = y_i p$$

$$x_i \gamma_i p_i^{\text{sat}} = y_i p$$

The above formula is commonly known as modified *Raoult's Law*.

Many models of solution thermodynamics are centered around obtaining good expressions and estimates for the activity coefficient.

The definition of the activity coefficient and its relationship to the excess chemical potential

$$\mu_i^{\text{E}}(T, p, x) = RT \ln \left[\frac{\hat{f}_i(T, p, x)}{x_i f_i(T, p)} \right] RT \ln \gamma_i$$

The chemical potential of species i in a mixture is also the partial molar Gibbs free energy of species i, so the excess chemical potential is the excess partial molar Gibbs free energy.

$$\mu_i^{\text{E}}(T, p, x) = RT \ln \gamma_i = \overline{G}_i^{\text{E}}$$

The chemical potential

$$\mu_i = \overline{G}_i = \left[\frac{\partial(n\underline{G})}{\partial n_i} \right]_{T, p, n_{j \neq i}}$$

The excess properties

$$\mu_i^{\text{E}} = \overline{G}_i^{\text{E}} = \left[\frac{\partial(n\underline{G}^{\text{E}})}{\partial n_i} \right]_{T, p, n_{j \neq i}}$$

The partial derivative is evaluated at constant T.

$$\frac{\mu_i^E}{RT} = \frac{\overline{G}_i^E}{RT} = \left[\frac{\partial(n\underline{G}^E/RT)}{\partial n_i}\right]_{T,p,n_{j\neq i}}$$

The natural logarithm of the activity coefficient is, in fact, a partial molar property of the excess molar Gibbs free energy divided by RT.

$$\ln\gamma_i = \frac{\mu_i^E}{RT} = \frac{\overline{G}_i^E}{RT} = \left[\frac{\partial(n\underline{G}^E/RT)}{\partial n_i}\right]_{T,p,n_{j\neq i}}$$

Since $\ln\gamma_i$ is a partial molar property of the excess molar Gibbs free energy, it allows us to use "summability" relationship for partial molar properties as

$$\frac{\underline{G}^E}{RT} = x_1\frac{\overline{G}_1^E}{RT} + x_2\frac{\overline{G}_2^E}{RT} = x_1\ln\gamma_1 + x_2\ln\gamma_2$$

The natural logarithm of the activity coefficient is a partial molar property of the excess molar Gibbs free energy in two ways: (1) we want to model real solutions using the activity coefficient and (2) there are useful relationships that apply to partial molar properties. Therefore, we exploit the fact that $\ln\gamma_i$ is a partial molar property in modeling approaches.

Excess Molar Gibbs Free Energy Models

In developing excess molar Gibbs free energy models, we have some built-in constraints that the models must satisfy.

$$x_1 \to 0, \ \frac{\underline{G}^E}{RT} = 0 \tag{5-131}$$

$$x_2 \to 0, \ \frac{\underline{G}^E}{RT} = 0 \tag{5-132}$$

A pure component is at both of those points (the pure components do a great job of validating the assumptions of an ideal solution-similar size and similar intermolecular interactions).

What would be the simplest expression that would satisfy these endpoints?

$$\frac{\underline{G}^E}{RT} = 0 \tag{5-133}$$

It is the ideal solution, since the excess molar Gibbs free energy is 0 at all compositions for the ideal solution.

What is the next simplest model that must satisfy those endpoint constraints(Fig.5-18)?

$$\frac{\underline{G}^E}{RT} = Ax_1x_2 \tag{5-134}$$

147

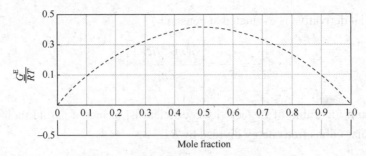

Fig.5-18 Excess molar Gibbs free energy models

The Gibbs-Duhem equation

$$x_1 d\bar{M}_1 + x_2 d\bar{M}_2 - \left(\frac{\partial M}{\partial T}\right)_{p,x_1,x_2} dT - \left(\frac{\partial M}{\partial p}\right)_{T,x_1,x_2} dp = 0 \tag{5-135}$$

Write this equation where $\dfrac{G^E}{RT}$ is our solution property ($\underline{M}$) and $\ln\gamma_1$ is its partical molar property.

$$x_1 d\ln\gamma_1 + x_2 d\ln\gamma_2 = 0 \tag{5-136}$$

A constraint that provides a relationship between activity coefficients that must be observed.

We can write this equation where G^E/RT is our solution property (M) and is its partial molar property.

If we consider this when evaluating p-xy data, we know our temperature is constant. Also, as we will show later in this chapter, G^E/RT does not change much with the pressure. Thus, the above equation provides a relationship between activity coefficients.

Do the activity coefficients from the 1-parameter Margules equation model obey this constraint?

$$\ln\gamma_i = \left[\frac{\partial\left(n\underline{G}^E/RT\right)}{\partial n_i}\right]_{T,p,n_{j\neq i}} \tag{5-137}$$

For a binary mixture,

$$\ln\gamma_1 = \left[\frac{\partial\left(n\underline{G}^E/RT\right)}{\partial n_1}\right]_{T,p,n_2} \tag{5-138}$$

$$\ln\gamma_2 = \left[\frac{\partial\left(n\underline{G}^E/RT\right)}{\partial n_2}\right]_{T,p,n_1} \tag{5-139}$$

$$\ln\gamma_1 = \left[\frac{\partial(nAx_1x_2)}{\partial n_1}\right]_{T,p,n_2} = \frac{\partial}{\partial n_1}\left[(n_1+n_2)A\frac{n_1 n_2}{(n_1+n_2)(n_1+n_2)}\right]_{T,p,n_2}$$

$$= \frac{\partial}{\partial n_1}\left(\frac{An_1 n_2}{n_1+n_2}\right)_{T,p,n_2} = An_2\frac{\partial}{\partial n_1}\left(\frac{n_1}{n_1+n_2}\right)_{T,p,n_2} = An_2\frac{\partial}{\partial n_1}\left[n_1(n_1+n_2)^{-1}\right]_{T,p,n_2} \tag{5-140}$$

$$\ln \gamma_1 = An_2 \frac{\partial}{\partial n_1}\left[n_1(n_1+n_2)^{-1}\right]_{T,p,n_2} = An_2\left[\frac{1}{n_1+n_2} - \frac{n_1}{(n_1+n_2)^2}\right] \quad (5\text{-}141)$$

$$= A\frac{n_2}{n_1+n_2}\left[\frac{n_1+n_2}{n_1+n_2} - \frac{n_1}{n_1+n_2}\right] = Ax_2(1-x_1) = Ax_2^2$$

Likewise,

$$\ln \gamma_2 = Ax_1^2 \quad (5\text{-}142)$$

Does the 1-parameter Margules equation satisfy the Gibbs-Duhem equation as given by $x_1 \mathrm{d}\ln\gamma_1 + x_2 \mathrm{d}\ln\gamma_2 = 0$?

Evaluate this differential relationship for small changes in $\mathrm{d}x_1$.

$$x_1 \frac{\mathrm{d}\ln\gamma_1}{\mathrm{d}x_1} + x_2 \frac{\mathrm{d}\ln\gamma_2}{\mathrm{d}x_1} = 0 \quad (5\text{-}143)$$

$$x_1 \frac{\mathrm{d}\ln\gamma_1}{\mathrm{d}x_1} + x_2 \frac{\mathrm{d}\ln\gamma_2}{\mathrm{d}x_1} = x_1 \frac{\mathrm{d}(Ax_2^2)}{\mathrm{d}x_1} + x_2 \frac{\mathrm{d}(Ax_1^2)}{\mathrm{d}x_1} = x_1 \frac{\mathrm{d}\left[A(1-x_1)^2\right]}{\mathrm{d}x_1} + x_2 \frac{\mathrm{d}(Ax_1^2)}{\mathrm{d}x_1} = 0 \quad (5\text{-}144)$$

If one develops an excess molar Gibbs free energy model that does not satisfy the Gibbs-Duhem equation, it is not thermodynamically correct and not a self-consistent model.

The activity coefficients from 1-parameter Margules equation are symmetric about $x_1 = 0.5$.

We see that the activity coefficients are symmetric about $x_1 = 0.5$. Therefore, the use of the 1-parameter Margules equation is limiting in that, while it considers the deviations from ideal solution, it does so in a way that does not distinguish between the components. System concentrated in one of the component A and dilute in the other component B will interact the same as a system concentrated in B and dilute in A, according to the 1-parameter Margules equation.

At this point, we have introduced the simplest excess molar Gibbs free energy model that can describe a non-ideal solution, the 1-parameter Margules equation. We have obtained the activity coefficient expressions for this model and have shown that the model does satisfy the Gibbs-Duhem equation. We also plotted the symmetric behavior of the activity coefficients from this model as a function of composition(Fig.5-19). However, rather than provide a list of different excess molar Gibbs free energy models (and there are several in use) we are going to focus on the utility of excess molar Gibbs free energy modeling approaches in solving problems. We will introduce the excess molar Gibbs free energy models in the context of the problems. Note that, once we develop the utility of a model, the same approach can be used for other excess molar Gibbs free energy models.

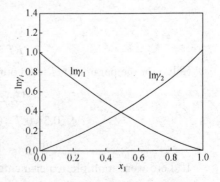

Fig.5-19 Activity coefficient symmetry of Margules equation

Excess Molar Gibbs Free Energy Models: Usage

EXAMPLE 5-7

A single vapor-liquid equilibrium points for the water (1) + ethanol (2) system is experimentally measured at 30℃.

The experiment provides the following information:

$$x_1 = 0.30, \quad y_1 = 0.23, \text{ and } p = 10.1 \text{ kPa}.$$

Use this information to estimate the system pressure and vapor-phase mole fraction when $x_1 = 0.80$.

$$p_1^{\text{sat}}(30℃) = 4.20 \text{ kPa}$$

$$p_2^{\text{sat}}(30℃) = 10.42 \text{ kPa}$$

SOLUTION

This system is at low pressure, so the ideal gas law is a good model for the vapor phase. However, water and ethanol are not ideal solutions.

$$\gamma_i x_i p_i^{\text{sat}} = y_i p$$

Our strategy is to use the known data at $x_1 = 0.3$ to obtain the parameters for an excess molar Gibbs free energy model. Then we can use this model at $x_1 = 0.8$ to estimate the activity coefficient and ultimately estimate the vapor phase composition and system pressure.

Calculating the activity coefficient from experimental data.

$$\gamma_1 = \frac{y_1 p}{x_1 p_1^{\text{sat}}} = \frac{0.23 \times 10.1}{0.3 \times 4.20} = 1.844$$

$$\gamma_2 = \frac{y_2 p}{x_2 p_2^{\text{sat}}} = \frac{0.77 \times 10.1}{0.7 \times 10.42} = 1.066$$

Excess molar Gibbs free energy model selection and parameterization.

$$\frac{G^{\text{E}}}{RT} = A x_1 x_2$$

$$\frac{G^{\text{E}}}{RT} = x_1 \ln \gamma_1 + x_2 \ln \gamma_2$$

Solve for the parameter A from our 1-parameter Margules equation:

$$A x_1 x_2 = x_1 \ln \gamma_1 + x_2 \ln \gamma_2$$

$$A \times 0.3 \times 0.7 = 0.3 \times \ln 1.844 + 0.7 \times \ln 1.066$$

$$A = 1.086$$

If there were multiple experimental data points, but one parameter, we would need to develop an objective function and minimize its value in order to determine the "best fit" parameter that would describe our data best over the range of data provided.

Predicting the activity coefficients at the new point.

$$\ln \gamma_1 = A x_2^2$$

$$\ln \gamma_2 = A x_1^2$$

For $x_1 = 0.8$

$$\gamma_1 = 1.044, \ \gamma_2 = 2.004$$

Predicting the vapor phase composition and system pressure.

$$\gamma_1 x_1 p_1^{sat} = y_1 p$$

$$\gamma_2 (1 - x_1) p_2^{sat} = (1 - y_1) p$$

These can be rearranged to solve consecutively as:

$$y_1 = \frac{\gamma_1 x_1 p_1^{sat}}{\gamma_1 x_1 p_1^{sat} + \gamma_2 x_2 p_2^{sat}} = 0.46$$

$$p = \gamma_1 x_1 p_1^{sat} + \gamma_2 x_2 p_2^{sat} = 7.69 \text{ kPa}$$

2-parameter Margules equation

A modification of the 1-parameter Margules equation is the 2-parameter Margules equation.

$$\frac{G^E}{RT} = x_1 x_2 (A_{21} x_1 + A_{12} x_2) \quad (5\text{-}145)$$

The activity coefficient expressions for the 2-parameter Margules equation

$$\ln \gamma_1 = x_2^2 \left[A_{12} + 2(A_{21} - A_{12}) x_1 \right] \quad (5\text{-}146)$$

$$\ln \gamma_2 = x_1^2 \left[A_{21} + 2(A_{12} - A_{21}) x_2 \right] \quad (5\text{-}147)$$

Unlike the 1-parameter Margules equation, the 2-parameter Margules equation is not necessarily symmetric.

EXAMPLE 5-8

Perform a reduction of the data in Table 5-5 for the di-isopropyl ether (1) + 1-propanol (2) system at 303.15 K using both the 1-parameter and 2-parameter Margules equations. Estimate the system pressure and vapor-phase mole fraction when $x_1 = 0.30$ and $x_1 = 0.80$.

Data reduction using the 1- and 2-parameter Margules Equation.

The experiment provides the following information:

$$x_1 = 0.4090, \ y_1 = 0.8524, \text{ and } p = 17.91 \text{ kPa}.$$

$$p_1^{sat}(30°C) = 4.20 \text{ kPa}$$

$$p_2^{sat}(30°C) = 10.42 \text{ kPa}$$

Table 5-5 Experimental p-xy data for the di-isopropyl ether (1) + 1-propanol (2) system at 303.15 K

p/kPa	x_1	y_1	p/kPa	x_1	y_1
3.77	0	0	19.51	0.5296	0.8774
5.05	0.0199	0.2671	20.23	0.5902	0.8890
6.15	0.0399	0.4090	20.71	0.6505	0.8974
7.22	0.0601	0.5061	21.35	0.7101	0.9093
8.29	0.0799	0.5783	21.92	0.7865	0.9209
10.60	0.1192	0.6847	22.62	0.8300	0.9372
12.16	0.1694	0.7346	23.20	0.8803	0.9521
14.07	0.2294	0.7822	23.59	0.9176	0.9637
15.62	0.2891	0.8133	23.80	0.9397	0.9709
16.81	0.3495	0.8343	23.99	0.9581	0.9785
17.91	0.4090	0.8524	24.19	0.9804	0.9885
18.77	0.4780	0.8659	24.36	1	1

At this point, it is important to note a few things. First, the activity coefficient for component 1 when $x_1 = 0.5$ does not exist (since component 1 is not present at this point). Likewise, the activity coefficient for component 2 when $x_1 = 0.5$ also does not exist. Thus, those entries are not included in Table 5-6.

Second, by definition, the excess molar Gibbs free energy when $x_1 = 0.5$ or $x_2 = 0.5$ is 0. Thus, we can put those endpoint entries into the table by hand.

Third, when a substance becomes very dilute in a mixture, its activity coefficient approaches a limiting value called the infinite dilution activity coefficient and is denoted as i. We introduced this earlier in the chapter when working with *Henry's Law*. Recall that i changes depending on which component is dilute and which component is concentrated. For example, in this problem, when 1-propanol is dilute in di-isopropyl ether, its infinite dilution activity coefficient would seem to be a little above 3.75, according to Table 5-6. On the other hand, when di-isopropyl ether is dilute in 1-propanol, its infinite dilution activity coefficient is likely a little above 2.78. Just by looking at the data, we can tell that the 1-parameter Margules equation will have trouble modeling this system, since it predicts a symmetric system (and the same infinite dilution activity coefficient for each compound). At any rate, we have the experimental data we need for modeling and the next step is to find the model parameters.

$$\gamma_i x_i p_i^{sat} = y_i p$$

$$\gamma_1 = ?; \gamma_2 = ?$$

$$\ln \gamma_1 = x_2^2 \left[A_{12} + 2(A_{21} - A_{12})x_1 \right]$$

$$\ln \gamma_2 = x_1^2 \left[A_{21} + 2(A_{12} - A_{21})x_2 \right]$$

$$A_{12} = ?; A_{21} = ?$$

$$x_1 = 0.8; x_2 = 0.2$$

$$\gamma_1 = ?; \gamma_2 = ?$$

$$\gamma_i x_i p_i^{sat} = y_i p$$

$$y_1 = ?; p = ?$$

$$\gamma_1 = \frac{y_1 p}{x_1 p_1^{sat}}$$

$$\gamma_2 = \frac{(1-y_1)p}{(1-x_1)p_2^{sat}}$$

$$\frac{G^E}{RT} = x_1 \ln \gamma_1 + x_2 \ln \gamma_2$$

$$\ln \gamma_1 = x_2^2 \left[A_{12} + 2(A_{21} - A_{12})x_1 \right]$$

$$\ln \gamma_2 = x_1^2 \left[A_{21} + 2(A_{12} - A_{21})x_2 \right]$$

Table 5-6 Experimental activity coefficient data for the di-isopropyl ether (1) + 1-propanol (2) system at 303.15 K

p/kPa	x_1	y_1	γ_1	γ_2	G^E/RT
3.77	0	0	-	1	0
5.05	0.0199	0.2671	2.782	1.001	0.0213
6.15	0.0399	0.4090	2.587	1.003	0.0408
7.22	0.0601	0.5061	2.499	1.005	0.0597
8.29	0.0799	0.5783	2.464	1.006	0.0776
10.60	0.1192	0.6847	2.498	1.005	0.1135
12.16	0.1694	0.7346	2.165	1.030	0.1554
14.07	0.2294	0.7822	1.970	1.053	0.1953
15.62	0.2891	0.8133	1.804	1.087	0.2299
16.81	0.3495	0.8343	1.648	1.134	0.2564
17.91	0.4090	0.8524	1.533	1.185	0.2751
18.77	0.4780	0.8659	1.417	1.260	0.2864
19.51	0.5296	0.8774	1.327	1.346	0.2896
20.23	0.5902	0.8890	1.251	1.452	0.2850
20.71	0.6505	0.8974	1.173	1.611	0.2705
21.35	0.7101	0.9093	1.123	1.770	0.2479
21.92	0.7685	0.9209	1.078	1.984	0.2163
22.62	0.8300	0.9372	1.049	2.214	0.1748
23.20	0.8803	0.9521	1.030	2.460	0.1338
23.59	0.9179	0.9637	1.017	2.764	0.0989
23.80	0.9397	0.9709	1.009	3.047	0.0756
23.99	0.9581	0.9785	1.006	3.259	0.0552
24.19	0.9804	0.9885	1.001	3.750	0.0269
24.36	1	1	1	-	0

Defining the objective function.

One approach is to try to minimize the deviation between the predicted value of $\frac{G^E}{RT}$ from the model and the experimental value for the 24 points.

$$\text{OBJ} = \frac{1}{n} \sum_{i=1}^{n} \left[\frac{\left(\frac{G^E}{RT}\right)_i^{\text{mod}} - \left(\frac{G^E}{RT}\right)_i^{\text{exp}}}{\left(\frac{G^E}{RT}\right)_i^{\text{exp}}} \right]^2$$

In Example 5-7, we had a single data point. Thus, we used that one data point to solve for the A parameter of the 1-parameter Margules equation. Here, we have 24 data points and we need to find the best value for A (1-parameter Margules equation) and the best value for A_{12} and A_{21} (2-parameter Margules equation) across all the data. Therefore, we need to define an objective function that, once minimized, will provide us with the best values for the parameters.

There are a few procedural issues that one must be aware of when performing minimizations of this kind. First, if you have a program that does minimization (such as Solver in MS Excel), it will try to minimize your objective function-all the way to negative infinity, if possible. Therefore, you need to define your objective function so that only positive terms exist. Second, it is important to make the terms in your objective function have the same order of magnitude (which is just good practice). This way, if fitting data values change across orders of magnitude, the minimization process isn't biased toward the larger values (whose actual deviations from the model will be a much smaller percentage than the smaller values).

Here, OBJ will always be positive, and the terms are scaled so that they are the same order of magnitude. Note that the endpoints $x_1 = 0$ and $x_1 = 1$ have been removed since the denominator of the objective function for both of those points will be 0 and, thus, the objective function will become undefined.

The experimental data is given by empty circles while the models are given by the solid lines: dark for the 1-parameter Margules equation and light for the 2-parameter Margules equation(Fig.5-20 and Fig.5-21).

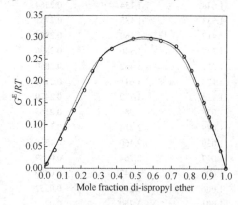

Fig.5-20 The excess molar Gibbs free energy for the di-isopropyl ether(1) + 1-propanol (2) system at 303.15 K

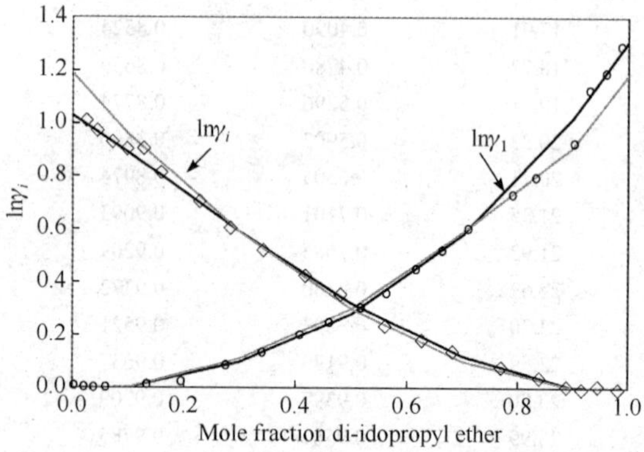

Fig.5-21 The natural logarithm of the activity coefficients for the di-isopropyl ether (1) +1-propanol (2) system at 303.15 K

In the next example, we introduce another excess molar Gibbs free energy model that can be used in order to take fitted parameters at one temperature and apply them to another temperature.

Margules expressed the excess Gibbs free energy of a binary liquid mixture as a power series of the mole fraction ions:

$$\frac{G^E}{RT} = x_1 x_2 (A_{21} x_1 + A_{12} x_2) + x_1^2 x_2^2 (B_{21} x_1 + B_{12} x_2) + \cdots + x_1^m x_2^m (M_{21} x_1 + M_{12} x_2)$$

5.4 Wilson and Van Laar Equation

5.4.1 Wilson Equation

A popular excess molar Gibbs free energy model, whose parameters depend on temperature, is called Wilson Equation.

$$\frac{G^E}{RT} = -x_1 \ln(x_1 + \Lambda_{12} x_2) - x_2 \ln(x_2 + \Lambda_{21} x_1) \tag{5-148}$$

The temperature dependence of the parameters Λ_{21} and Λ_{21} is given

$$\Lambda_{12} = \frac{V_2}{V_1} \exp\left(\frac{-\alpha_{12}}{RT}\right) \tag{5-149}$$

$$\Lambda_{21} = \frac{V_1}{V_2} \exp\left(\frac{-\alpha_{21}}{RT}\right) \tag{5-150}$$

The activity coefficient of this model is:

$$\ln \gamma_1 = -\ln(x_1 + \Lambda_{12} x_2) + x_2 \left(\frac{\Lambda_{12}}{x_1 + \Lambda_{12} x_2} - \frac{\Lambda_{21}}{x_2 + \Lambda_{21} x_1}\right) \tag{5-151}$$

$$\ln \gamma_2 = -\ln(x_2 + \Lambda_{21} x_1) - x_1 \left(\frac{\Lambda_{12}}{x_1 + \Lambda_{12} x_2} - \frac{\Lambda_{21}}{x_2 + \Lambda_{21} x_1}\right) \tag{5-152}$$

EXAMPLE 5-9

When evaluating the separation of ethanol (1) + butyl methyl ether (2) mixture, the VLE of the system at 323.15 K was measured(Table 5-7). Then, the size of the distillation column required for the system measurement of vapor-liquid equilibrium at 338.15 K needs to be estimated. It is better to estimate the VLE at 338.15 K using the data at 323.15 K than the experiment at 338.15 K. The data of this system at 323.15 K is shown in the table.

Table 5-7 A p-xy data of ethanol (1) + butyl methyl ether (2) system at 323.15 K

x_1	p/kPa	y_1
0	50.69	0
0.0316	52.37	0.0592
0.1067	53.95	0.1597
0.1899	55.83	0.2426
0.3393	56.73	0.2990
0.4304	56.10	0.3317
0.6225	53.75	0.4030
0.6649	52.74	0.4334
0.7352	50.52	0.4902
0.8124	47.49	0.5284
0.8525	45.52	0.6058
0.9091	41.41	0.6677

		Continued
x_1	p/kPa	y_1
0.9542	36.75	0.7751
0.9820	32.49	0.8956
1	29.34	1

The activity coefficient and excess molar Gibbs free energy at 323.15 K were obtained (Table 5-8).

$$\gamma_1 = \frac{y_1 p}{x_1 p_1^{sat}}$$

$$\gamma_2 = \frac{(1-y_1)p}{(1-x_1)p_2^{sat}}$$

$$\frac{G^E}{RT} = x_1 \ln \gamma_1 + x_2 \ln \gamma_2$$

Table 5-8 Activity coefficient of ethanol (1) + butyl methyl ether (2) system at 323.15 K

x_1	p/kPa	y_1	γ_1	γ_2	G^E/RT
0	50.69	0		1	0
0.0316	52.37	0.0592	3.3924	1.0204	0.0582
0.1067	53.95	0.1597	2.7839	1.0182	0.1254
0.1899	55-83	0.2426	2.4537	1.0479	0.2084
0.3393	56.73	0.2990	1.7176	1.2089	0.3089
0.4303	56.10	0.3317	1.4847	1.3222	0.3292
0.6225	53.75	0.4030	1.1929	1.7093	0.3122
0.6649	52.74	0.4334	1.1778	1.7939	0.3046
0.7352	50.52	0.4902	1.1529	1.9583	0.2826
0.8124	47.49	0.5284	1.0566	2.4052	0.2094
0.8525	45.52	0.6058	1.1053	2.4545	0.2178
0.9091	41.41	0.6677	1.0385	3.0582	0.1360
0.9542	36.75	0.7751	1.0183	3.6548	0.0767
0.9820	32.49	0.8956	1.0101	3.8290	0.0340
1	29.34	1	1	1	0

Fitting parameters at 323.15 K.

$$\Lambda_{12} = 0.5045$$

$$\Lambda_{21} = 0.3031$$

$$OBJ = 0.145$$

Get Wilson parameter.

$$\underline{V}_1 = 58.67 \text{ cm}^3/\text{mol}$$

$$\underline{V}_2 = 119.25 \text{ cm}^3/\text{mol}$$

$$\frac{\alpha_{12}}{R} = 450.2789 \text{ K}$$

$$\frac{\alpha_{21}}{R} = 156.5022 \text{ K}$$

At 338.15 K, the new values of Λ_{12} and Λ_{21} are obtained.

$$\Lambda_{12} = 0.5367$$

$$\Lambda_{21} = 0.3097$$

Predicting the VLE at 338.15 K(Fig.5-22).

$$x_1 = 0.2 \quad x_1 = 0.4 \quad x_1 = 0.6 \quad x_1 = 0.8$$

$$y_1 = \frac{\gamma_1 x_1 p_1^{\text{sat}}}{\gamma_1 x_1 p_1^{\text{sat}} + \gamma_2 x_2 p_2^{\text{sat}}}$$

$$p = \gamma_1 x_1 p_1^{\text{sat}} + \gamma_2 x_2 p_2^{\text{sat}}$$

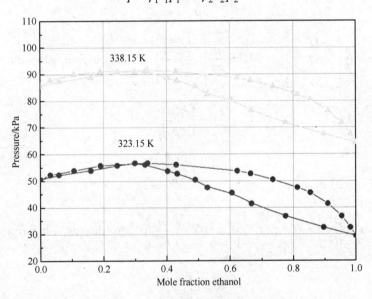

Fig.5-22 A p-xy diagram of ethanol (1) + butyl methyl ether (2) system at two different temperatures

5.4.2 Relationship between Activity Coefficient and Temperature and Pressure

The basic properties of M varying with system variables.

$$d(n\underline{M}) = \left[\frac{\partial(n\underline{M})}{\partial T}\right]_{p,n_1,n_2} dT + \left[\frac{\partial(n\underline{M})}{\partial p}\right]_{T,n_1,n_2} dp + \left[\frac{\partial(n\underline{M})}{\partial(n_1)}\right]_{T,p,n_2} dn_1 + \left[\frac{\partial(n\underline{M})}{\partial(n_2)}\right]_{T,p,n_1} dn_2$$

(5-153)

So far, we have not discussed the effect of temperature and pressure on the activity coefficient itself and all jobs are stable.

Temperature and the implication (not stated) was that the pressure effects on the excess molar Gibbs free energy (and, in turn, the activity coefficients) did not strongly impact the results. Indeed, none of the models were explicit in the pressure.

Before moving further, let's explore this notion in this subsection as it also sets the stage for something we will need in a later chapter.

$$d\left(n\frac{G^E}{RT}\right) = \left[\frac{\partial\left(n\frac{G^E}{RT}\right)}{\partial T}\right]_{p,n_1,n_2} dT + \left[\frac{\partial\left(n\frac{G^E}{RT}\right)}{\partial p}\right]_{T,n_1,n_2} dp + \left[\frac{\partial\left(n\frac{G^E}{RT}\right)}{\partial(n_1)}\right]_{T,p,n_2} dn_1$$

$$+ \left[\frac{\partial\left(n\frac{G^E}{RT}\right)}{\partial(n_2)}\right]_{T,p,n_1} dn_2 \qquad (5\text{-}154)$$

$$d\left(n\frac{G^E}{RT}\right) = \left[\frac{\partial\left(n\frac{G^E}{RT}\right)}{\partial T}\right]_{p,n_1,n_2} dT + \left[\frac{\partial\left(n\frac{G^E}{RT}\right)}{\partial p}\right]_{T,n_1,n_2} dp + \ln\gamma_1 dn_1 + \ln\gamma_2 dn_2$$

$$= \frac{n}{R}\left[\frac{\partial\left(\frac{G^E}{T}\right)}{\partial T}\right]_{p,n_1,n_2} dT + \frac{n}{RT}\left[\frac{\partial(G^E)}{\partial p}\right]_{T,n_1,n_2} dp + \ln\gamma_1 dn_1 + \ln\gamma_2 dn_2 \qquad (5\text{-}155)$$

$$\underline{V} = \left(\frac{\partial \underline{G}}{\partial p}\right)_T \qquad (5\text{-}156)$$

$$\underline{V}^E = \left(\frac{\partial \underline{G}^E}{\partial p}\right)_T \qquad (5\text{-}157)$$

$$d\left(n\frac{G^E}{RT}\right) = \frac{n}{R}\left[\frac{\partial\left(\frac{G^E}{T}\right)}{\partial T}\right]_{p,n_1,n_2} dT + \left(\frac{n}{RT}\right)\underline{V}^E dp + \ln\gamma_1 dn_1 + \ln\gamma_2 dn_2 \qquad (5\text{-}158)$$

$$\underline{S} = -\left(\frac{\partial \underline{G}}{\partial T}\right)_p \qquad (5\text{-}159)$$

$$\underline{S}^E = -\left(\frac{\partial \underline{G}^E}{\partial T}\right)_p \qquad (5\text{-}160)$$

$$\left[\frac{\partial\left(\frac{G^E}{T}\right)}{\partial T}\right]_{p,n_1,n_2} = \frac{T\left[\frac{\partial(G^E)}{\partial T}\right]_{p,n_1,n_2} - G^E}{T^2} \qquad (5\text{-}161)$$

$$\left[\frac{\partial\left(\frac{G^E}{T}\right)}{\partial T}\right]_{p,n_1,n_2} = \frac{-T\underline{S}^E - \underline{G}^E}{T^2} \qquad (5\text{-}162)$$

Equation is an expression we can use to answer our question about the impact of temperature and pressure on the excess molar Gibbs free energy.

$$\underline{H} = \underline{G} + T\underline{S} \tag{5-163}$$

$$\underline{H}^E = \underline{G}^E + T\underline{S}^E \tag{5-164}$$

$$\left[\frac{\partial\left(\frac{\underline{G}^E}{T}\right)}{\partial T}\right]_{p,n_1,n_2} = \frac{-T\underline{S}^E - \underline{G}^E}{T^2} = \frac{-\underline{H}^E}{T^2} \tag{5-165}$$

$$d\left(\frac{n\underline{G}^E}{RT}\right) = -\left(\frac{n\underline{H}^E}{RT^2}\right)dT + \left(\frac{n\underline{V}^E}{RT}\right)dp + (\ln\gamma_1)dn_1 + (\ln\gamma_2)dn_2 \tag{5-166}$$

$$d\left(\frac{n\underline{G}^E}{RT}\right) = -\left(\frac{n\underline{H}^E}{RT^2}\right)dT + \left(\frac{n\underline{V}^E}{RT}\right)dp + (\ln\gamma_1)dn_1 + (\ln\gamma_2)dn_2 \tag{5-167}$$

$$\frac{\partial}{\partial T}\left(\frac{n\underline{G}^E}{RT}\right)_{p,n_1,n_2} = -\left(\frac{n\underline{H}^E}{RT^2}\right)\left(\frac{\partial T}{\partial T}\right)_{p,n_1,n_2} + \left(\frac{n\underline{V}^E}{RT}\right)\left(\frac{\partial p}{\partial T}\right)_{p,n_1,n_2} + (\ln\gamma_1)\left(\frac{\partial n_1}{\partial T}\right)_{p,n_1,n_2}$$

$$+ (\ln\gamma_2)\left(\frac{\partial n_2}{\partial T}\right)_{p,n_1,n_2} \frac{\partial}{\partial T}\left(\frac{n\underline{G}^E}{RT}\right)_{p,n_1,n_2} = -\left(\frac{n\underline{H}^E}{RT^2}\right) \tag{5-168}$$

To explore the effect of temperature, we divide by a differential temperature at constant pressure and number of moles of each component to obtain the required partial derivative.

$$\frac{\partial}{\partial T}\left(\frac{\underline{G}^E}{RT}\right)_{p,n_1,n_2} = -\left(\frac{\underline{H}^E}{RT^2}\right) \tag{5-169}$$

Examine the impact of temperature on the excess molar Gibbs free energy and the activity coefficients.

The influence of temperature on excess molar Gibbs free energy is investigated. Thus,

$$d\left(\frac{n\underline{G}^E}{RT}\right) = -\left(\frac{n\underline{H}^E}{RT^2}\right)dT + \left(\frac{n\underline{V}^E}{RT}\right)dp + \ln\gamma_1 dn_1 + \ln\gamma_2 dn_2 \tag{5-170}$$

$$\frac{\partial}{\partial p}\left(\frac{n\underline{G}^E}{RT}\right)_{T,n_1,n_2} = -\left(\frac{n\underline{H}^E}{RT^2}\right)\left(\frac{\partial T}{\partial p}\right)_{T,n_1,n_2} + \left(\frac{n\underline{V}^E}{RT}\right)\left(\frac{\partial p}{\partial p}\right)_{T,n_1,n_2} + \ln\gamma_1\left(\frac{\partial n_1}{\partial p}\right)_{T,n_1,n_2}$$

$$+ \ln\gamma_2\left(\frac{\partial n_2}{\partial p}\right)_{T,n_1,n_2} \tag{5-171}$$

Similarly, we divide equation by a differential pressure at constant temperature and number of moles of each component.

$$\frac{\partial}{\partial p}\left(\frac{n\underline{G}^E}{RT}\right)_{T,n_1,n_2} = -\left(\frac{n\underline{V}^E}{RT}\right) \tag{5-172}$$

$$\frac{\partial}{\partial p}\left(\frac{\underline{G}^E}{RT}\right)_{T,n_1,n_2} = -\left(\frac{\underline{V}^E}{RT}\right) \tag{5-173}$$

Examine the impact of pressure, respectively on the excess molar Gibbs free energy and the activity coefficients.

For example, consider the system tetrahydrofuran (1) + 2,2,2-trifluoroethanol (2) at 298.15 K. The measured excess molar enthalpy of the equimolar mixture is about 22.0 kJ/mol, while the measured excess molar volume of the equimolar mixture is about 1 cm³/mol.

$$\frac{\partial}{\partial T}\left(\frac{G^E}{RT}\right)_{p,n_1,n_2} = 2.7 \times 10^{-3} \text{ K}^{-1} \tag{5-174}$$

$$\frac{\partial}{\partial p}\left(\frac{G^E}{RT}\right)_{T,n_1,n_2} = 4.03 \times 10^{-10} \text{ Pa}^{-1} \tag{5-175}$$

The impact of a change in temperature far exceeds that of pressure for the excess molar Gibbs free energy. For the derivative with respect to temperature, we see that there is a change on the order of 1023 per change in 1 Kelvin. For the derivative with respect to pressure, we see that there is a change on the order of 1025 for every change in 100 kPa. Thus, the impact of a change in temperature far exceeds that of pressure for the excess molar Gibbs free energy. Therefore, models for the excess molar Gibbs free energy will not normally include the effect of pressure, accordingly, activity coefficients are often assumed to be independent of pressure.

5.4.3 Van Laar Equation and Regular Solution Theory

The excess property is the mixing property minus it if the mixture is modeled by an ideal solution.

$$\underline{V}^E = \underline{V} - \underline{V}^{im} \tag{5-176}$$

$$\underline{U}^E = \underline{U} - \underline{U}^{im} \tag{5-177}$$

$$\underline{S}^E = \underline{S} - \underline{S}^{im} \tag{5-178}$$

$$\underline{G} = \underline{U} + p\underline{V} - T\underline{S} \tag{5-179}$$

$$\underline{G}^E = \underline{U}^E + p\underline{V}^E - T\underline{S}^E \tag{5-180}$$

First, assume that the excess molar volume and excess molar entropy of the system are 0, which is called "regular solution theory."

$$\underline{S}^E = 0 \tag{5-181}$$

$$\underline{V}^E = 0 \tag{5-182}$$

Secondly, assume that both pure components and mixtures conform to van der Waals equation of state. The latter hypothesis may not come as a surprise, because van Laar is a doctoral student at Van der Waals college.

The first assumption of van Laar isolates the excess internal energy as the key to calculating the excess molar Gibbs free energy (since VE and SE are set to 0). Van Laar's strategy is to create a hypothetical process with three steps. The first step is to decompress both pure liquids to an ideal gas in an isothermal manner, then use the van der Waals equation of state to calculate this internal energy change. The second step is to mix these ideal gases in an isothermal manner to form a mixture. Recall that the internal energy of mixing for an ideal gas is 0, so this step contributes nothing to the calculation. Finally, he would recompress this ideal gas mixture to a liquid at the pressure of interest (again in an isothermal manner) and calculate this internal energy change via the van der Waals equation of state.

$$\Delta \underline{U} = \underline{U}^E = \underline{U} - \left(x_1 \underline{U}_1 + x_2 \underline{U}_2\right) \tag{5-183}$$

$$\Delta \underline{U}^{ig} = 0 \tag{5-184}$$

$$\Delta \underline{U} = \underline{U} - \left(x_1 \underline{U}_1 + x_2 \underline{U}_2\right) - \left[\underline{U}^{ig} - \left(x_1 \underline{U}_1^{ig} + x_2 \underline{U}_2^{ig}\right)\right] \tag{5-185}$$

$$\Delta \underline{U} = \left(\underline{U} - \underline{U}^{ig}\right) - x_1\left(\underline{U}_1 - \underline{U}_1^{ig}\right) - x_2\left(\underline{U}_2 - \underline{U}_2^{ig}\right) \tag{5-186}$$

Eq. (5-185) describes what van Laar proposed, namely to calculate the molar internal energies it needed referenced back to the ideal gas state, and then describe those residual values through the van der Waals equation. In other words, each of the three residuals in Eq. (5-186) is described by the van der Waals equation, the former is for the mixture and the latter two are for the pure components.

While we have calculated pure-component parameters for the van der Waals equation in Chapter 5, we have yet to calculate mixture parameters. In order to do this, we introduce the van der Waals one-fluid mixing rules here.

5.4.4 Van Der Waals One-Fluid Mixing Rules

For the parameters a and b in van der Waals equation, we use the quadratic mixing rule and express them as mixing parameters with subscript m.

$$a_m = x_1 x_1 a_{11} + x_1 x_2 a_{12} + x_2 x_1 a_{21} + x_2 x_2 a_{22} \tag{5-187}$$

$$b_m = x_1 x_1 b_{11} + x_1 x_2 b_{12} + x_2 x_1 b_{21} + x_2 x_2 b_{22} \tag{5-188}$$

The two subscripts on parameters a and b represent the components involved. For those with the same subscript, the pure component is implied.

$$a_{11} = a_1, b_{11} = b_1, a_{22} = a_2, b_{22} = b_2 \tag{5-189}$$

Although we have calculated pure-component parameters for the van der Waals equation in Chapter 5, we have yet to calculate mixture parameters. In order to do this, we introduce the van der Waals one-fluid mixing rules here.

Parameters with different subscripts are called cross parameters. They are defined as follows. For the cross parameter on a, it is a normal method.

It is a geometric mean combination rule:

$$a_{12} = a_{21} = \sqrt{a_{11} a_{22}} = \sqrt{a_1 a_2} \tag{5-190}$$

For b, the crossover parameter is defined by arithmetic mean combination rule:

161

$$b_{12} = b_{21} = \frac{b_{11}+b_{22}}{2} = \frac{b_1+b_2}{2} \tag{5-191}$$

Hybrid rules b_m will be simplified.

$$b_m = x_1 b_1 + x_2 b_2 \tag{5-192}$$

If one follows the strategy described above in calculating the excess molar internal energy via Eq. (5-193) through use of the van der Waals equation of state, the result is an excess molar Gibbs free energy model.

$$\frac{G^E}{RT} = x_1 x_2 \frac{\tau_{12}\tau_{21}}{\tau_{12}x_1 + \tau_{21}x_2} \tag{5-193}$$

Where the activity coefficients, as obtained via Eq. (5-194), are

$$\ln \gamma_1 = \tau_{12}\left(1 + \frac{\tau_{12}x_1}{\tau_{21}x_2}\right)^{-2} \tag{5-194}$$

$$\ln \gamma_2 = \tau_{21}\left(1 + \frac{\tau_{21}x_2}{\tau_{12}x_1}\right)^{-2} \tag{5-195}$$

Here, τ_{12} and τ_{21} are given in terms of the van der Waals parameters as

$$\tau_{12} = \frac{b_1}{RT}\left(\frac{\sqrt{a_1}}{b_1} - \frac{\sqrt{a_2}}{b_2}\right)^2 \tag{5-196}$$

$$\tau_{21} = \frac{b_2}{RT}\left(\frac{\sqrt{a_1}}{b_1} - \frac{\sqrt{a_2}}{b_2}\right)^2 \tag{5-197}$$

If one follows the strategy described above in calculating the excess molar internal energy via Eq. (5-197) through use of the van der Waals equation of state, the result is an excess molar Gibbs free energy model. The van der Waals equation of state is not very accurate from a predictive standpoint.

Scatchard-Hildebrand approach, George Scarchard and Joel Henry Hildebrand (each) improved the van der method using experimental pure components.

Instead of van der Waals parameter.

$$\frac{G^E}{RT} = x_1 x_2 \frac{M_{12}M_{21}}{M_{12}x_1 + M_{21}x_2} \tag{5-198}$$

$$\ln \gamma_1 = M_{12}\left(1 + \frac{M_{12}x_1}{M_{21}x_2}\right)^{-2} \tag{5-199}$$

$$\ln \gamma_2 = M_{21}\left(1 + \frac{M_{21}x_2}{M_{12}x_1}\right)^{-2} \tag{5-200}$$

They use the experimental molar internal energy of vaporization (normally calculated at 258℃) as well as the molar volumes of the pure liquids in order to estimate both the mixture internal energy of vaporization and mixture molar volume. With George Scatchard, Hildebrand developed an equation for excess molar volumes in mixtures.

$$M_{12} = \frac{V_1}{RT}(\delta_1 - \delta_2)^2 \tag{5-201}$$

$$M_{21} = \frac{V_2}{RT}(\delta_1 - \delta_2)^2 \tag{5-202}$$

In this case, δ_1 and δ_2 are known as solubility parameters and given as:

$$\delta_1 = \sqrt{\frac{\Delta \underline{U}_1^{vap}}{\underline{V}_1}} \tag{5-203}$$

$$\delta_2 = \sqrt{\frac{\Delta \underline{U}_2^{vap}}{\underline{V}_2}} \tag{5-204}$$

As evidenced by the definitions for M_{12} and M_{21}, the solubility parameters from this modeling approach provide a quick estimate of a mixture's deviation from ideal solution. When the solubility parameters of two compounds are the same, the mixture is predicted to behave as an ideal solution. The larger the difference of the solubility parameters from each other is, the more non-ideal the mixture is predicted to behave. However, systems that best validate the assumptions inherent in the approach, namely that excess molar volume and excess molar entropy are 0, are where predictive results provide the most benefit. This is normally for non-polar or slightly polar systems with small excess molar volumes. We demonstrate this in Example 5-10.

EXAMPLE 5-10

The vapor-liquid equilibrium of n-pentane (1) + benzene (2) system at 313.15 K was estimated by van Laar equation.

The first step for VLE problems is to decide the modeling approach needed. Should we try *Raoult's Law*? From the liquid point of view, the two molecules, while non-polar, are not similar in size and shape (pentane is a chain and benzene are a ringed structure). Thus, we will need an excess molar Gibbs free energy model for the liquid (as provided in the problem statement, we should use the van Laar model). What about the vapor point of view? Do we need to account for vapor phase non-idealities? A quick check for this would be to look at the vapor pressure of the pure components at the temperature of interest.

The Antoine equation can help us find a good estimate of the vapor pressure of both compounds at 313.15 K. Using the Antoine parameters for the two components, then

$$p_1^{sat}(313.15 \text{ K}) = 119 \text{ kPa}$$

$$p_1^{sat}(313.15 \text{ K}) = 24 \text{ kPa}$$

Unless this mixture produces very large, positive deviations from *Raoult's Law*, the pressure of the mixture is not likely to go well above 119 kPa. And since these are two non-polar compounds, the ideal gas is a reasonable assumption in the absence of experimental data. Thus, assuming an ideal gas for the vapor phase is quite reasonable here(Fig.5-23). Our modeling approach will be modified by *Raoult's Law*:

$$T_{c1} = 496.7 \text{ K} \quad p_{c1} = 3370 \text{ kPa}$$

$$T_{c2} = 562.05 \text{ K} \quad p_{c2} = 4895 \text{ kPa}$$

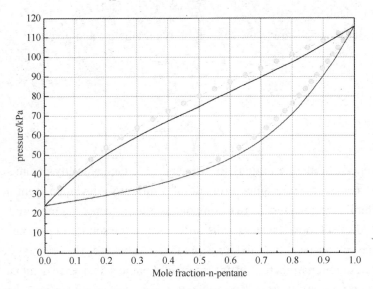

Fig.5-23 A *p-xy* diagram of n-pentane (1) and benzene (2) system at 313.15 K

Thermodynamic Consistency

Gibbs-Duhem equation is the relationship between thermodynamic variables in the system, which can be used to measure all independent thermodynamic variables of the system, and then check whether it is true (within some reasonable experimental errors).

First, if you measure all but one of the independent thermodynamic variables in the system, the Gibbs-Duhem equation can be used to "solve" for the one variable that is not measured. This is similar to the material balance example above, where we measure the two mass flow rates of the input streams and solve for the mass flow rate of the output stream. Relative to the modeling of VLE, the Gibbs-Duhem equation, for example, can be used to solve for the vapor-phase composition in a binary mixture without having to measure it experimentally.

If the data have large deviations from the Gibbs-Duhem equation, this will cause the user to have an increasing level of skepticism as to the accuracy and utility of the data.

$$x_1 d\bar{M}_1 + x_2 d\bar{M}_2 - \left(\frac{\partial \underline{M}}{\partial T}\right)_{p, x_1, x_2} dT - \left(\frac{\partial \underline{M}}{\partial p}\right)_{T, x_1, x_2} dp = 0 \qquad (5\text{-}205)$$

Once again, for example purposes, we assume we are evaluating *p-xy* data and that means the temperature is constant. Also, as we have shown in Section 11.7.1, activity coefficients are only slight functions of pressure. Thus, under these assumptions:

$$x_1 d\ln\gamma_1 + x_2 d\ln\gamma_2 = 0 \tag{5-206}$$

$$x_1 \frac{d\ln\gamma_1}{dx_1} + x_2 \frac{d\ln\gamma_2}{dx_1} = 0 \tag{5-207}$$

This is a convenient form of Gibbs-Duhem equation because it contains the relationship between activity coefficients.

The Gibbs-Duhem equation.

$$x_1 d\bar{M}_1 + x_2 d\bar{M}_2 - \left(\frac{\partial M}{\partial T}\right)_{p,x_1,x_2} dT - \left(\frac{\partial M}{\partial p}\right)_{T,x_1,x_2} dp = 0 \tag{5-208}$$

$$x_1 d\ln\gamma_1 + x_2 d\ln\gamma_2 = 0 \tag{5-209}$$

Once again, we can evaluate this differential relationship for small changes in dx_1. Thus,

$$x_1 \frac{d\ln\gamma_1}{dx_1} + x_2 \frac{d\ln\gamma_2}{dx_1} = 0 \tag{5-210}$$

This is a convenient form of Gibbs-Dihem equation because it contains the relationship between activity coefficients.

Integral (Area) Test

Let's assume that there is an experimental thermodynamic data for a binary mixture at a constant temperature, which means that p-xy data exists at a given temperature.

The experimental activity coefficient can be solved. The experimental excess molar Gibbs free energy can be calculated from these experimental activity coefficients.

$$\frac{G^{E^*}}{RT} = x_1 \ln(\gamma_1)^* + x_2 \ln(\gamma_2)^* \tag{5-211}$$

In order to discriminate between the experimental and model values in what follows, we assign an asterisk to the experimental values.

$$\frac{G^{E^*}}{RT} = x_1 \ln(\gamma_1)^* + (1-x_1)\ln(\gamma_2)^* \tag{5-212}$$

$$\frac{d}{dx_1}\left(\frac{G^E}{RT}\right)^* = \frac{d}{dx_1}\left[x_1 \ln(\gamma_1)^*\right] + \frac{d}{dx_1}\left[(1-x_1)\ln(\gamma_2)^*\right]$$

$$= x_1 \frac{d}{dx_1}\left[\ln(\gamma_1)^*\right] + \ln(\gamma_1)^* + (1-x_1)\frac{d}{dx_1}\left[\ln(\gamma_2)^*\right] - \ln(\gamma_2)^*$$

$$= x_1 \frac{d}{dx_1}\left[\ln(\gamma_1)^*\right] + x_2 \frac{d}{dx_1}\left[\ln(\gamma_2)^*\right] + \ln\left(\frac{\gamma_1}{\gamma_2}\right)^* \tag{5-213}$$

If the experimental data conform to Gibbs-Duhem equation,

$$x_1 \frac{d}{dx_1}\left[\ln(\gamma_1)^*\right] + x_2 \frac{d}{dx_1}\left[\ln(\gamma_2)^*\right] = 0 \tag{5-214}$$

$$\frac{d}{dx_1}\left(\frac{G^E}{RT}\right)^* = x_1 \frac{d}{dx_1}\left[\ln(\gamma_1)^*\right] + x_2 \frac{d}{dx_1}\left[\ln(\gamma_2)^*\right] + \ln\left(\frac{\gamma_1}{\gamma_2}\right)^* \tag{5-215}$$

$$\frac{d}{dx_1}\left(\frac{G^E}{RT}\right)^* = \ln\left(\frac{\gamma_1}{\gamma_2}\right)^* \tag{5-216}$$

What we see is that the first two terms on the right-hand side of Eq. (5-216) are the Gibbs-Duhem equation. Accordingly, if the experimental data is consistent with the Gibbs-Duhem equation, those terms must sum to 0.

$$d\left(\frac{G^E}{RT}\right)^* = \ln\left(\frac{\gamma_1}{\gamma_2}\right)^* dx_1 \tag{5-217}$$

Or alternatively,

$$\int_{x_1=0}^{x_1=1} d\left(\frac{G^E}{RT}\right)^* = \int_{x_1=0}^{x_1=1} \ln\left(\frac{\gamma_1}{\gamma_2}\right)^* dx_1 \tag{5-218}$$

To make use of this relationship, we can integrate both sides from $x_1=0.5$ to $x_1=1$, as

$$\frac{G^E}{RT}(x_1=1) - \frac{G^E}{RT}(x_1=0) = \int_{x_1=0}^{x_1=1} \ln\left(\frac{\gamma_1}{\gamma_2}\right)^* dx_1 \tag{5-219}$$

$$0 = \int_{x_1=0}^{x_1=1} \ln\left(\frac{\gamma_1}{\gamma_2}\right)^* dx_1 \tag{5-220}$$

Each of the terms on the left-hand side are equal to 0 by themselves, since they define the pure component endpoints ($x_1=0$ and $x_1=0.5$), both of which have an excess molar Gibbs free energy equal to 0.

What Eq. (5-220) shows is that, if we plot the natural logarithm of the ratio of the activity coefficients and obtain the area under the curve from $x_1=0$ to $x_1=0.5$, this area should be 0 when the experimental data satisfy the Gibbs-Duhem equation.

5.5 Supplementary Simulation Examples

5.5.1 Vapor-Liquid Equilibrium Calculations Using Activity Coefficient Models

Aspen Plus (Advanced System for Process Engineering, Aspen for short) is a large general process simulation system, which is a suitable process simulation software introduced to the market by American Aspen Tech in the 1980s.

The process simulation software is based on the material and heat balance, phase equilibrium, chemical equilibrium and reaction kinetics, and usually provides the physical property database, strict thermodynamic estimation model library and rich process unit model library.

Among them, the process unit model library includes common chemical operation units, such as pressure change (such as pump, compressor, expander and valve), heat exchange, separation (flash evaporator, tower, etc.), reactor, etc.

(1) Define the simulation process

The definition of simulation flow is to choose the appropriate unit process module and connect it with material flow and energy flow, but it is not the same with the actual process flow, and its essence is the process of establishing mathematical model for the actual production situation.

(2) Set process simulation parameters

This part includes the setting of chemical composition, thermodynamic method, flow information, module parameters and calculation method of simulation. Its operation can be completed under the guidance of the next navigation button of the software, but the selection of appropriate physical property method is the key to the success of simulation.

(3) Simulation analysis tools

When the basic simulation process is completed, the software simulation analysis tools (design requirements, sensitivity analysis, optimization, etc.) can be used to complete the design requirements, operational performance analysis, economic optimization and other purposes.

(4) Output simulation results

After the simulation is completed, the results can be viewed in multiple places, and different output methods can be used for different purposes.

Vapor-Liquid Equilibrium (VLE) is a very important knowledge point during Chemical Thermodynamics, and it is an important physical property data of mixture distillation and separation. One of the difficulties of this course is to use appropriate methods to regress and analyze VLE data. Textbooks usually introduce the iterative optimization process based on a certain equation, which is not only tedious but also error-prone if students work out the iterative optimization by hand. Therefore, using computer software to complete this kind of physical property data regression can not only improve the calculation efficiency, but also improve the students' application ability.

The software Aspen Plus has a powerful regression function of vapor-liquid equilibrium data, and different thermodynamic physical property methods can be selected to bring the vapor-liquid equilibrium experimental data into the software for regression processing, and the best physical property methods can be selected according to the accuracy of regression for subsequent process simulation calculation. The practical design mode combining the Aspen Plus software with the "Vapor Liquid Equilibrium Data Regression Analysis" the course of Chemical Thermodynamics can accurately process the data.

EXAMPLE 5-11

Aspen Plus software are used to regression the vapor liquid equilibrium data of ETHAN-01 and N-HEX-01 binary system as an example, compared with the traditional manual iterative optimization calculation, the calculation amount can be greatly reduced and the accuracy can be improved(Fig.5-24 and Fig.5-25).

The way Aspen Plus is used generally falls into four parts.

Property Analysis Method.

Low pressure vapor-liquid equilibria of the ethanol-hexane mixture.

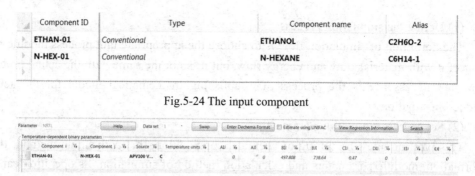

Fig.5-24 The input component

Fig.5-25 View binary interaction parameters

Since the mixture contains a polar component and is at low pressure, an activity coefficient model should be used.

If user parameters are used, for example, as a result of regressing your own data or a data set obtained from NIST TDE as described later in this chapter, the source would then be listed as USER(Fig.5-26 to Fig.5-29).

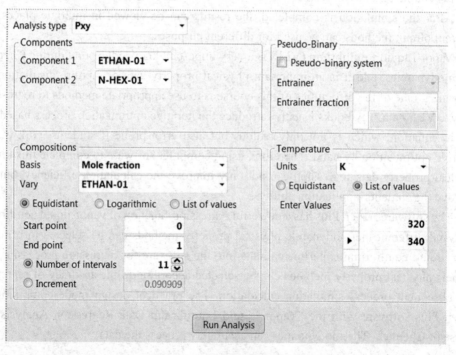

Fig.5-26 Set up the binary analysis interface

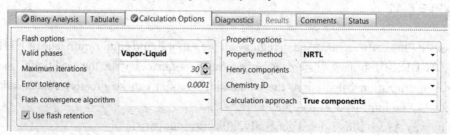

Fig.5-27 Choose property options

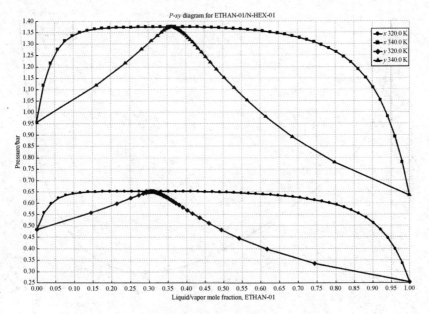

Fig.5-28 *p-xy* diagram

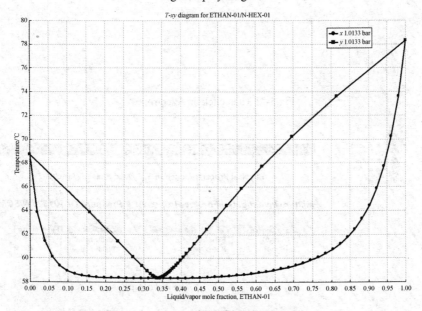

Fig.5-29 *T-xy* diagram

All the correlative models (NRTL, UNIQUAC, and Wilson) give very similar results.
It is not always true that the different activity coefficient models will give similar results.
Change the type of calculation from Properties to Simulation.
Go to Azeotrope Search in the Analysis.
The simulation method.
Flash block, consider an equimolar mixture of ethanol and hexane at 380 K and 1000 kPa, described by the NRTL model, it is adiabatically flashed to 100 kPa(Fig.5-30 to Fig.5-34).

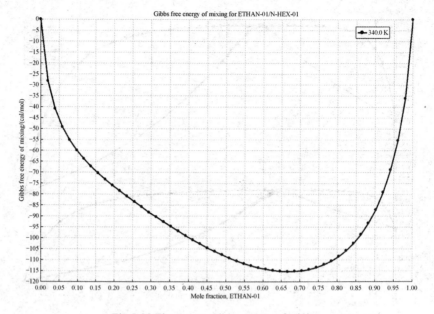

Fig.5-30 The excess Gibbs energy of mixing

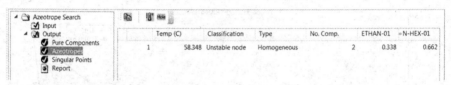

Fig.5-31 Then click on Azeotropes

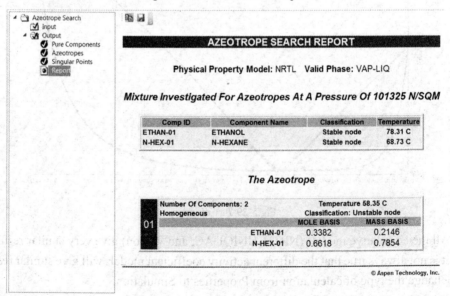

Fig.5-32 Report

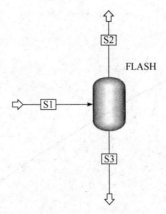

Fig.5-33 Two outlet flash evaporator process

Material | Heat | Load | Work | Vol.% Curves | Wt. % Curves | Petroleum | Polymers | Solids

	Units	S1	S2	S3
Molar Enthalpy	cal/mol	-53073.7	-44726.8	-56530.4
Mass Enthalpy	cal/gm	-802.65	-622.548	-886.702
Molar Entropy	cal/mol-K	-108.055	-100.778	-110.167
Mass Entropy	cal/gm-K	-1.63415	-1.40272	-1.72802
Molar Density	mol/cc	0.00933603	3.62915e-05	0.0107141
Mass Density	gm/cc	0.617327	0.00260736	0.683063
Enthalpy Flow	cal/sec	-14742.7	-3638.44	-11104.3
Average MW		66.1231	71.8448	63.7535
+ Mole Flows	**kmol/hr**	**1**	**0.292853**	**0.707147**
+ Mole Fractions				
+ Mass Flows	**kg/hr**	**66.1231**	**21.04**	**45.0831**
+ Mass Fractions				

Fig.5-34 Results

Calculate the bubble point of a benzene-decane equimolar mixture at 350 K using the NRTL model. The results here are obtained by physical property analysis(Fig.5-35 and Fig.5-36).

TEMP	MOLEFRAC N-DEC-01	TOTAL PRES	TOTAL KVL BENZE-01	TOTAL KVL N-DEC-01	LIQUID1 GAMMA BENZE-01	LIQUID1 GAMMA N-DEC-01	LIQUID2 GAMMA BENZE-01	L GA
K		bar						
350	0.22	0.747189	1.26696	0.0535013	1.03474	1.14426		
350	0.24	0.73234	1.29883	0.0537216	1.03968	1.12614		
350	0.26	0.717353	1.33236	0.0540571	1.0447	1.10998		
350	0.28	0.702201	1.36769	0.0545073	1.04975	1.09558		
350	0.3	0.686861	1.40497	0.0550726	1.05481	1.08276		
350	0.32	0.671316	1.44435	0.0557546	1.05983	1.07136		
350	0.34	0.655549	1.48602	0.056556	1.0648	1.06124		
350	0.36	0.639551	1.53017	0.0574805	1.06967	1.05227		
350	0.38	0.623312	1.57703	0.0585331	1.07444	1.04433		
350	0.4	0.606826	1.62685	0.05972	1.07907	1.03732		
350	0.42	0.59009	1.67993	0.0610489	1.08354	1.03116		
350	0.44	0.573104	1.73658	0.0625291	1.08784	1.02576		
350	0.46	0.555868	1.79719	0.0641715	1.09195	1.02104		
350	0.48	0.538387	1.86216	0.0659894	1.09585	1.01695		
350	0.5	0.520665	1.932	0.067998	1.09952	1.01341		
350	0.52	0.502709	2.00727	0.0702157	1.10296	1.01037		
350	0.54	0.484527	2.08861	0.0726641	1.10615	1.00778		

Fig.5-35 View binary analysis results

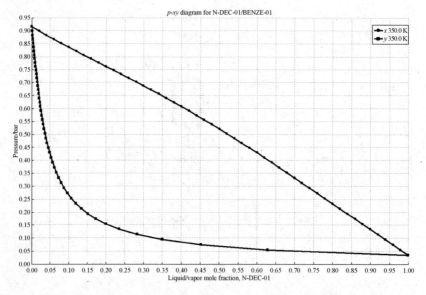

Fig.5-36 The results of physical property analysis were compared with experimental data

EXAMPLE 5-12

Regression of Binary VLE Data with Activity Coefficient Models I. Brown and F. Smith, Aust. J. Chem. 1954, 7, 264 the components are ethanol and benzene, and the VLE data at 45℃, regression of p-xy data at constant temperature to obtain parameters in several activity coefficient models(Table 5-9).

As shown in Fig.5-37 to Fig.5-39.

Table 5-9 VLE data at 45℃

x_{eth}	y_{eth}	p/kPa	x_{eth}	y_{eth}	p/kPa
0.0000	0.0000	29.39	0.5284	0.4101	40.93
0.0374	0.1965	36.19	0.6155	0.4343	40.28
0.0972	0.2895	39.53	0.7087	0.4751	38.91
0.2183	0.3370	40.88	0.8102	0.5456	36.16
0.3141	0.3625	41.24	0.9193	0.7078	30.36
0.4150	0.3842	41.28	0.9591	0.8201	27.11
0.5199	4.4065	41.00	1.0000	1.0000	23.21

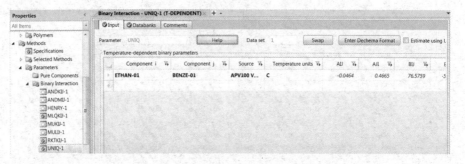

Fig.5-37 View binary interaction parameters

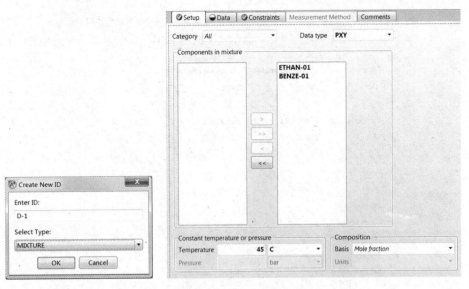

Fig.5-38 Create a new data type

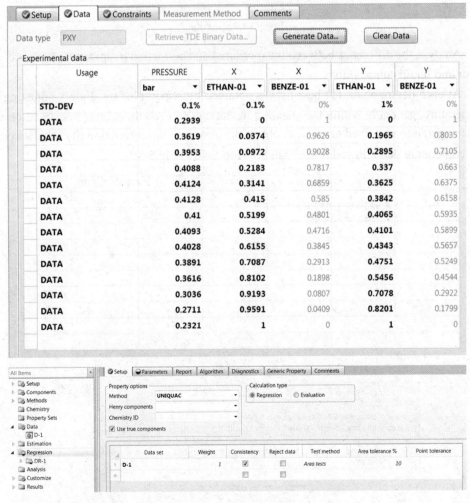

Fig.5-39

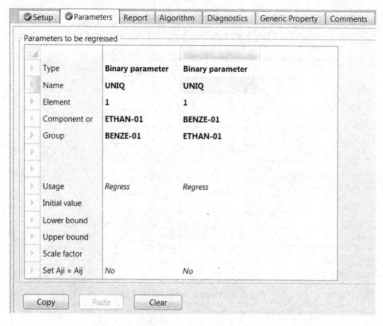

Fig.5-39 Regression

Analyze the difference between the measured and correlated temperature, pressure, and vapor and liquid compositions.

At each datum point, the best fit was obtained by Aspen Plus by slightly adjusting the temperature, generally within the standard deviation. So even though the specification was that temperature was fixed at 45 ℃, to obtain the best fit in the correlation of each data point, the temperature was allowed to vary slightly(Fig.5-40 to Fig.5-49).

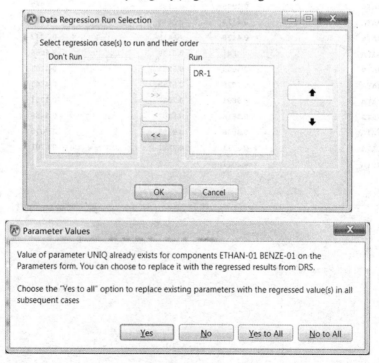

```
->Processing input specifications ...


->Data Regression begins ...


     BEGIN CASE DR-1                                    TIME =    0.01
     STARTING POINT                                      TIME:    0.01
     DM-ITER:   1  SSQ: 0.134322E+03  REL CHANGE: 0.287    TIME:    0.01
     DM-ITER:   2  SSQ: 0.132559E+03  REL CHANGE: 0.149E-01 TIME:    0.01
     DM-ITER:   3  SSQ: 0.132546E+03  REL CHANGE: 0.719E-03 TIME:    0.01
     BL-ITER:   4  SSQ: 0.132545E+03  REL CHANGE: 0.357E-04 TIME:    0.01
     BL-ITER:   5  SSQ: 0.133476E+03  REL CHANGE: 0.158E-02 TIME:    0.01
     BL-ITER:   6  SSQ: 0.133475E+03  REL CHANGE: 0.104E-03 TIME:    0.01
     FINAL SSQ:   133.475

->Data regression completed
```

Fig.5-40 Run-Yes

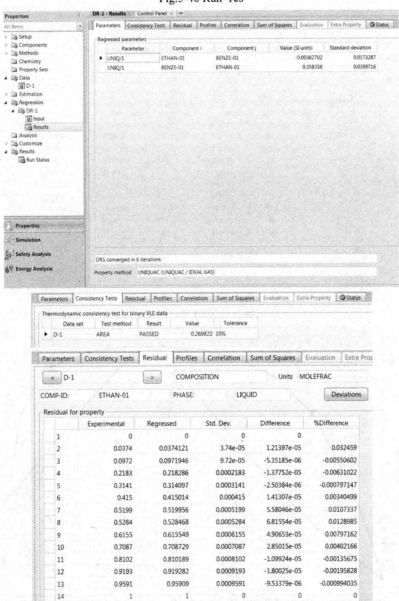

Fig.5-41

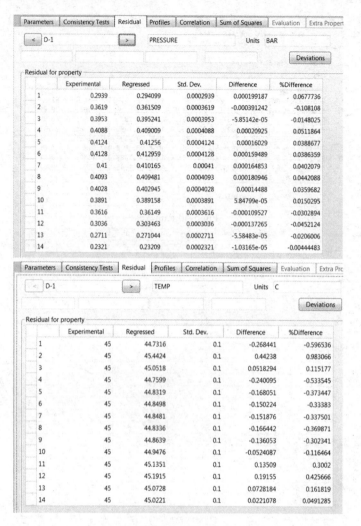

Fig.5-41 View the results

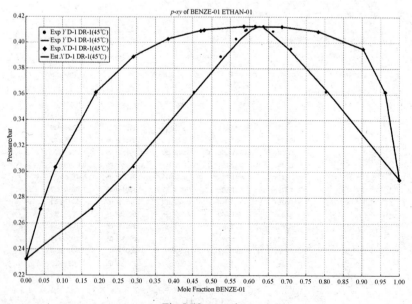

Fig.5-42 *p-xy* plot

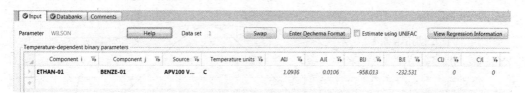

Fig.5-43 View binary interaction parameters

Fig.5-44 Selecting physical methods Wilson

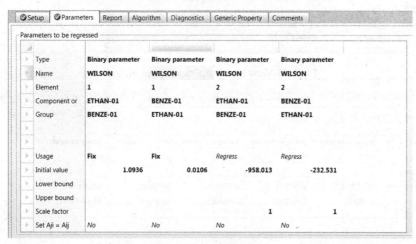

Fig.5-45 Regression

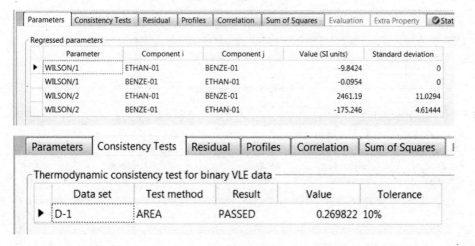

Fig.5-46

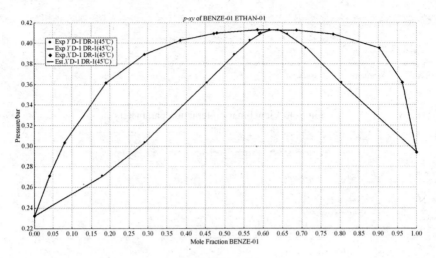

Fig.5-46 Results

NRTL

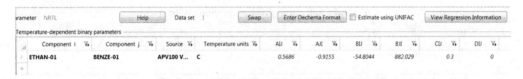

Fig.5-47 View binary interaction parameters

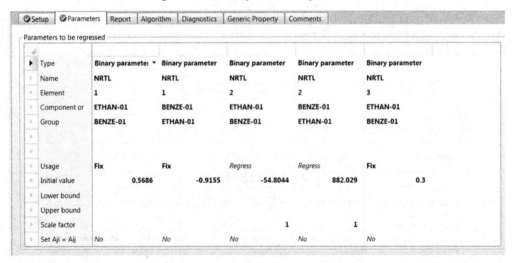

Fig.5-48 Regression

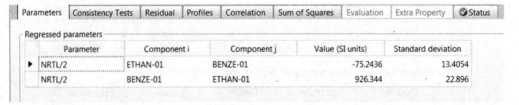

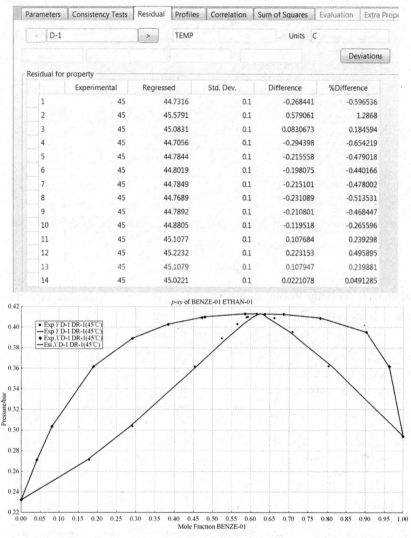

Fig.5-49 Results

5.5.2 Vapor-Liquid Equilibrium Calculations Using an Equation of State

EXAMPLE 5-13

VLE of the benzene + n-decane system

5.5.2.1 The Property Analysis Method

As shown in Fig.5-50 to Fig.5-53.

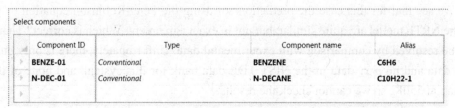

Fig.5-50 Components

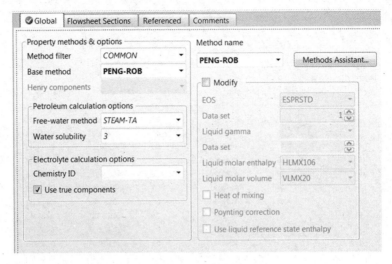

Fig.5-51 Methods

Note, there is no Peng-Robinson binary parameter available for the benzene + decane binary system in the Aspen Plus database.

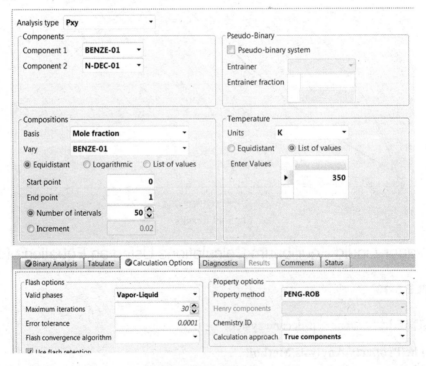

Fig.5-52 Analysis p-xy

We see that for this case the predictions of using the Peng-Robinson equation of state and the NRTL model are quite similar, but not in exact agreement. Which is correct? That can only be resolved by comparison with experimental data. Unfortunately, there is only bubble point data and no p-xy data in the NIST TDE data bank for this system, and none of those data are at 350K, so we cannot check the results.

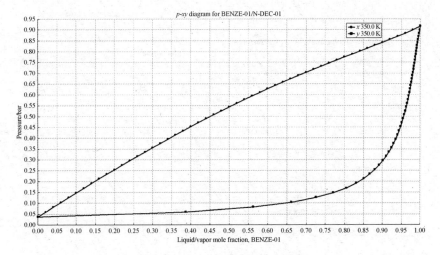

Fig.5-53 *p-xy*

5.5.2.2 The Simulation Method

As shown in Fig.5-54 to Fig.5-57.

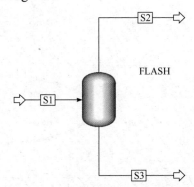

Fig.5-54 Blocks

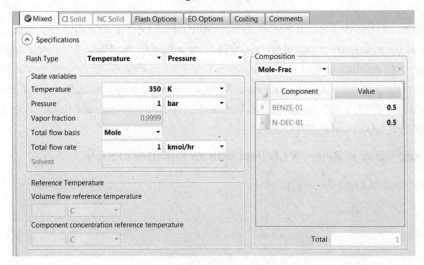

Fig.5-55 Input blocks parameters

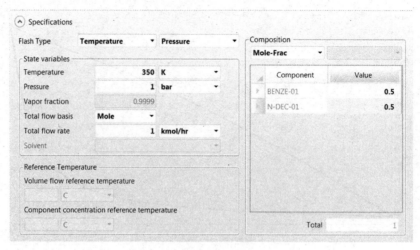

Fig.5-56 Input stream parameters

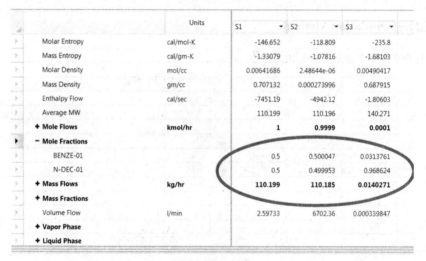

Fig.5-57 Results

Therefore, the dew point pressure is predicted to be 7.2 kPa and the benzene mole fraction of the equilibrium liquid to be 0.0314 according to the Peng-Robinson equation for the equal mole mixture at 350 K.

The results of the two calculations for this system are close with dew point pressure of 0.071 atm (PR EOS) and 0.067 atm (NRTL), and the dew point benzene liquid mole fractions of 0.0314 (PR EOS) and 0.334 (NRTL).

5.5.2.3 Regression of Binary VLE Data with an Equation of State

As shown in Fig.5-58 to Fig.5-64.

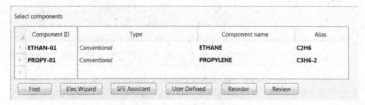

Fig.5-58 Ethane-propylene mixture

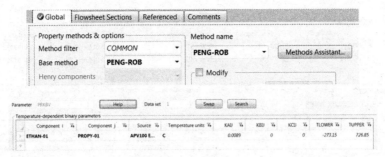

Fig.5-59 Methods

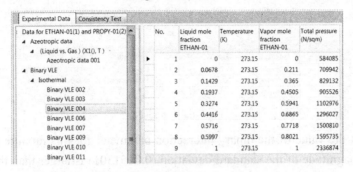

Fig.5-60 Data binary VLE 004

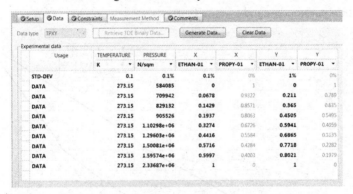

Fig.5-61 Save data

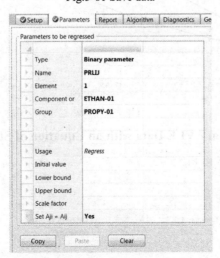

Fig.5-62 Regression

Regressed parameters				
Parameter	Component i	Component j	Value (SI units)	Standard deviation
PRLIJ/1	ETHAN-01	PROPY-01	0.0149992	0.0157052
PRLIJ/1	PROPY-01	ETHAN-01	0.0149992	0.0157052

Summary of regression results

Exp Val TEMP	Est Val TEMP	Exp Val PRES	Est Val PRES	Exp Val MOLEFRAC X ETHAN-01	Est Val MOLEFRAC X ETHAN-01	Exp Val MOLEFRAC X PROPY-01	Est Val MOLEFRAC X PROPY-01
C	C	bar	bar				
0	0.0218999	5.84085	5.84042	0	0	1	
0	0.79872	7.09942	7.07749	0.0678	0.0678843	0.9322	0.9321
0	0.53524	8.29132	8.2741	0.1429	0.143024	0.8571	0.8569
0	0.186119	9.05526	9.04828	0.1937	0.193792	0.8063	0.8062
0	-0.689751	11.0298	11.0602	0.3274	0.326922	0.6726	0.6730
0	-0.620853	12.9603	12.9951	0.4416	0.440643	0.5584	0.5593
0	-0.961652	15.0081	15.0722	0.5716	0.569173	0.4284	0.4308
0	-0.0108226	15.9574	15.9598	0.5997	0.59933	0.4003	0.400
0	-0.947855	23.3687	23.4596	1	1	0	

Fig.5-63 Run results

Note that the value of the binary interaction parameter for this mixture is quite small, and that the magnitude of the standard deviation (0.0151301) is larger than the value of the binary interaction parameter (0.0091725). Consequently, for this system the value of the binary interaction parameter can be set to 0.

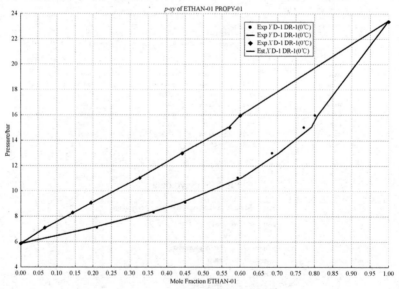

Fig.5-64 p-xy

5.5.2.4 Regression of Binary VLE Data with an Equation of State

As shown in Fig.5-65 to Fig.5-72.

Component ID	Type	Component name	Alias
CARBO-01	Conventional	CARBON-DIOXIDE	CO2
N-HEX-01	Conventional	N-HEXANE	C6H14-1

Fig.5-65 Carbon dioxide-hexane system

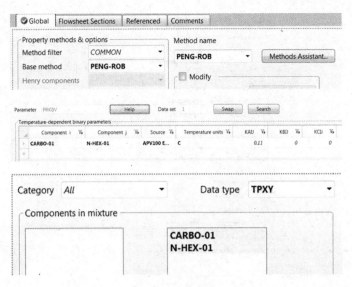

Global
Flowsheet Sections
Referenced
Comments
Property methods & options
Method filter
COMMON
Base method
PENG-ROB
Henry components
Method name
PENG-ROB
Methods Assistant...
Modify
Parameter PRKBV
Help
Data set 1
Swap
Search
Temperature-dependent binary parameters
Component i
Component j
Source
Temperature units
KAIJ
KBIJ
KCIJ
CARBO-01 N-HEX-01 APV100 E... C 0.11 0 0
Category All
Data type TPXY
Components in mixture
CARBO-01
N-HEX-01

Fig.5-66 Methods

Experimental Data
Consistency Test
Data for CARBO-01(1) and N-HEX-01(2)
Binary VLE
Isothermal
Binary VLE 001
Binary VLE 004
Binary VLE 005
Binary VLE 007
Binary VLE 008
Binary VLE 009
Binary VLE 011
Binary VLE 012
Binary VLE 013
Binary VLE 014
Binary VLE 016
Binary VLE 017
Binary VLE 018
Binary VLE 020
Binary VLE 021
Binary VLE 022
Binary VLE 025
No. | Liquid mole fraction CARBO-01 | Temperature (K) | Vapor mole fraction CARBO-01 | Total pressure (N/sqm)
1 0.0495 298.09438 0.9423 443700
2 0.1185 298.09438 0.9685 971300
3 0.1893 298.09438 0.9777 1471600
4 0.2683 298.09438 0.9793 1920300
5 0.3506 298.09438 0.983 2530900
6 0.3675 298.09438 0.9828 2676400
7 0.4251 298.09438 0.9843 3051100
8 0.5263 298.09438 0.9852 3649300
9 0.7255 298.09438 0.9872 4458400
10 0.8856 298.09438 0.9904 5205700
Ohgaki, K.; Katayama, T. J. Chem. Eng. Data, 1976, 21, 53-5 Isothermal Vapor-Liquid Equilibrium Data for Binary Systems Containg Carbon
Display Uncertainty
Save Data
Data Regression
Help
TDE version: 10.0, Dat

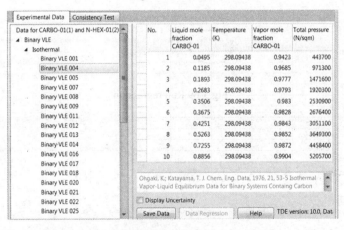

Fig.5-67 Data Binary VLE 004

(Ohgaki K, Katayama T J. Isothermal Vapor-Liquid Equilibrium Data for Binary Systems Containg Carbon Dioxide at High Pressures: Methanol-Carbon Dioxide, n-Hexane-Carbon Dioxide and Benzene-Carbon Dioxide Systems. Journal of Chemical and Engineering Data, 1976.)

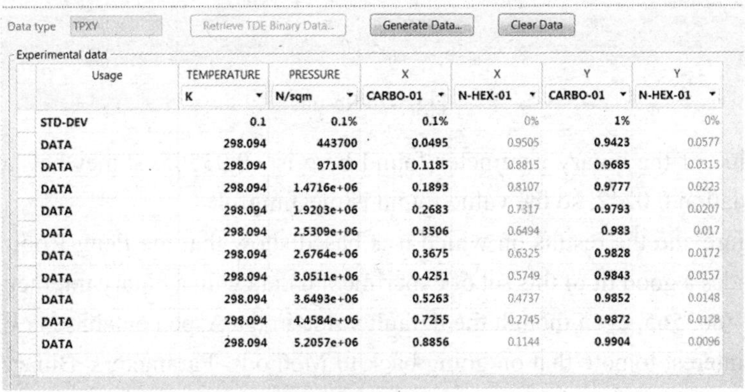

Data type TPXY
Retrieve TDE Binary Data...
Generate Data...
Clear Data
Experimental data
Usage | TEMPERATURE K | PRESSURE N/sqm | X CARBO-01 | X N-HEX-01 | Y CARBO-01 | Y N-HEX-01
STD-DEV 0.1 0.1% 0.1% 0% 1% 0%
DATA 298.094 443700 0.0495 0.9505 0.9423 0.0577
DATA 298.094 971300 0.1185 0.8815 0.9685 0.0315
DATA 298.094 1.4716e+06 0.1893 0.8107 0.9777 0.0223
DATA 298.094 1.9203e+06 0.2683 0.7317 0.9793 0.0207
DATA 298.094 2.5309e+06 0.3506 0.6494 0.983 0.017
DATA 298.094 2.6764e+06 0.3675 0.6325 0.9828 0.0172
DATA 298.094 3.0511e+06 0.4251 0.5749 0.9843 0.0157
DATA 298.094 3.6493e+06 0.5263 0.4737 0.9852 0.0148
DATA 298.094 4.4584e+06 0.7255 0.2745 0.9872 0.0128
DATA 298.094 5.2057e+06 0.8856 0.1144 0.9904 0.0096

Fig.5-68 Save Data

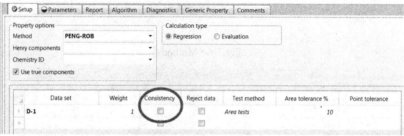

Fig.5-69 Run regression

Fig.5-70 Results

The value of the binary parameter found here is −0.03565, somewhat larger than the standard deviation 0.0242, so the value found is meaningful.

This figure and the results on which it is based show that the Peng-Robinson equation of state provides a good fit of this set of experimental data with a binary interaction parameter value of $k_{ij} = 0.03565$, even though the default value in the Aspen database is $k_{ij} = 0.11$.

It is of interest to note that on going back to Methods>Parameters>Binary Interaction> PRLIJ-1 the value of the binary parameter that now appears is the one just obtained by regression and the source is listed as regression R-1.

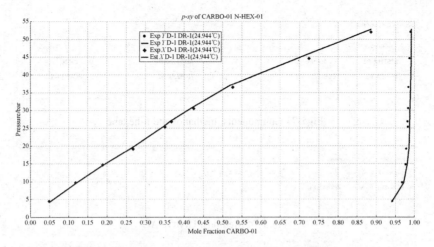

Fig.5-71 *p-xy*

Fig.5-72 View binary interaction parameters

5.5.2.5 Problems

The following vapor-liquid equilibrium data(Table 5-10) are available for ethane-ethylene mixtures at −0.01 ℃.

Table 5-10 The vapor-liquid equilibrium data of ethane-ethylene at −0.01 ℃

Pressure/kPa	Liquid/%	Vapor/%
780	6.2	39.73
2280	19.7	37.07
3030	25.5	35~60
5900	53.1	32.13
8900	85.4	25.45

Find the value of the binary interaction parameter in the Peng-Robinson equation of state with the van der Waals one-fluid mixing rules that best fits these data.

Regression of binary liquid-liquid equilibrium (LLE) data to obtain values of the parameters in activity models.

The calculation of liquid-liquid equilibrium is very sensitive to the thermodynamic models used, the values of the activity coefficients, and their accuracy over the whole concentration range.

Since LLE and VLLE only occur in very nonideal mixtures, generally involving very different chemical species, activity coefficient models are used in this chapter(Fig.5-73 to Fig.5-87).

Select components

Component ID	Type	Component name	Alias
METHA-01	*Conventional*	METHANOL	CH4O
N-HEX-01	*Conventional*	N-HEXANE	C6H14-1

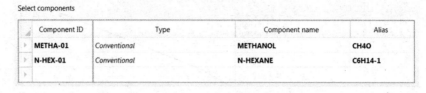

Fig5-73 Components: methanol and n-hexane

ırameter UNIQ Help Data set 1 Swap Enter Dechema Format Estimate using UNIFAC View Regression Info

Temperature-dependent binary parameters

Component i	Component j	Source	Temp. units	AIJ	AJI	BIJ	BJI	CIJ	CJI	DIJ	DJI
METHA-01	N-HEX-01	APV100 LLE-LIT	C	*0*	*0*	*5.0185*	*-610.14*	*0*	*0*	*0*	*0*

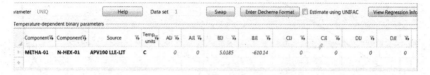

Fig.5-74 Methods: UNIQUAC

Experimental Data | Consistency Test

Data for METHA-01(1) and N-HEX-01(2)
- Binary LLE
  - (Liquid 1 vs. Liquid 2) (T, P, X1(L1))
    - Binary LLE 001
    - Binary LLE 002
    - Binary LLE 003
    - Binary LLE 004
    - Binary LLE 005
    - Binary LLE 006
    - Binary LLE 007
    - Binary LLE 008
    - Binary LLE 009
    - Binary LLE 010

No.	Liquid1 mole fraction METHA-01	Temperature (K)	Pressure (N/sqm)	Liquid2 mole fraction METHA-01
1	0.822	293.19536	101300	0.193
2	0.808	295.69486	101300	0.217
3	0.792	298.19436	101300	0.245
4	0.771	300.69379	101300	0.28
5	0.743	303.19304	101300	0.324
6	0.696	305.69229	101300	0.391
7	0.553	307.69169	101300	

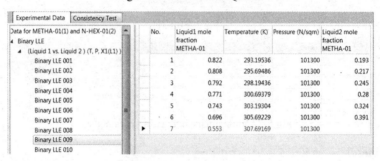

Fig.5-75 Data type: TPXX

(Schmidt R, Werner G, Schuberth H Z. Determination of heloroazeotropic properties of the systems n-hexane - methanol and methylcyclohexane - methanol with a new equilibrium apparatus. Phys. Chem. (Leipzig), 1969, 242: 381-390)

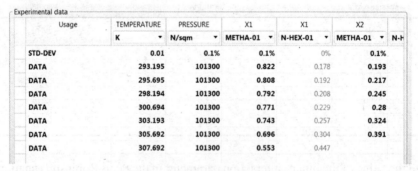

Experimental data

Usage	TEMPERATURE	PRESSURE	X1	X1	X2	
	K	N/sqm	METHA-01	N-HEX-01	METHA-01	N-H
STD-DEV	0.01	0.1%	0.1%	0%	0.1%	
DATA	293.195	101300	0.822	0.178	0.193	
DATA	295.695	101300	0.808	0.192	0.217	
DATA	298.194	101300	0.792	0.208	0.245	
DATA	300.694	101300	0.771	0.229	0.28	
DATA	303.193	101300	0.743	0.257	0.324	
DATA	305.692	101300	0.696	0.304	0.391	
DATA	307.692	101300	0.553	0.447		

Fig.5-76 Data

Experimental data

Usage	TEMPERATURE	PRESSURE	X1	X1	X2	
	K	N/sqm	METHA-01	N-HEX-01	METHA-01	N-H
STD-DEV	0.01	0.1%	0.1%	0%	0.1%	
DATA	293.195	101300	0.822	0.178	0.193	
DATA	295.695	101300	0.808	0.192	0.217	
DATA	298.194	101300	0.792	0.208	0.245	
DATA	300.694	101300	0.771	0.229	0.28	
DATA	303.193	101300	0.743	0.257	0.324	
DATA	305.692	101300	0.696	0.304	0.391	

Fig.5-77 Delete Data

To simplify the regression, the last (highest temperature) data point that is in the one-phase region will be eliminated.

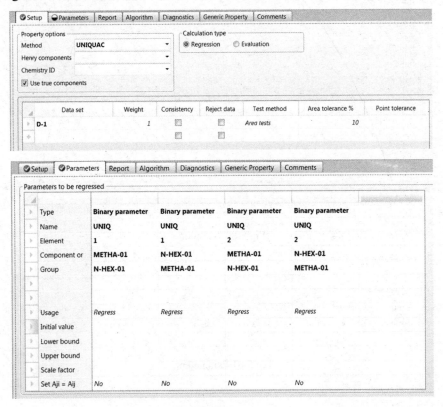

Fig.5-78 Regression

```
->Data Regression begins ...


   BEGIN CASE DR-1                              TIME =      0.00
   STARTING POINT                                     TIME:   0.01
   DM-ITER:   1  SSQ: 0.337464E+07  REL CHANGE: 0.266  TIME:   0.01
   DM-ITER:   2  SSQ: 0.286158E+06  REL CHANGE: 0.495  TIME:   0.01
   DM-ITER:   3  SSQ: 0.134468E+04  REL CHANGE: 1.26   TIME:   0.01
   DM-ITER:   4  SSQ: 0.836173E+03  REL CHANGE: 0.202E-01 TIME: 0.01
   BL-ITER:   5  SSQ: 0.836194E+03  REL CHANGE: 0.248E-03 TIME: 0.01
   BL-ITER:   6  SSQ: 0.829022E+03  REL CHANGE: 0.210E-03 TIME: 0.01
   BL-ITER:   7  SSQ: 0.829003E+03  REL CHANGE: 0.137E-04 TIME: 0.01
   FINAL SSQ:    829.002

->Data regression completed
```

Parameter	Component i	Component j	Value (SI units)	Standard deviation
UNIQ/1	METHA-01	N-HEX-01	0.314908	0.165736
UNIQ/1	N-HEX-01	METHA-01	4.38309	0.497891
UNIQ/2	METHA-01	N-HEX-01	-92.8991	49.5059
UNIQ/2	N-HEX-01	METHA-01	-1888.94	148.781

Fig.5-79 Results

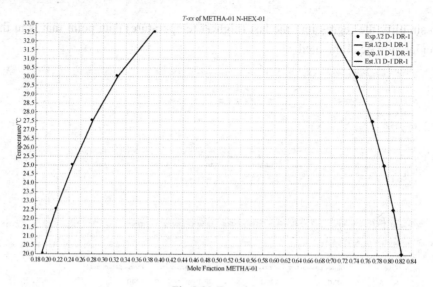

Fig.5-80 *T-xx* plot

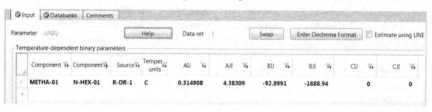

Fig.5-81 Parameters

Fig.5-82 Binary analysis

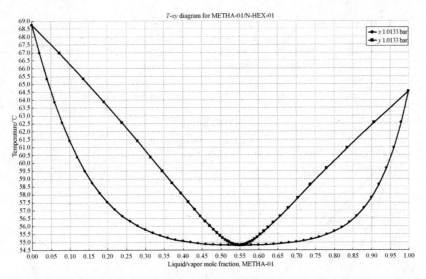

Fig.5-83 Run analysis

So even though the parameters were obtained by correlating liquid-liquid equilibrium data at 101.35 kPa, if we allow for a vapor phase in the calculations, only vapor-liquid equilibria are predicted with the correlated parameters over the whole composition range. That liquid-liquid equilibrium is not observed because the default temperature range is above the liquid-liquid coexistence temperature.

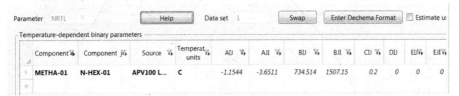

Fig.5-84 NRTL model

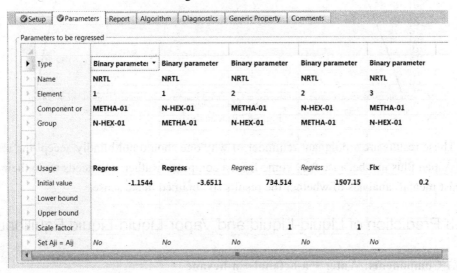

Fig.5-85 Regression

Regressed parameters

	Parameter	Component i	Component j	Value (SI units)	Standard deviation
▶	NRTL/1	METHA-01	N-HEX-01	3.06692	0.901451
	NRTL/1	N-HEX-01	METHA-01	-8.95857	0.888557
	NRTL/2	METHA-01	N-HEX-01	-475.608	269.171
	NRTL/2	N-HEX-01	METHA-01	3006.66	265.46

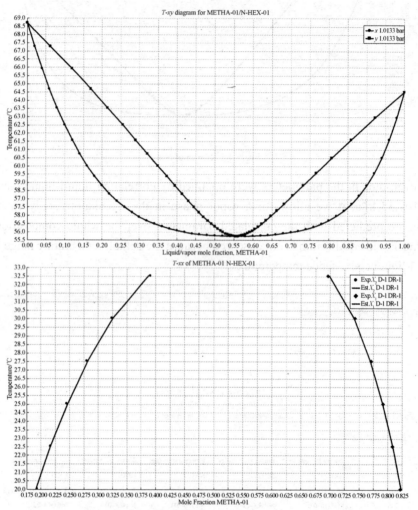

Fig.5-86 Results

Fig.5-87 Run analysis

These results are a poignant reminder of why one should not blindly accept the results from Aspen Plus just because they come from a computer. Rather, one needs to do their own (at least mental) analysis of whether the results so obtained make sense.

5.5.3 Prediction of Liquid-Liquid and Vapor-Liquid-Liquid Equilibrium

5.5.3.1 Components: Water + n-octanol + n-hexane

Gibbs phase rule: $F = C - P - R + 2$. In a binary system, where $C = 2$, there is only one degree of freedom, $F = 1$.

That is, at each fixed composition (the one degree of freedom) there is only one temperature-pressure point at which three phases coexist(Fig.5-88 to Fig.5-92).

Tools>Analysis>Property>Ternary

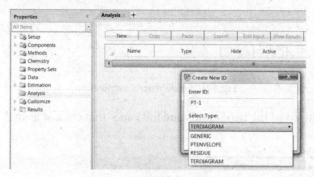

Fig.5-88 Analysis>Ter diagram

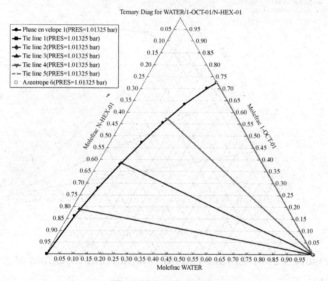

Fig.5-89 Run Analysis

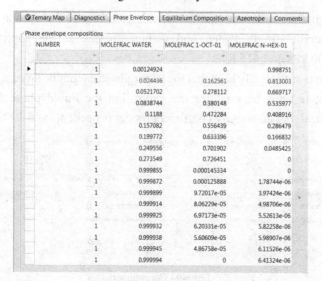

Fig.5-90 Phase envelop

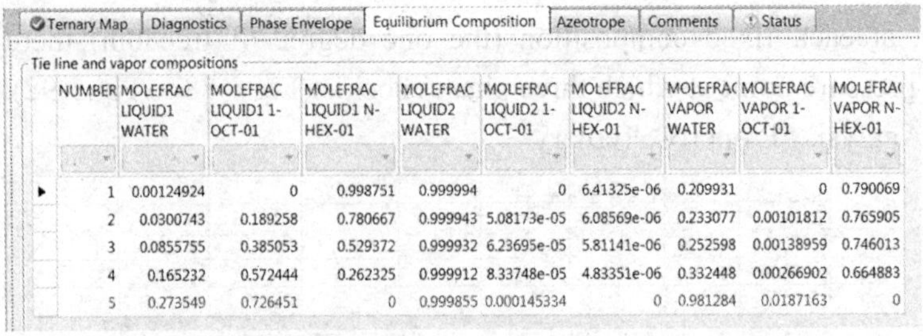

NUMBER	MOLEFRAC LIQUID1 WATER	MOLEFRAC LIQUID1 1-OCT-01	MOLEFRAC LIQUID1 N-HEX-01	MOLEFRAC LIQUID2 WATER	MOLEFRAC LIQUID2 1-OCT-01	MOLEFRAC LIQUID2 N-HEX-01	MOLEFRAC VAPOR WATER	MOLEFRAC VAPOR 1-OCT-01	MOLEFRAC VAPOR N-HEX-01
1	0.00124924	0	0.998751	0.999994	0	6.41325e-06	0.209931	0	0.790069
2	0.0300743	0.189258	0.780667	0.999943	5.08173e-05	6.08569e-06	0.233077	0.00101812	0.765905
3	0.0855755	0.385053	0.529372	0.999932	6.23695e-05	5.81141e-06	0.252598	0.00138959	0.746013
4	0.165232	0.572444	0.262325	0.999912	8.33748e-05	4.83351e-06	0.332448	0.00266902	0.664883
5	0.273549	0.726451	0	0.999855	0.000145334	0	0.981284	0.0187163	0

Fig.5-91 Equilibrium composition

The compositions of the two liquids and the vapor that coexist at each tie line(Fig.5-92).

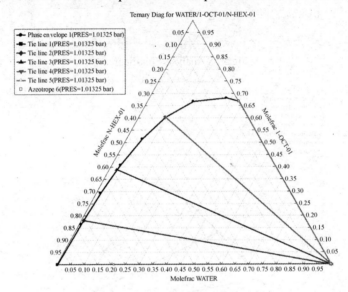

Fig.5-92 Azeotrope

5.5.3.2 High Pressure Vapor-Liquid-Liquid Equilibrium Equation of State Activity Coefficient

As an example, consider an equimolar mixture of each of n-decane (to represent the hydrocarbon reservoir fluid), water, and carbon dioxide at the approximate reservoir conditions of 7500 kPa and 415 K.

The pressure is sufficiently high that the vapor phase cannot be an ideal gas. In this case, an equation of state must be used for the vapor, but the liquid phase is, or phases are, sufficiently nonideal that an activity coefficient model is needed(Fig.5-93 to Fig.5-101).

Component ID	Type	Component name	Alias
N-DEC-01	Conventional	N-DECANE	C10H22-1
WATER	Conventional	WATER	H2O
CARBO-01	Conventional	CARBON-DIOXIDE	CO2

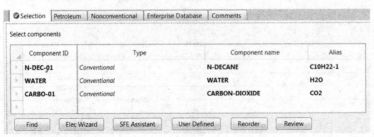

Fig.5-93 Components

Method: UNIQUAC

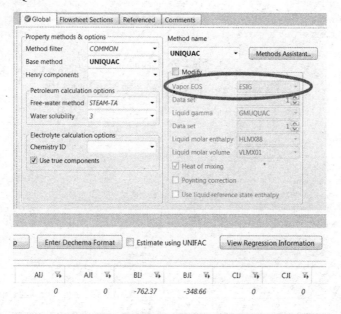

Fig.5-94 UNIQUAC

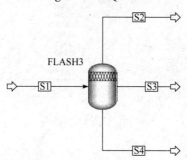

Fig.5-95 Flowsheet

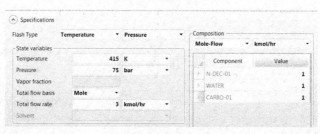

Fig.5-96 Streams

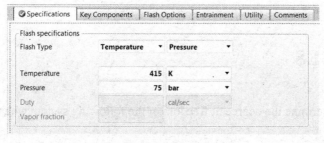

Fig.5-97 Blocks

- Mole Fractions				
N-DEC-01	0.333333	0.00344554	0.636588	2.78807e-05
WATER	0.333333	0.046031	0.0659891	0.902198
CARBO-01	0.333333	0.950523	0.297423	0.0977745

Fig.5-98 Results

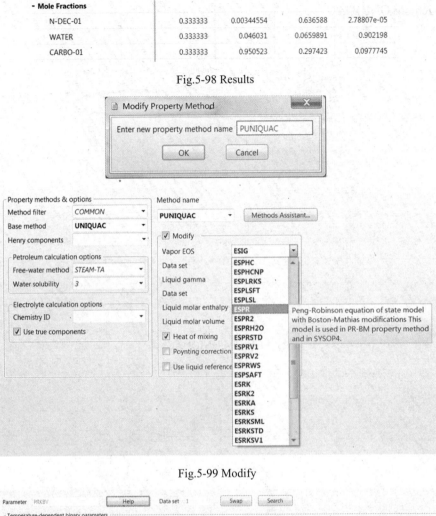

Fig.5-99 Modify

Fig.5-100 Parameter

- Mole Fractions				
N-DEC-01	0.333333	0.0135681	0.676056	2.1777e-05
WATER	0.333333	0.0687124	0.0628781	0.918289
CARBO-01	0.333333	0.917719	0.261065	0.0816891

Fig.5-101 Results

EXERCISES

1. Try to determine the degree of freedom of the following system when it reaches phase equilibrium:

(1) Triple point of water;
(2) Water-vapor equilibrium;
(3) Water-water-inert gas.

2. For a binary solution composed of components A and B, the vapor phase can be regarded as an ideal gas, while the liquid phase is a non-ideal solution. The correlation between the activity coefficient of the liquid phase and its composition is expressed as follows.

$$\ln \gamma_A = 0.5 x_B^2, \quad \ln \gamma_B = 0.5 x_A^2$$

At 80℃, the saturated vapor pressures of components A and B are respectively

$$p_A^s = 1.2 \times 10^5 \text{ Pa}, \quad p_B^s = 0.8 \times 10^5 \text{ Pa}$$

Does the vapor-liquid equilibrium of this solution form azeotrope at 80℃? If so, what is the boiling pressure and boiling composition?

3. The methanol-methyl ethyl ketone binary system forms an azeotrope containing 84.2% (mole fraction) methanol at a temperature of 64.3℃ and a pressure of 1.013×10^5 Pa. The Wilson equation is used to calculate the parameters of the constant boiling point data and the t-x_1-y_1 data of the binary system at a pressure of 1.013×10^5 Pa.

The Antoine constant of y is given as follows

component		A	B	C
methyl alcohol	(1)	11.9673	3626.55	−34.29
methyl ethyl ketone	(2)	9.9784	3150.42	−36.65

$$\text{Antoine equation} \quad \ln p_i^s = A - \frac{B}{T+C}$$

4. It is known that 2, 4-dimethylpentane (1) and benzene (2) form the maximum pressure azeotrope at 60℃, which is now separated by extractive distillation. It is known that 2-methyl amyl 2-4 diol is a suitable third component. How much of the third component must be added so that the relative volatility α_{12} of the original azeotrope is never less than 1, that is, the minimum α_{12} occurs at $x_2 = 0$. When $x_2 \to 0, \alpha_{12} = 1$, what is the concentration of 2-methylpentyl 2-4-diol?

Known data: saturated vapor pressure of two components at 60℃

$$p_1^s = 52.39 \text{ kPa}, \quad p_2^s = 52.26 \text{ kPa}$$

The infinite dilution activity coefficient of the system at this temperature is
binary system 1-2, $\gamma_1^\infty = 1.96, \gamma_2^\infty = 1.48$
binary system 1-3, $\gamma_1^\infty = 3.55, \gamma_3^\infty = 15.1$
binary system 2-3, $\gamma_2^\infty = 2.04, \gamma_3^\infty = 3.89$

5. At 330.3 K, the liquid mixture of acetone (A) and methanol is balanced at 101325 Pa. The equilibrium composition is liquid phase $x_A = 0.400$ and vapor phase $y_A = 0.400$. It is known that the vapor pressure of 330.3K pure component $p_A = 104791$ Pa, $p_E = 73460$ Pa.

Try to explain whether the liquid mixture is an ideal liquid mixture and why? If it is not an ideal liquid mixture, calculate the activity and activity factor of each component. (Pure liquid is the standard state)

6. In the Vapor-liquid balance system of a binary mixture, the infinite dilution $\gamma_1^\infty = \gamma_2^\infty = 1.648$ at 80℃ was measured at 101.3 kPa, and the activity coefficient model $\ln\gamma_1 = \beta x_2^2, \ln\gamma_2 = \beta x_1^2, p_1^S = 120$ kPa, $p_2^S = 80$ kPa was obtained. The azeotropic composition and azeotropic pressure of the system at 80℃ were obtained.

7. Try using H_2O-$NaNO_3$ system phase diagram for analysis of constant temperature evaporation process of $NaNO_3$ solution. And at 60℃, 100 kg 30% $NaNO_3$ solution, constant temperature evaporation to the system composition of 85%, precipitation and evaporation of water quality.

8. At a given T and p, the excess enthalpy model of a binary mixture solution is $\dfrac{G^E}{RT} = (-1.5x_1 - 1.8x_2)x_1x_2$. In equation (A) x is the mole fraction, try for:

(1) The expression of $\ln\gamma_1$ and $\ln\gamma_2$;

(2) The value of the $\ln\gamma_1^\infty$ and $\ln\gamma_2^\infty$;

(3) Combine the expression obtained by (1) with the formula $\dfrac{G^E}{RT} = \sum(x_i \ln\gamma_i)$, prove that equation (A) is reproducible.

9. The enthalpy of a binary mixture at a certain T and p can be expressed as follows: $H = x_1(a_1 - b_1x_1) + x_2(a_2 + b_2x_2)$. Where a and b are constants, try to find the expression of partial molar enthalpy $\bar{H}_i$ of component 1.

10. At a particular temperature, the excess enthalpy of a binary solution is expressed as $G^E/RT = Bx_1x_2$, If the binary system is considered as low pressure gas-liquid equilibrium, when $0 < x_1 < 1$, why does not phase splitting occur when B is valued?

11. At 25℃, binary solution A and B are in vapor-liquid-liquid three-phase equilibrium, and the composition of saturated liquid phase.

$$x_A^\alpha = 0.02, x_B^\alpha = 0.98$$

$$x_A^\beta = 0.98, x_B^\beta = 0.02$$

At 25℃, the saturated vapor pressure of substance A and substance B:

$$p_A^s = 0.01 \text{ MPa}, p_B^s = 0.1013 \text{ MPa},$$

Try to calculate the pressure and vapor phase composition of three phase coexistence equilibrium.

REFERENCES

[1] Amer H H, Paxton R R, Van Winkle M. Methanol ethanol-Acetone[J]. Ind. Eng. Chem., 1956, 48: 142-157.

[2] Benson G C. Molar excess Gibbs free energies of benzene-m-xylene mixtures[J]. Can. J. Chem., 1969, 47: 539-542.

[3] Bobbo S, Fedele L, Scattolini M, Camporese R. Isothermal VLE measurements for the binary mixtures HFC-134a + HFC-245fa and HC-600 + HFC-245fa[J]. Fluid Phase Equil., 2001, 185: 255-276.

[4] Gmehling J, Onken U. Vapor-liquid equilibrium data collection. DECHEMA. 1977.

[5] Mueller C R, Kearns E R. Thermodynamic studies of the system: acetone + chloroform[J]. J. Phys. Chem. 1958, 62, 1441-1457.

[6] Piacentini A, Stein F. An experimental and correlative study of the vapor-liquid equilibria of the tetrafluoromethane system[J]. Chem. Eng. Prog. Symp. Ser. 1967, 63: 28-39.

[7] Schuberth H. Phasengleichgewichtsmessungen. II. Vorausberechnung von gleichgewichtsdaten dampfformig-lussig idealer binrer systeme und prüfung derselben mittels einer neuen gleichgewichtsapparatur[J]. J. Prakt. Chem., 1958, 06: 129-142.

[8] Sinor J E, Weber J H. Vapor-liquid equilibria at atmospheric pressure. systems containing ethyl alcohol, n-Hexane, benzene, and methylcyclopentane[J]. J. Chem. Eng. Data, 1960, 05: 243-256.

[9] Stein F, Proust P. Vapor-liquid equilibriums of the trifluoromethane-trifluorochloromethane system[J]. J. Chem. Eng. Data, 1971, 16: 389-395.

[10] Wilsak R, A Campbell S W, Thodos, G. Vapor-liquid equilibrium measurements for the n-pentanemethanol system at 372.7, 397.7 and 422.6 K[J]. Fluid Phase Equil, 1987, 33: 157-168.

Chapter 6

Energy Analysis of Chemical Process

Learning Objectives

- Examine the performance of engineering devices in light of the second law of thermodynamics. Use the Poynting correction to estimate the fugacity of liquids and solids at elevated pressures.
- Define exergy, which is the maximum useful work that could be obtained from the system at a given state in a specified environment.
- Define reversible work, which is the maximum useful work that can be obtained as a system undergoes a process between two specified states.
- Define the exergy destruction, which is the wasted work potential during a process as a result of irreversibilities.
- Define the second-law efficiency.
- Develop the exergy balance relation.
- Apply exergy balance to closed systems and control volumes.

6.1 The Definition of Entropy Exergy

The first law of thermodynamics deals with the quantity of energy and asserts that energy cannot be created or destroyed. The second law of thermodynamics deals with the quality of energy. A property to determine the useful work potential of a given amount of energy at some specified state. This property is exergy, which is also called the availability or available energy.

Exergy is the maximum useful work that could be obtained from the system at a given state in a specified environment. Exergy is the part of energy that can be converted into any other form of energy theoretically when the system changes reversibly from any state to a state that is equilibrium with a given environment. So, the ideal work done by the system as it changes from its state to the reference state is the exergy (Y A Cengel, 2014). The atmosphere contains a tremendous amount of energy, but no exergy. A system is said to be in the dead-state when it is in thermodynamic equilibrium with the environment it is in.

Unless specified situations, the dead-state temperature and pressure are taken to be $T_0 = 25\ ℃$ and $p_0 = 1.01325 \times 10^5$ Pa.

A system has zero exergy at the dead-state. A system must go to the dead-state at the end of the process to maximize the work output. A system delivers the maximum possible work as it undergoes a reversible process from the specified initial state to the state of its environment (the dead-state). This represents the useful work potential of the system at the specified state and is called exergy. Exergy is a property of the system-environment combination and not of the system alone. The term availability was made popular in the United States by the M.I.T. School of Engineering in the 1940s. Today, an equivalent term, exergy, introduced in Europe in the 1950s, has found global acceptance partly because it is shorter, it is different from energy and entropy, and it can be adapted without requiring translation. In this text the preferred term is exergy.

6.2 Exergy (Work Potential) Associated with Kinetic and Potential Energy

Kinetic energy, which can be completely converted into work, is one of the ways in which mechanical energy exists (Zhu Ziqiang, 2012). Therefore, no matter how the temperature and pressure in the environment change, the potential energy or exergy of the kinetic energy of the system is always the same as the kinetic energy itself.

Exergy of kinetic energy can be shown as:

$$X_{ke} = ke = \frac{V^2}{2} (\text{kJ/kg}) \tag{6-1}$$

The speed of the system relative to the environment is denoted by V.

In the same way, potential energy is a form of mechanical energy that can be completely converted into work. Thus, regardless of the ambient temperature and pressure, the potential energy or exergy of the system is equal to the potential energy itself (Fig 6-1).

Exergy of potential energy is shown as:

$$X_{pe} = pe = gz (\text{kJ/kg}) \tag{6-2}$$

In the equation, g represents the acceleration of gravity, and z represents the height difference between the system and the specified reference level in the environment.

Therefore, kinetic energy and potential energy exergy are equal to themselves, which can be completely converted into work. However, the internal energy U and enthalpy H of the system, on the contrary, are not fully used for work, as will be discussed later in this book.

EXAMPLE 6-1

Maximum Power Generation by a Wind Turbine.

A wind turbine with a 10-m-diameter rotor, as shown in Fig.6-2, is to be installed at a location where the wind is blowing steadily at an average velocity of 15 m/s. Determine the maximum power that can be generated by the wind turbine.

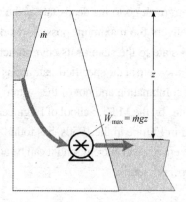
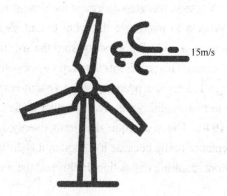

Fig.6-1 The work potential or exergy of potential energy is equal to the potential energy itself

Fig.6-2 The schematic of Example 6-1

SOLUTION

The location of wind turbines needs to be reasonably discussed. And the maximum power that wind turbines can produce is uncertain.

Assumption that in 1 atm and 25℃ standard condition, the density of air is 1.18 kg/m³. Air in motion has the same characteristics as air at rest except that it has speed and some kinetic energy. When air reaches a state of rest, it will reach a dead-state. Therefore, the kinetic energy of the flowing air is exergy.

$$ke = \frac{V^2}{2} = \frac{(15 \text{ m/s})^2}{2} \times \frac{1 \text{ kJ/kg}}{1000 \text{ m}^2/\text{s}^2} = 0.1125 \text{ kJ/kg}$$

That is to say, the work potential of air per unit mass flowing at 15 meters per second is 0.1125 kJ/kg. In other words, an ideal wind turbine would be able to freeze the air completely and capture a potential of 0.1125 kJ/kg. To get the maximum power, we need to take the amount of air passing through the wind turbine per unit time and the mass flow rate. It can be determined as:

$$m = \rho AV = \rho \frac{\pi D^2}{4} V = 1.18 \text{ kg/m}^3 \times \frac{\pi \times (10 \text{ m})^2}{4} \times 15 \text{ m/s} = 1390 \text{ kg/s}$$

Thus,

Maximum power = $m \times 0.1125$ kJ/kg = 1390 kg/s × 0.1125 kJ/kg = 156.38 kW

This is the maximum power that can be utilized by wind turbines.

Assuming a conversion efficiency of 30%, an ideal wind turbine can convert 20.0 kilowatts of electricity. But the important thing to note is that the work potential in this case is equal to the kinetic energy of the entire air.

It is important to note that although all the kinetic energy of the wind can be used to generate electricity, *Baez's law* dictates that the output of a wind turbine is at its maximum when the wind slows down to about a third of its initial speed. Thus, for maximum power (and minimum cost per installed power), the maximum efficiency of a wind turbine is about 59%. In practical life applications, the actual efficiency ranges from 20% to 40%, with many wind turbines achieving an actual efficiency of around 35%.

Wind power is generated when the average wind speed is at least 6 meters per second (or 13 miles per hour) of steady wind. Recent improvements in wind turbine design have made wind power around 5 cents per kilowatt-hour, making it competitive with electricity from other sources.

6.3 Reversible Work and Irreversibility

In this section, we describe two quantities that related to the actual initial and final states of processes and serve as valuable tools in the thermodynamic analysis of components or systems. These two quantities are the reversible work and irreversibility (or exergy destruction). But first we examine the surroundings work, which is the work done by or against the surroundings during a process (SI Sandler, 2015).

The work done by work-producing devices is not always entirely in a usable form. Part of the work done by the gas is used to push the atmospheric air out of the way of the piston.

$$W_{surr} = p_0 (V_2 - V_1) \tag{6-3}$$

The difference between the actual work W and the surroundings work W_{surr} is called the useful work W_u:

$$W_u = W - W_{surr} = W - p_0 (V_2 - V_1) \tag{6-4}$$

Any difference between the reversible work W_{surr} and the useful work W_u is due to the irreversibilities present during the process, and this difference is called irreversibility.

The irreversibility is equivalent to the exergy destroyed. For a totally reversible process, the actual and reversible work terms are identical, and thus the irreversibility is zero. This is expected since totally reversible processes generate no entropy. Irreversibility is a positive quantity for all actual (irreversible) processes since $W_{rev} \geq W_u$ for work-producing devices and $W_{rev} \leq W_u$ for work-consuming devices. Irreversibility can be viewed as the wasted work potential or the lost opportunity to do work.

EXAMPLE 6-2

The Rate of Irreversibility of a Heat Engine.

A heat engine receives heat from a source at 1500 K and rate of 600 kJ/s, rejects the waste heat to a medium at 303.15 K. The power output of the heat engine is 220 kW. Determine the reversible power and the irreversibility rate for this process.

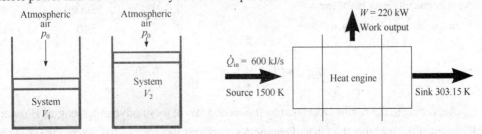

Fig.6-3 As a closed system expands, some work needs to be done to push the atmospheric air out of the way (W_{surr})

Fig.6-4 Schematic for Example 6-2

SOLUTION

The reversible power for this process is the amount of power that a reversible heat engine, such as a *Carnot* heat engine, would produce when operating between the same temperature limits and is determined to be:

$$\dot{W}_{\text{rev,out}} = \eta_{\text{th,rev}} \dot{Q}_{\text{in}} = \left(1 - \frac{T_{\text{sink}}}{T_{\text{source}}}\right) \dot{Q}_{\text{in}} = \left(1 - \frac{303.15}{1500}\right) \times 600 \text{ kW} = 478.74 \text{ kW}$$

The irreversibility rate is the difference between the reversible power and the useful power output:

$$\dot{I} = \dot{W}_{\text{rev,out}} - \dot{W}_{\text{u,out}} = (478.74 - 220) \text{ kW} = 267.74 \text{ kW}$$

Note that 175 kW of power potential is wasted during this process as a result of irreversibility. Also, the $(600 - 478.74)\text{kW} = 121.26$ kW of heat rejected to the sink is not available for converting to work and thus is not part of the irreversibility.

6.4 Second-law Efficiency

We have introduced the first law of thermodynamics in Section 6.1, but it could not determine the most efficient performance based on the first law of thermodynamics, which may lead to misleading performance selection (Ma Peisheng et al., 2009). As shown as(Fig.6-5):

There are two engines, both of which have a thermodynamic efficiency of 30% and both release energy to the medium at a temperature of 300 K. The heat input to one engine (engine A) is 600 K, and the heat input to the other engine (engine B) is 1000 K. It seems that the two engines convert the input energy in the same proportion. However, if we use the second law of thermodynamics as the standard, we get the opposite result. Calculated by the second law of thermodynamics, the efficiency of these engines would be:

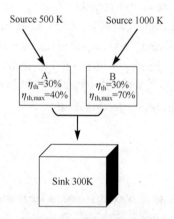

Fig.6-5 Two heat engines that have the same thermal efficiency but different maximum thermal efficiencies

$$\eta_{\text{rev,A}} = \left(1 - \frac{T_L}{T_H}\right)_A = 1 - \frac{300}{500} = 0.40$$

$$\eta_{\text{rev,B}} = \left(1 - \frac{T_L}{T_H}\right)_B = 1 - \frac{300}{1000} = 0.70$$

According to the results calculated by the second law of thermodynamics, engine B's working efficiency of 0.7 is higher than that of engine A's working efficiency of 0.4, so engine B has greater working potential. Therefore, although the two engines have the same thermal efficiency, we can say that engine B is superior to engine A.

As this example shows, the first law of thermodynamics is not a practical measure of equipment performance. To overcome this shortcoming, we define the second law of thermodynamics efficiency η_{II} as the ratio of the actual thermal efficiency to the maximum thermal efficiency under the same conditions.

$$\eta_{\mathrm{II}} = \frac{\eta_{\mathrm{th}}}{\eta_{\mathrm{th,rev}}} \tag{6-5}$$

According to this definition of efficiency by the second law of thermodynamics, the efficiencies of the two engines are:

$$\eta_{\mathrm{II,A}} = \frac{0.3}{0.4} = 0.75$$

$$\eta_{\mathrm{II,B}} = \frac{0.3}{0.7} = 0.43$$

In other words, engine A can convert 75 percent of its maximum power to output power. For engine B, the ratio is only 43 percent. Therefore, the second law of thermodynamics efficiency is the ratio of the effective work to the maximum work:

$$\eta_{\mathrm{II}} = \frac{W_{\mathrm{u}}}{W_{\mathrm{rev}}} (\text{work-producing devices}) \tag{6-6}$$

This equation is more widely used because it can be applied to other processes involving cycles. For work-consuming devices that do not involve cycles, we can also define the second law of thermodynamics as the ratio of the minimum input work to the effective input work.

$$\eta_{\mathrm{II}} = \frac{W_{\mathrm{u}}}{W_{\mathrm{rev}}} (\text{work-consuming devices}) \tag{6-7}$$

For work-consuming devices involving cycles, the second law of thermodynamics is defined as:

$$\eta_{\mathrm{II}} = \frac{\mathrm{COP}}{\mathrm{COP}_{\mathrm{rev}}} (\text{work-consuming devices}) \tag{6-8}$$

There are limitations to the efficiency of the two second laws of thermodynamics, such as their inapplicability to devices that do not require external work to be done and cannot be done externally. There are also some differences in the definition of the second law of thermodynamics efficiency, so we may encounter different second law efficiency calculations for the same device. Considering that the purpose of the second law efficiency is as a parameter to measure the exergy efficiency, we can define the second law efficiency of thermodynamics in a process as:

$$\eta_{\mathrm{II}} = \frac{\text{Exergy recovered}}{\text{Exergy expended}} = 1 - \frac{\text{Exergy destroyed}}{\text{Exergy expended}} \tag{6-9}$$

According to this equation, before calculating the efficiency of the second law of thermodynamics, we first need to determine how much energy is consumed in the process. In the reversible process, all the exergy consumed should be recovered. In this case, the maximum value of the second efficiency is 1. When we cannot recover the energy consumed by the system, the second law of thermodynamics reaches its lowest efficiency, which is zero. It is important to note that exergy can be recovered in various forms, such as heat, work, kinetic energy, potential energy, and enthalpy. Because there are many different interpretations of exergy, this will lead to different definitions of the second law of thermodynamics. But in any case, the sum of the recovered and wasted exergy must equal the exergy consumed. Therefore, we must define the system precisely in order to calculate the energy exchange between the system and the environment correctly.

In the case of an engine or heat pump, the wasted exergy is equational to the difference between the heat pump supplied to the machine and the heat or work output from the machine. The net output is equal to the heat that can be recovered. For a heat exchanger with two fluids, the exergy usually consumed is the reduction of exergy of the high-temperature fluid, and the exergy recovered is the increase of the low-temperature fluid. The wasted exergy is the difference between the reduced exergy of the hot fluid and the increased exergy of the cold fluid.

6.5 Exergy Change of a System

Energy is the work potential of the system in a given environment, which represents the maximum work that can be obtained when the system reaches equilibrium with the environment. The exergy of a system in equilibrium with the outside world is zero (Chen Xinzhi, 2015). Thermodynamically, a system is said to be "dead-state" when it reaches a state in which no work is done.

In this section we are only dealing with thermal-mechanical (exergy), so any mixing and chemical reactions are ignored. Thus, if a system has the same temperature and pressure with the environment and has no kinetic or potential energy relative to the environment, the system is called "confined dead-state". However, its chemical composition may differ from the environment (Zheng Danxing, 2010).

Exergy related to different chemical compositions and chemical reactions will be discussed in a later chapter. We can derive a relationship between exergy and exergy changes in fixed mass and flowing flow.

6.5.1 Exergy of a Fixed Mass: Nonflow (or Closed System) Exergy

Generally speaking, internal energy consists of manifest energy, potential energy, chemical energy and nuclear energy. However, in the absence of any chemical or nuclear reaction, the chemical and nuclear energy can be ignored, and the internal energy can be considered by the only sensible heat and potential energy that can be transferred from a system to the thermal system boundary whenever there is a temperature difference. The second law of thermodynamics denotes that heat cannot be completely converted into work, so the work internal potential of energy must be less than internal energy itself. But how much less?

The exergy of a given mass and state is the useful work that can be produced by a reversible process of the mass towards the state of the environment. The exergy in the initial state of this process (Fig. 6-6) is the useful work of the system at the time of its delivery. That is, the final temperature of the system is T_0 and the pressure is p_0.

Consider a piston-cylinder assembly containing a fluid of mass m, temperature T, and pressure p. The system (the mass inside the cylinder) has a volume V, internal energy U, and entropy S.

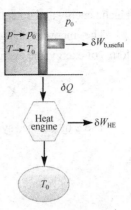

The system is now allowed to make a differential change of state in which the volume change is the micro component dV and heat is transferred as the micro component δQ. The energy balance of the system can be expressed in a differential process in which the direction of heat and work transfer (output heat and work) of the system is taken as an example:

$$\underbrace{\delta E_{in} - \delta E_{out}}_{\substack{\text{Net energy transfer} \\ \text{by heat, work, and mass}}} = \underbrace{dE_{system}}_{\substack{\text{Change in internal, kinetic,} \\ \text{potential, etc, energies}}} \qquad (6\text{-}10)$$

Fig.6-6 The exergy of a given mass and state is the useful work that can be produced as the mass undergoes a reversible process to the state of the environment

$$-\delta Q - \delta W = dU$$

Since internal energy is the only form of energy contained in the system, heat and work are the only forms of energy transfer involved in a fixed mass energy. Also, the only form of work done in a reversible process by a simple compressible system is boundary work. When the direction of work is towards the system, boundary work is $\Delta W = pdV \, (\text{or} - pdV)$, the pressure p in the pdV expression is the absolute pressure measured from absolute zero, and any available piston-cylinder device that can be done is operating at a pressure above the atmospheric level.

As a result,

$$\delta W = pdV \qquad (6\text{-}11)$$

Any useful work done by piston-cylinder units is due to pressures above atmospheric pressure:

$$\delta W = pdV = (p - p_0)dV + p_0 dV = \delta W_{b,useful} + p_0 dV \qquad (6\text{-}12)$$

Because of this heat transfer, the differential work done by the engine is:

$$\delta W_{HE} = \left(1 - \frac{T_0}{T}\right)\delta Q = \delta Q - \frac{T_0}{T}\delta Q = \delta Q - (-T_0 dS)$$

$$\delta Q = \delta W_{HE} - T_0 dS \qquad (6\text{-}13)$$

Rearranging

$$\delta W_{total,useful} = \delta W_{HE} + \delta W_{b,useful} = -dU - p_0 dV + T_0 dS \qquad (6\text{-}14)$$

The integral from a given state to a dead-state

$$W_{total,useful} = U - U_0 + p_0(V - V_0) - T_0(S - S_0) \qquad (6\text{-}15)$$

$W_{total,useful}$ is the total useful work delivered as the system undergoes a reversible process from the given state to the dead-state, which is exergy by definition.

In general, a closed system can have both kinetic energy and potential energy, and the sum of kinetic energy and potential energy within a closed system is its total energy.

The important thing to note here is that kinetic energy and potential energy themselves are forms of exergy, the exergy of a closed system of mass m is:

$$X = U - U_0 + p_0(V - V_0) - T_0(S - S_0) + \frac{mV^2}{2} + mgz \tag{6-16}$$

The φ of a closed system (or non-flow) on a unit mass basis is shown as:

$$\begin{aligned} \phi &= (u - u_0) + p_0(V - V_0) - T_0(S - S_0) + \frac{V^2}{2} + gz \\ &= (e - e_0) + p_0(V - V_0) - T_0(S - S_0) \end{aligned} \tag{6-17}$$

Where u_0, U_0, and S_0 are the properties of the system evaluated at the dead-state. Note that the exergy of a system is zero at the dead-state since $e=e_0$, $V=V_0$, and $S=S_0$ at that state.

The difference between the final exergy and the initial exergy of a system is the exergy change of a closed system in a process. The value of the exergy attribute does not change unless the state changes.

$$\begin{aligned} \Delta X = X_2 - X_1 &= m(\phi_2 - \phi_1) = (E_2 - E_1) + p_0(V_2 - V_1) - T_0(S_2 - S_1) \\ &= (U_2 - U_1) + p_0(V_2 - V_1) - T_0(S_2 - S_1) + m\frac{V_2^2 - V_1^2}{2} + mg(z_2 - z_1) \end{aligned} \tag{6-18}$$

or, on a unit mass basis,

$$\begin{aligned} \Delta\phi = \phi_2 - \phi_1 &= (u_2 - u_1) + p_0(V_2 - V_1) + T_0(S_2 - S_1) + \frac{V_2^L - V_1^2}{2} + g(z_2 - z_1) \\ &= (e_2 - e_1) + p_0(V_2 - V_1) - T_0(S_2 - S_1) \end{aligned} \tag{6-19}$$

In a fixed closed system, the kinetic and potential terms need to be omitted.

However, when there are different properties of the system, exergy of the system can be shown as:

$$X_{\text{system}} = \int \phi \delta m = \int_V \phi \rho dV \tag{6-20}$$

Where V is the volume of the system, ρ is the density of the system. It should be noted that exergy, as a kind of attribute, has a value that is related to the state. As the state changes, the value of exergy changes. Therefore, if the state or environment of the system does not change during the process, the exergy change of the system will be zero. For example, when nozzles, compressors, turbines, pumps and heat exchangers and other stationary flow equipment are in stable operation, exergy change under a given environment is zero.

There is no negative exergy value for a closed system. Even if the medium in the low temperature, and/or lower pressures contains exergy, because the cold medium can neutralize the heat generated by the heat engine and absorb heat from the environment at T_0, while the vacuum condition enables atmospheric pressure to move the piston and do useful work.

6.5.2 Exergy of a Flow Stream: Flow (or Stream) Exergy

Previous content show that the flowing liquid has an additional form of energy, called energy flow, which is the energy needed to maintain the transportation in the pipeline or pipe. This is expressed as $W_{flow} = pV$, where V is the specific volume of the liquid, which is equivalent to the volume change per unit mass of the fluid when it is displaced in the flow (Shi Yunhai et al., 2013) (Fig.6-7). The nature of boundary work is that the fluid flowing done to downstream of the fluid, so the exergy-related work of the flow is equivalent to the exergy-related work of the boundary, that is, which is the boundary work in excess of the work done against the atmospheric air at p_0 to displace it by a volume V (Chen Guangjin, 2006)(Fig.6-8).

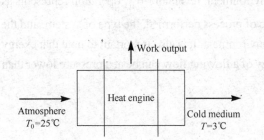

Fig.6-7 The exergy of a cold medium is also a positive quantity since work can be produced by transferring heat to it

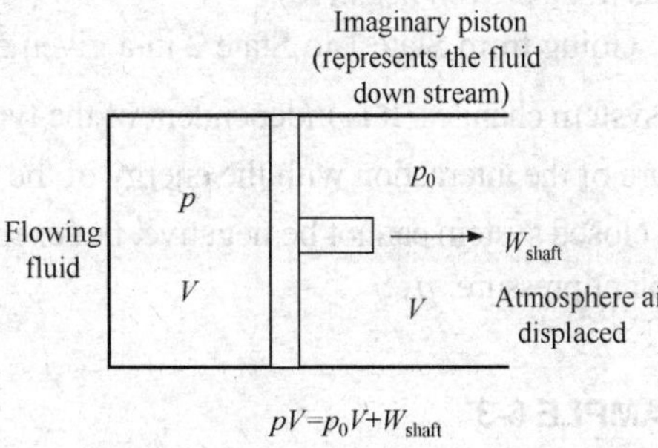

Fig.6-8 The exergy associated with flow energy is the useful work that would be delivered by an imaginary piston in the flow section

Note that the flow work is pV and the work done to the atmosphere is p_0V. Exergy related to flow energy can be shown as:

$$X_{flow} = pV - p_0V = (p - p_0)V \qquad (6\text{-}21)$$

Therefore, exergy for the fluid related to flow energy could be replaced by the pressure p that the pressure higher than the atmospheric pressure: $p-p_0$.

For non-flowing fluids, the exergy of the flow can be determined only by adding the above exergy relation to the exergy relation Eq. (6-22).

$$\begin{aligned}X_{flowing\ fluid} &= X_{nonflowing\ fluid} + X_{flow}\\ &= (u-u_0)+p_0(V-V_0)-T_0(S-S_0)\\ &\quad +\frac{V^2}{2}+gz+(p-p_0)V\\ &= (u-pV)-(u_0+p_0V_0)-T_0(S-S_0)\\ &\quad +\frac{V^2}{2}+gz\\ &= (h-h_0)-T_0(S-S_0)+\frac{V^2}{2}+gz \qquad (6\text{-}22)\end{aligned}$$

Energy:
$e = u + \frac{V^2}{2} + gz$ — Fixed mass

Exergy:
$\phi = (u-u_0)+p_0(V-V_0)-T_0(S-S_0)+\frac{V^2}{2}+gz$

(a) A fixed mass (nonflowing)

Energy:
$\theta = u + \frac{V^2}{2} + gz$ — Fluid stream

Exergy:
$\psi = (H-H_0)-T_0(S-S_0)+\frac{V^2}{2}+gz$

(b) A fluid stream (flowing)

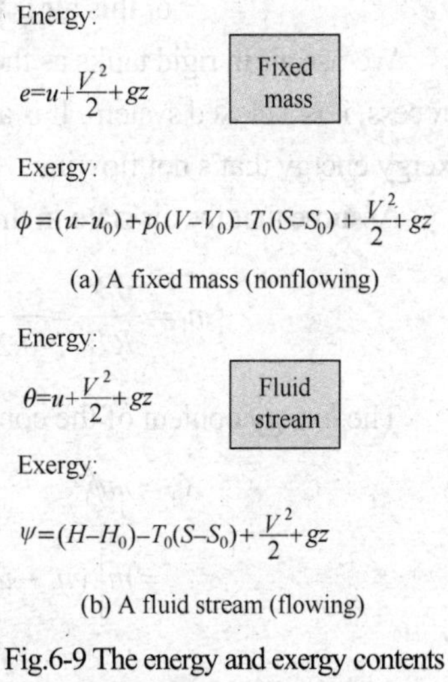

Fig.6-9 The energy and exergy contents of (a) a fixed mass and (b) a fluid stream

The resulting expression is in the form of exergy and is represented by ψ.

$$\psi = (H-H_0)-T_0(S-S_0)+\frac{V^2}{2}+gz \qquad (6\text{-}23)$$

Then, after the process from State 1 to State 2, the exergy change of the fluid is:

$$\Delta \psi = \psi_2 - \psi_1 = (H_2 - H_1) - T_0(S_2 - S_1) + \frac{V_2^2 - V_1^2}{2} + g(z_2 - z_1) \qquad (6\text{-}24)$$

For fluids where kinetic energy and potential energy are negligible, kinetic energy and potential energy terms do not exist. Note that changes in exergy or flow represent the usefulness of the work that is the maximum amount of work that can be done (or if it is the least useful work, it needs to be proven negative).

Going from State 1 to State 2 in a given environment, reversible W_{rev} operation represents as the system changes. It is independent of the type of process performed, the type of system, and the nature of the interaction with the energy of the environment. It is also important to note that exergy of a closed system cannot be negative, but exergy of a flowing flow can be at a pressure lower than ambient pressure p_0.

EXAMPLE 6-3

Work Potential of Compressed Air in a Tank.

A 300 m³ rigid tank contains 1 MPa and 303.15 K of compressed air. Determine how much work the air can do when the environmental conditions are 1 MPa and 303.15 K(Fig.6-10).

Assumptions: ① Air is an ideal gas. ② Kinetic and potential energies are negligible.

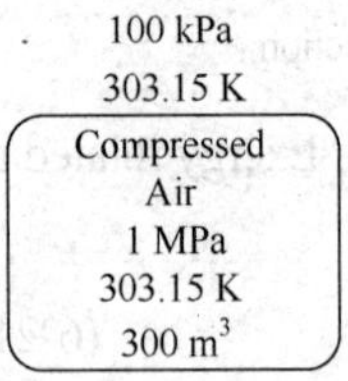

Fig.6-10 Schematic for Example 6-3

SOLUTION

Compressed air stored in a large tank is considered. The work potential of this air is to be determined.

We use air in rigid tanks as the system, since no mass crosses the system boundary during this process, it is a closed system. It is a problem here. By definition, work potential of a fixed mass, it's exergy energy that's not flowing.

Assume that the air state in the tank is 1, $T_1 = T_0 = 303.15$ K, and the air mass in the tank is:

$$m_1 = \frac{p_1 V}{RT_1} = \frac{1000 \text{ kPa} \times 300 \text{ m}^3}{0.287 \text{ kPa} \cdot \text{m}^3/(\text{kg} \cdot \text{K}) \times 303.15 \text{ K}} = 3448 \text{ kg}$$

The exergy content of the compressed air can be determined from:

$$X_1 = m\phi_1$$
$$= m\left[(u_1 - u_0) + p_0(V_1 - V_0) - \left|T_0(S_1 - S_0) + \frac{V_1^2}{2} + gz_1\right|\right]$$

$$(u_1 - u_0) = 0, \frac{V_1^2}{2} = 0, gz_1 = 0$$

$$X_1 = m\left[p_0(V_1 - V_0) - T_0(S_1 - S_0)\right]$$

We note that:

$$p_0(V_1 - V_0) = p_0\left(\frac{RT_1}{p_1} - \frac{RT_0}{p_0}\right) = RT_0\left(\frac{p_0}{p_1} - 1\right) \text{(since } T_1 = T_0\text{)}$$

$$T_0(S_2 - S_0) = T_0\left(c_p \ln\frac{T_1}{T_0} - R\ln\frac{p_1}{p_0}\right) = -RT_0 \ln\frac{p_1}{p_0} \text{(since } T_1 = T_0\text{)}$$

Therefore,

$$\phi_1 = RT_0\left(\frac{p_0}{p_1} - 1\right) + RT_0 \ln\frac{p_1}{p_0} = RT_0\left(\ln\frac{p_1}{p_0} + \frac{p_0}{p_1} - 1\right)$$

$$= 0.287 \text{ kJ/(kg·K)} \times 303.15 \text{ K} \times \left(\ln\frac{1000 \text{ kPa}}{100 \text{ kPa}} + \frac{100 \text{ kPa}}{1000 \text{ kPa}} - 1\right)$$

$$= 122.02 \text{ kJ/kg}$$

And

$$X_1 = m\phi_1 = 3448 \text{kg} \times 122.02 \text{ kJ/kg} = 420724 \text{ kJ} \cong 421 \text{ MJ}$$

Compressed air stored in a large tank is considered. The work potential of this air is to be determined.

EXAMPLE 6-4

Exergy Change During a Compression Process.

The refrigerant was stable compressed from 0.14 MPa and -10°C to 0.8 MPa and 50°C by the compressor. Determine how the exergy change of the refrigerant in the process and the minimum work input per unit mass of refrigerant to the compressor under the environmental conditions of 20°C and 95 kPa(Fig.6-11).

Assumptions: ① Steady operating conditions exist. ② The kinetic and potential energies are negligible.

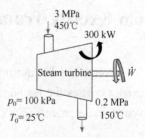

Fig.6-11 Schematic for Example 6-4

The properties of the refrigerant in the inlet and outlet states are:

$$p_1 = 0.14 \text{ MPa}, \quad T_1 = -10^\circ\text{C}$$

$$\hat{H}_1 = 246.37 \text{ kJ/kg}, \quad \hat{S}_1 = 0.9724 \text{ kJ/(kg·K)}$$

$$p_2 = 0.8 \text{ MPa}, \quad T_2 = 50^\circ\text{C}$$

$$\hat{H}_2 = 286.71 \text{ kJ/kg}, \quad \hat{S}_2 = 0.9803 \text{ kJ/(kg·K)}$$

SOLUTION

$$\Delta\psi = \psi_2 - \psi_1 = (H_2 - H_1) - T_0(S_2 - S_1) + \frac{V_2^2 - V_1^2}{2} + g(z_2 - z_1)$$

In there

$$\frac{V_2^2 - V_1^2}{2} = 0$$

$$g(z_2 - z_1) = 0$$

Thus

$$\Delta \psi = (H_2 - H_1) - T_0(S_2 - S_1)$$

$$\Delta \psi = \left[(286.71 - 246.37) - 293(0.9803 - 0.9724)\right] \text{kJ/kg}$$
$$= 38.0 \text{ kJ/kg}$$

The exergy of the refrigerant during compression is increased by 38.0 kJ/kg.

An exergy changes in a system under a given environment represents the reversible work under that environment. It is the minimum input of work consumed by working device, such as a compressor.

Thus, the addition of refrigerant (exergy) is equal to the minimum work needed to supply the compressor:

$$W_{\text{in,min}} = \hat{X}_2 - \hat{X}_1 = 38.0 \text{ kJ/kg} \qquad (6\text{-}25)$$

Note that if the compressed refrigerant at 0.8 MPa and 50℃ were to be expanded to 0.14 MPa and −10℃ in a turbine at the same environment in a reversible manner, 38.0 kJ · kg^{-1} of work would be produced.

6.6 Exergy Transfer by heat, work, and mass

Exergy can be transmitted to the environment in three forms: heat, work and mass. When exergy crosses the boundary of the environment, it represents the exergy gained or lost by the system in the process (Hu Ying, 2002).

6.6.1 Exergy Transfer by Heat, Q

Heat is a form of disordered energy, so it cannot be completely converted into work, as dictated by the second law of thermodynamics. We can get some work by transferring heat at high temperature to the engine, and then transferring waste heat from the engine to the environment (Prausnitz waited, 2006). Because the heat transfer efficiency is not higher than 100 percent, so the heat transfer is always accompanied by the ignition transfer. It can be expressed by the following equation:

$$X_{\text{heat}} = \left(1 - \frac{T_0}{T}\right) Q \text{ (kJ)} \qquad (6\text{-}26)$$

This equation represents the exergy transfer associated with the transfer of heat Q for temperature T not equal to T_0. When $T > T_0$, the heat transfer from the environment to the system increases the exergy of the system, while the heat transfer from the system to the environment reduces the exergy of the system; but when $T < T_0$, heat transfer Q at T_0 is the heat denied to the cold medium, so it cannot be mixed with the heat provided by the environment at T_0. If $T = T_0$, then no heat transfer will occur, so exergy with heat transfer will be zero (S I Sandler, 2006).

When $T > T_0$, which means that exergy and heat transfer are in the same direction, which means that both exergy and energy of the heat transfer medium have increased. When $T < T_0$, the direction of exergy is opposite to that of heat transfer, so the energy of the cold medium will increase due to heat transfer, but the exergy will decrease. When the system and the ambient temperature are the same, the exergy of the cold medium is zero. Eq. (6-27) can also be regarded as the relationship between thermal energy and exergy at a certain temperature.

When the heat transfer temperature is not constant, the exergy transfer accompanying the heat transfer can be determined by the next integral calculation:

$$X_{\text{heat}} = \int \left(1 - \frac{T_0}{T}\right) \delta Q \tag{6-27}$$

It should be noted that heat transfer under a certain temperature difference is irreversible, so the heat transfer process will generate a certain amount of entropy. The creation of entropy is always accompanied by a certain amount of exergy loss. And the heat transfer Q at the temperature T is always accompanied by the entropy transfer of Q/T and the exergy transfer of $\left(1 - \frac{T_0}{T}\right) Q$.

6.6.2 Exergy Transfer from Work, X_{work}

Exergy is a useful transmission of useful work, and exergy transmitted by work can be expressed by the following equation:

$$X_{\text{work}} = \begin{cases} W - W_{\text{surr}} & \text{(for boundary work)} \\ W & \text{(for other forms of work)} \end{cases} \tag{6-28}$$

Where $W_{\text{surr}} = p_0(V_2 - V_1)$, p_0 is atmospheric pressure, and V_1 and V_2 are the initial and final volumes of the system. Therefore, with the shaft work, electricians and other work of exergy transfer is equal to the work itself. When designing boundary systems, the work done to the environment during expansion cannot be transferred, so it must be subtracted. In addition, part of the work in the compression process is done by the atmosphere, so less work is needed from the outside (Gao Guanghua, 2010).

To further illustrate this point, suppose a cylinder in which the pressure of the gas is always one atmosphere, regardless of the gravity and friction of the piston. When heat is added to the system, according to the ideal gas equation, the volume of the gas in the cylinder is going to increase, so the piston is going to go up, and the system is going to be doing work to the environment, but that's not going to make any sense, because the work that the system is doing to the environment is just pushing the air out of the atmosphere. Similarly, when the temperature of the system is transferred to the environment to cool the gas, the volume of the gas decreases, and the piston will drop. This is a process in which the environment does the work for the system, but does not require external work from the system. So, we can conclude that the work done by the atmosphere or done to the atmosphere is not being used valuably and can therefore be excluded from the available work (He Liming, 2004).

6.6.3 Exergy Transfer by Mass, m

The mass is the carrier of exergy, energy and entropy, so the content of exergy, energy and entropy in the system is proportional to the mass, and it can be deduced that the energy transfer rate of exergy, energy and entropy in the system is proportional to the mass flow rate (Cui Keqing, 2004). The flow rate of mass is a parameter introduced to measure exergy, energy, and entropy input or output. The exergy transmitted through mass when a stream of mass m enters or leaves a system is:

$$X_{\text{mass}} = m\psi \tag{6-29}$$

In the equation

$$\psi = (H - H_0) - T_0(S - S_0) + \frac{V^2}{2} + gz \tag{6-30}$$

Therefore, the exergy of a system increases by $m\psi$ when mass in the amount of m enters, and it decreases by the same amount when the same amount of mass at the same state leaves the system.

Exergy flow associated with a fluid stream when the fluid properties are variable can be determined by integration from:

$$X_{\text{mass}} = \int_{A_c} \psi \rho V_n \mathrm{d}A_c$$

$$X_{\text{mass}} = \int \psi \delta m = \int_{\Delta t} X_{\text{mass}} \mathrm{d}t \tag{6-31}$$

Where A_c is the cross-sectional area of the flow and V_n is the local velocity normal to $\mathrm{d}A_c$.

Note that exergy transfer by heat X_{heat} is zero for adiabatic systems, and the exergy transfer by mass X_{mass} is zero for systems that involve no mass flow across their boundaries (i.e., closed systems). The total exergy transfer is zero for isolated systems since they involve no heat, work, or mass transfer.

6.7 The Decrease of Exergy Principle and Exergy Destruction

Energy cannot be created or destroyed in a process, which is the conservation of energy principle that we introduced, one of the expressions of the second law of thermodynamics is the increase of entropy principle, and we establish such a rule. We generally think of entropy as something that can be created but cannot be destroyed. That is, entropy generation must be positive or zero (reversible process), but it cannot be negative. Another version of the second law of thermodynamics is called the decrease of exergy principle, this corresponds to the entropy principle. Conservation of energy principle includes increase of entropy principle and decrease of exergy principle (Feng Xin et al., 2009).

Consider the isolated system shown in Fig. 6-12, there's no transfer of energy or entropy. Then the energy and entropy equilibrium in the isolated system can be shown as:

Energy equilibrium:

$$E_{in} - E_{out} = \Delta E_{system}$$

$$0 = E_2 - E_1$$

Entropy equilibrium:

$$S_{in} - S_{out} + S_{gen} = \Delta S_{system}$$

$$S_{gen} = S_2 - S_1$$

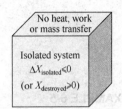

Fig.6-12 The isolated system considered in the development of the decrease of exergy principle

The second relationship times T_0, and then subtracting that from the first relationship, you get:

$$-T_0 S_{gen} = E_2 - E_1 - T_0 (S_2 - S_1) \tag{6-32}$$

$$\begin{aligned} X_2 - X_1 &= (E_2 - E_1) + p_0(V_2 - V_1) - T_0(S_2 - S_1) \\ &= (E_2 - E_1) = -T_0(S_2 - S_1) \end{aligned} \tag{6-33}$$

Because for the isolated system $V_2 = V_1$ (which does not contain any moving boundaries and therefore does not contain any boundary work).

$$X_2 - X_1 = -T_0 S_{gen} \leqslant 0 \tag{6-34}$$

$$\Delta X_{isolated} = (X_2 - X_1)_{isolated} \leqslant 0 \tag{6-35}$$

In the limiting condition of a reversible process, this equation can be expressed as the exergy of an isolated system which is always decreasing or remains constant in a process. In other words, it never increases, which is the so-called decrease of exergy principle.

The reduction of exergy equals exergy destroyed, which can be applied to isolated systems.

Anything that produces entropy destroys the energy state. Irreversibilities such as friction, mixing, chemical reactions, unconstrained expansion, nonequilibrium compression, or expansion always produces entropy, and anything that produces entropy always destroys energy. The exergy destroyed is directly proportional to the entropy generated, which is shown as:

$$X_{destroyed} = T_0 S_{gen} \geqslant 0 \tag{6-36}$$

Note that the exergy consumed by any real process is positive and zero for a reversible process. The work potential lost is generally expressed in terms of the exergy of destruction, also known as irreversibility or work lost.

There is no real process that is truly reversible, and the exergy that can be thought of as an isolated system is diminishing. The more irreversible a process is, the greater the exergy damage is in the process. There is no exergy damage during a reversible process.

The reduction of the exergy principle does not mean that the exergy of the system cannot be increased. In a process, the exergy changes to the system can be positive or negative valve, but the destructive exergy cannot be negative.

The decrease of exergy principle can be summarized as follows:

$$X_{\text{destroyed}} \begin{cases} > 0 \text{ irreversible process} \\ = 0 \text{ reversible process} \\ < 0 \text{ impossible process} \end{cases} \quad (6-37)$$

EXAMPLE 6-5

Exergy Analysis of Heating a Room with a Radiator.

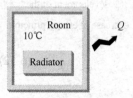

Fig.6-13 Schematic for Example 6-5

A 60 L electrical radiator containing heating oil is placed in a well-sealed 85 m³ room (Fig. 6-13). Both the air in the room and the oil in the radiator are initially at the environment temperature of 10 ℃. The electronic radiator is rated at 2.4 kilowatts. Heat is also lost from the room at an average rate of 0.75 kW. The heater is turned off after some time when the temperatures of the room air and oil are measured to be 25 ℃ and 60 ℃, respectively. Taking the density and the specific heat of oil to be 1000 kg/m³ and 2.8 kJ/kg, determine (a) how long the heater is kept on, (b) the second-law efficiency for this process.

SOLUTION

An electrical radiator is placed in a room and it is turned on for a period of time. The time period for which the heater was on, the exergy destruction, and the second-law efficiency are to be determined. The properties of air at room temperature are $R = 0.287$ kPa·m³/(kg·K), $c_P = 1.005$ kJ/(kg·K), $c_V = 0.718$ kJ/(kg·K). The properties of oil are given to be $\rho = 950$ kg/m³, $c = 2.2$ kJ/(kg·K).
Analysis:

(a) The masses of air and oil are:

$$m_a = \frac{p_1 V}{RT_1} = \frac{101.3 \text{ kPa} \times 85 \text{ m}^3}{0.287 \text{ kPa}\cdot\text{m}^3/(\text{kg}\cdot\text{K}) \times (10+273)\text{K}} = 106.01 \text{ kg}$$

$$m_{\text{oil}} = \rho V_{\text{oil}} = 950 \text{ kg/m}^3 \times 0.060 \text{ m}^3 = 57 \text{ kg}$$

An energy balance on the system can be used to determine time period for which the heater was kept on.

$$\left(W_{\text{in}} - Q_{\text{out}}\right)\Delta t = \left[mc_V(T_2 - T_1)\right]_a + \left[mc(T_2 - T_1)\right]_{\text{oil}}$$

$$(2.4 - 0.75)\text{kW}\Delta t = 106.01 \text{ kg} \times 0.718 \text{ kJ/(kg}\cdot\text{℃)} \times (25-10)\text{℃}$$
$$+ 57 \text{kg} \times 2.2 \text{ kJ/(kg}\cdot\text{℃)} \times (60-10)\text{℃}$$

$$\Delta t = 4491.96 \text{ s} = 74.9 \text{ min}$$

(b) The pressure of the air at the final state is:

$$p_{a_2} = \frac{m_a R T_{a_2}}{V} = \frac{106.01 \text{ kg} \times 0.287 \text{ kPa} \cdot \text{m}^3/(\text{kg} \cdot \text{K}) \times (25+273)\text{K}}{85 \text{ m}^3} = 106.67 \text{ kPa}$$

The amount of heat transfer to the surroundings is:

$$Q'_{out} = Q_{out} \Delta t = 0.75 \text{ kJ/s} \times 4491.96 \text{ s} = 3368.97 \text{ kJ}$$

The entropy generation is the sum of the entropy changes of air, oil, and the surroundings.

$$\Delta S_a = m\left(c_p \ln\frac{T_2}{T_1} - R \ln\frac{p_2}{p_1}\right)$$

$$= 94.88 \text{ kg} \times 1.005 \text{ kJ/(kg} \cdot \text{K)} \times \ln\frac{(20+273)\text{K}}{(6+273)\text{K}} - 0.287 \text{ kJ/(kg} \cdot \text{K)} \times \ln\frac{106.4 \text{ kPa}}{101.3 \text{ kPa}} = 5.525 \text{ kJ/K}$$

$$\Delta S_{oil} = mc \ln\frac{T_2}{T_1} = 57 \text{ kg} \times 2.8 \text{ kJ/K} \times \ln\frac{(60+273)\text{K}}{(10+273)\text{K}} = 25.97 \text{ kJ/K}$$

$$\Delta S_{surr} = \frac{Q'_{out}}{T_{surr}} = \frac{3368.97 \text{ kJ}}{(10+273)\text{K}} = 11.9 \text{ kJ/K}$$

(c) The second-law efficiency may be defined in this case as the ratio of the exergy covered to the exergy expended. That is,

$$\Delta X_a = m\left[c_V(T_2 - T_1)\right] - T_0 \Delta S_a$$

$$= 106.01 \text{ kg} \times 0.718 \text{ kJ/(kg} \cdot \text{K)} \times (25-10)\text{K} - (10+273)\text{K} \times 5.525 \text{ kJ/K}$$

$$= -421.85 \text{ kJ}$$

$$\Delta X_{oil} = m\left[c(T_2 - T_1)\right] - T_0 \Delta S_{oil}$$

$$= 57 \text{ kg} \times 2.8 \text{ kJ/(kg} \cdot \text{K)} \times (60-10)\text{K} - (10+273)\text{K} \times 11.9 \text{ kJ/K}$$

$$= 4612.3 \text{ kJ}$$

$$\eta_{II} = \frac{X_{recovered}}{X_{expended}} = \frac{\Delta X_a + \Delta X_{oil}}{W_{in}\Delta t} \times \frac{(4612.3 - 421.85)\text{kJ}}{2.8 \text{ kJ/s} \times 4491.96 \text{ s}} = 0.333 \text{ or } 33.3\%$$

DISCUSSION

This is a highly irreversible process since the most valuable form of energy, work, is used to heat the room air with a corresponding second-law efficiency of 33.3 percent.

EXAMPLE 6-6

Work Potential of Heat Transfer Between Two Tanks.

Two constant-volume tanks, each filled with 40 kg of air, have temperatures of 900 K and 300 K (Fig. 6-14). A heat engine placed between the two tanks extracts heat from the high-temperature

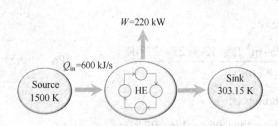

Fig.6-14 Schematic for Example 6-6

tank, produces work, and transmits heat to the low-temperature tank. Determine the maximum work that can be produced by the heat engine and the final temperatures of the tanks. Assume constant specific heats at room temperature.

SOLUTION

A heat engine operates between two tanks filled with air at different temperatures. The maximum work that can be produced and the final temperature of the tanks are to be determined.

Assumptions that air is an ideal gas with constant specific heats at room temperature. The gas constant of air is $0.287 \text{ kPa} \cdot \text{m}^3/(\text{kg} \cdot \text{K})$. The constant volume specific heat of air at room temperature is $c_V = 0.718 \text{ kJ}/(\text{kg} \cdot \text{K})$.

For maximum work production, the process must be reversible, and thus the entropy generation must be zero. We take the two tanks (the heat source and heat sink) and the heat engine as the system. Noting that the system involves no heat and mass transfer and that the entropy change for cyclic devices is zero, the entropy balance can be shown as:

$$\underbrace{S_{\text{in}} - S_{\text{out}}}_{\substack{\text{let entropy transfer} \\ \text{by heat and mass}}} + \underbrace{S'_{\text{gen}}}_{\substack{\text{Entropy} \\ \text{generation}}} = \underbrace{\Delta S_{\text{system}}}_{\substack{\text{Change} \\ \text{in entropy}}}$$

$$0 + S'_{\text{gen}} \ 0 = \Delta S_{\text{tank,source}} + \Delta S_{\text{tank,sink}} + \Delta S'_{\text{heat engine}}$$

$$0 = \Delta S_{\text{tank,source}} + \Delta S_{\text{tank,sink}}$$

$$\left(mc_V \ln \frac{T_2}{T_1} + mR \ln \frac{V'_2}{V_1} \right)_{\text{source}} + \left(mc_V \ln \frac{T_2}{T_1} + mR \ln \frac{V'_2}{V_1} \right)_{\text{sink}} = 0$$

$$\ln \frac{T_2 T_2}{T_{1,A} T_{1,B}} = 0 \to T_2^2 = T_{1,A} T_{1,B}$$

$$\ln \frac{T_2 T_2}{T_{1,A} T_{1,B}} = 0 \to T_2^2 = T_{1,A} T_{1,B}$$

Where $T_{1,A}$, and $T_{1,B}$ are the initial temperatures of the source and the sink, respectively, and T_2 is the common final temperature. Therefore, the final temperature of the tanks for maximum power production.

$$T_2 = \sqrt{T_{1,A} T_{1,B}} = \sqrt{900 \text{ K} \times 300 \text{ K}} = 519.6 \text{ K}$$

The energy balance $E_{\text{in}} - E_{\text{out}} = \Delta E_{\text{system}}$ for the source and sink can be shown as:

$$-Q_{\text{source, out}} = \Delta U = mc_V (T_2 - T_{1,A})$$

$$Q_{\text{source,out}} = mc_V (T_{1,A} - T_2) = 40 \text{ kg} \times 0.718 \text{ kJ}/(\text{kg} \cdot \text{K}) \times (900 - 519.6) \text{ K} = 10925 \text{ kJ}$$

$$Q_{\text{sink,in}} = mc_V\left(T_2 - T_{1,B}\right) = 40\text{ kg} \times 0.718\text{ kJ/(kg·K)} \times (519.6 - 300)\text{K} = 6307\text{ kJ}$$

Then the work produced in this case becomes:

$$W_{\text{max,out}} = Q_H - Q_L = Q_{\text{source,out}} - Q_{\text{sink,in}} = (10925 - 6307)\text{kJ} = 4618\text{ kJ}$$

DISCUSSION

Note that 6307 kJ of the 10925 kJ heat transferred from the source can be converted to work, and this is the best that can be done. This corresponds to a first law efficiency of 6307/10925 = 0.577 or 57.7% but to a second-law efficiency of 100 percent since the process involves no entropy generation and thus no exergy destruction.

6.8 Exergy Balance: Closed Systems

Exergy possesses properties completely different from those of entropy. Exergy can be destroyed but cannot be directly generated. Therefore, in the process, the exergy change of the system is smaller than the exergy transfer consumed in the process within the system boundary. (Fig.6-15) The reduced form of exergy principle can be shown as:

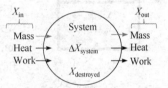

Fig.6-15 Mechanisms of exergy transfer

$$\begin{pmatrix}\text{Total}\\\text{exergy}\\\text{entering}\end{pmatrix} - \begin{pmatrix}\text{Total}\\\text{exergy}\\\text{leaving}\end{pmatrix} - \begin{pmatrix}\text{Total}\\\text{exergy}\\\text{destroyed}\end{pmatrix} = \begin{pmatrix}\text{Change in the}\\\text{total exergy}\\\text{of the system}\end{pmatrix}$$

$$X_{\text{in}} - X_{\text{out}} - X_{\text{destroyed}} = \Delta X_{\text{system}} \tag{6-38}$$

The exergy equilibrium relationship can be shown in the form that the exergy changes in the process of the system which is equal to the irreversibility of net exergy transfer through the system boundary and exergy destruction within the system boundary.

As mentioned earlier, exergy passes into or out of the system in the form of heat, work, and mass transfer. Therefore, in any process, any system experiencing exergy balance can be explicitly shown as:

General:

$$\underbrace{X_{\text{in}} - X_{\text{out}}}_{\substack{\text{Net exergy transfer}\\\text{by heat, work, and mass}}} - \underbrace{X_{\text{destroyed}}}_{\substack{\text{Exergy}\\\text{destruction}}} = \underbrace{\Delta X_{\text{system}}}_{\substack{\text{Change}\\\text{in exergy}}} \tag{6-39}$$

Or, in the rate form, as

$$\underbrace{\dot{X}_{\text{in}} - \dot{X}_{\text{out}}}_{\substack{\text{Rate of net exergy transfer}\\\text{by heat, work, and mass}}} - \underbrace{\dot{X}_{\text{destroyed}}}_{\substack{\text{Rate of exergy}\\\text{destruction}}} = \underbrace{\frac{dX_{\text{system}}}{dt}}_{\substack{\text{Rate of change}\\\text{in exergy}}} \text{ (kW)} \tag{6-40}$$

Where the rates of exergy transfer by heat, work, and mass are shown as, $\dot{X}_{heat} = (1 - T_0/T)\dot{Q}$, $\dot{X}_{mass} = \dot{m}\psi$ and $\dot{X}_{work} = \dot{W}_{useful}$ respectively. The exergy balance can also be shown per unit mass as:

$$(X_{in} - X_{out}) - X_{destroyed} = \Delta X_{system} \; (kJ/kg) \tag{6-41}$$

All quantities involved in the system can be expressed in terms of unit mass. Note that for a reversible process, exergy damage item $X_{destroyed}$ is removed from the above relationships. In addition, it is usually more convenient to find the entropy-producing S_{gen} firstly and then calculate the destruction exergy directly from Eq. (6-42).

$$X_{destroyed} = T_0 S_{gen} \tag{6-42}$$

When the environmental conditions p_0 and T_0 and the final state of the system are determined, no matter how the process proceeds, exergy change of the system can be directly measured in equation $\Delta X_{system} = X_2 - X_1$. However, knowledge of these interactions is required to determine exergy transfer through heat, work, and mass. The closed system does not involve any mass flow, and therefore does not involve any exergy transfer related to mass flow.

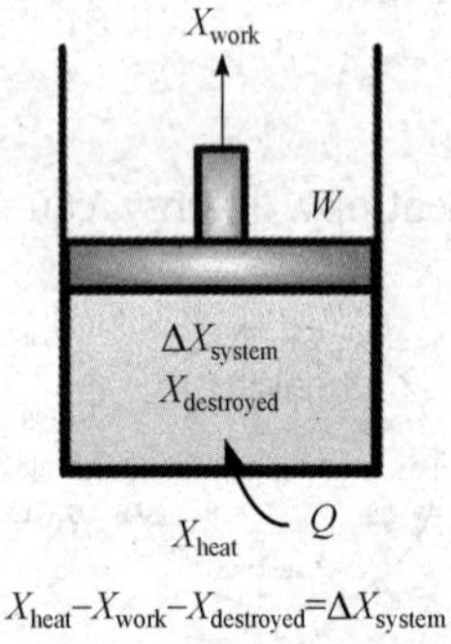

Exergy balance of the closed system can be expressed more clearly if the exergy balance is oriented towards the positive direction of heat transfer and the positive direction of work transfer to the system (Fig. 6-16).

Closed system:

$$X_{heat} - X_{work} - X_{destroyed} = \Delta X_{system} \tag{6-43}$$

$X_{heat} - X_{work} - X_{destroyed} = \Delta X_{system}$

Fig.6-16 Exergy balance for a closed system when the direction of heat transfer is taken to be to the system and the direction of work from the system

Or:

$$\sum\left(1 - \frac{T_0}{T_k}\right)Q_k - \left[W - p_0(V_2 - V_1)\right] - T_0 S_{gen} = X_2 - X_1 \tag{6-44}$$

Where Q_k is the heat transfer through the boundary at temperature T_k. Dividing the previous equation by the time interval Δt and taking the limit as $\Delta t \to 0$ gives the rate form of the exergy balance for a closed system.

Rate form:

$$\sum\left(1 - \frac{T_0}{T_k}\right)\dot{Q}_k - \left(\dot{W} - p_0\frac{dV_{system}}{dt}\right) - T_0 \dot{S}_{gen} = \frac{dX_{system}}{dt} \tag{6-45}$$

Note that for a closed system, the relationship above is obtained by setting the heat transfer and work done to the system as positive quantities. Therefore, the heat transfer of the system and the work done to the system should be considered negative when these relationships are applied.

The exergy destruction item was set to zero, and the reversible work W_{rew} was determined by the above exergy balance relationship. In this case, W becomes reversible work. Where $X_{destroyed} = T_0$,

$S_{gen} = 0$, $W = W_{rew}$.

Note that $X_{destroyed}$ only represents broken exergy within the system boundaries, not broken exergy that may occur outside the system boundaries during the process due to external irretrievability. Therefore, the process $X_{destroyed} = 0$ is internally reversible, but not necessarily completely reversible. Total exergy damage during the process can be determined by exergy balance on the extended system (both the system itself and the immediate environment where external irreversibility may occur.) (Fig. 6-16). Also, the sum of the exergy changes of the system and the exergy change of the exposed environment is equal to the exergy change in this situation. It is worth mentioning that under stable conditions, there is no change in the state or exergy of the exposed environment "buffer" at any point in time throughout the process, so the exergy change of the exposed environment is zero. When evaluating exergy transfer between the extended system and the environment, the ambient temperature, T_0, can simply be set to the boundary temperature of the extended system.

In the reversible process, there is no entropy generation and exergy consumption. In this case, exergy balance relationship is the same as energy balance relationship. In other words, exergy change of the system is equal to exergy transfer. Note that the energy transfer of any process is equivalent to the energy change of the system, but the exergy change of the system is only equivalent to the exergy transfer of the reversible process. In practice, according to the first law, the amount of energy is always constant, but known by the second law, mass must decrease. The decrease of mass will lead to the increase of entropy and the decrease of exergy. For example, when 10 kilojoules of heat are transferred from a hot medium to a cold medium, even after we get to the end of the process, we still have 10 kilojoules of energy in the system, but at lower temperatures, we can do work with lower mass and lower potential energy.

EXAMPLE 6-7

General Exergy Balance for Closed Systems.

Starting with energy and entropy balances, derive the general exergy balance relation for a closed system.

SOLUTION

Starting with energy and entropy balance relations, a general relation for exergy balance for a closed system is to be obtained. We have designed a general mass fixed closed system that freely exchanges heat and work with the surrounding environment (Fig. 6-17). The system went from state one to state two. Set the direction of heat transfer to the system as the positive direction and the direction of work done to the system as the positive direction, then the energy and entropy balance of the closed system can be shown as:

Energy balance:

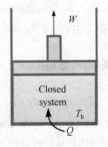

Fig.6-17 A general closed system considered in Example 6-7

$$E_{in} - E_{out} = \Delta E_{system} \to Q - W = E_2 - E_1$$

Entropy:

$$S_{in} - S_{out} + S_{gen} = \Delta S_{system} \to \int_1^2 \left(\frac{\delta Q}{T}\right)_{balance} + S_{gen} = S_2 - S_1$$

The second relationship is multiplied by T_0, and then subtracted from the first relationship:

$$Q - T_0 \int_1^2 \left(\frac{\delta Q}{T}\right)_{boundary} - W - T_0 S_{gen} = E_2 - E_1 - T_0(S_2 - S_1)$$

The heat transfer of process 1-2 can be shown as:

$$Q = \int_1^2 \delta Q$$

and the right side of the preceding equation is from:

$$(X_2 - X_1) - p_0(V_2 - V_1)$$

thus

$$\int_1^2 \delta Q - T_0 \int_1^2 \left(\frac{\delta Q}{T}\right)_{boundary} - W - T_0 S_{gen} = X_2 - X_1 - p_0(V_2 - V_1)$$

Letting T_b denote the boundary temperature and rearranging give:

$$\int_1^2 \left(1 - \frac{T_0}{T_b}\right) \delta Q - [W - p_0(V_2 - V_1)] - T_0 S_{gen} = X_2 - X_1$$

For exergy balance, it is equivalent to Eq. (6-43), and for convenience, the integral expression is replaced by the sum in the equation which completes the proof.

DISCUSSION

It should be noted that exergy balance is obtained by the addition of energy balance and entropy balance, so the equation obtained is not an independent one. However, in exergy analysis, it can be used to replace the entropy balance relation with the expression of the second law.

EXAMPLE 6-8

Exergy destruction during expansion of steam. At 1 MPa and 300 ℃, the piston-cylinder unit contains 0.05 kg of steam. The steam expands to its final state of 200 kilopascals, 150 degrees Celsius, and it does some work. In this process, assuming an ambient temperature of $T_0 = 25$ ℃ and $p_0 = 100$ kPa, the heat loss from the system to the environment is estimated to be 2 kJ. Then determine (a) the exergy of the steam in its initial and final state, (b) the exergy change of the steam, (c) exergy destruction, and (d) the second law efficiency of the process.

SOLUTION

The steam in a piston-cylinder device expands to a specified state. Exergy, exergy change, exergy destruction and exergy second law efficiency of steam in the initial and final states in the process are measured.

Note that the kinetic and potential energies are negligible. We take the steam contained within the piston-cylinder device as the system (Fig. 6-18). This is a closed system since no mass crosses the system boundary during the process. We note that boundary work is done by the system and heat is lost from the system during the process.

(a) First, we determine the properties of the steam at the initial and final states as well as the state of the surroundings:

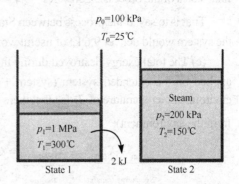

Fig.6-18 Schematic for Example 6-8

State 1: $\left.\begin{array}{l} p_1 = 1\text{ MPa} \\ T_1 = 300°C \end{array}\right\}$ $\quad \begin{array}{l} u_1 = 2793.7 \text{ kJ/kg} \\ U_1 = 0.25799 \text{ m}^3/\text{kg} \\ S_1 = 7.1246 \text{ kJ/(kg} \cdot \text{K)} \end{array}$

State 2: $\left.\begin{array}{l} p_2 = 200\text{ kPa} \\ T_2 = 150°C \end{array}\right\}$ $\quad \begin{array}{l} u_2 = 2577.1 \text{ kJ/kg} \\ U_2 = 0.95986 \text{ m}^3/\text{kg} \\ S_2 = 7.2810 \text{ kJ/(kg} \cdot \text{K)} \end{array}$

Dead-state: $\left.\begin{array}{l} p_0 = 100\text{ kPa} \\ T_0 = 25°C \end{array}\right\}$ $\quad \begin{array}{l} u_0 \cong u_{f=25°C} = 104.83 \text{ kJ/kg} \\ U_0 \cong U_{f=25°C} = 0.00103 \text{ m}^3/\text{kg} \\ S_0 \cong S_{f=25°C} = 0.3672 \text{ kJ/(kg} \cdot \text{K)} \end{array}$

The exergies of the system at the initial state X_1 and the final state X_2 are determined from Eq. (6-13).

$$X_1 = m\left[(u_1 - u_0) - T_0(S_1 - S_0) + p_0(U_1 - U_0)\right]$$
$$= 0.05 \text{ kg} \times [(2793.7 - 104.83)\text{kJ/kg}$$
$$- 298 \text{ K} \times (7.1246 - 0.3672)\text{kJ/(kg} \cdot \text{K)}$$
$$+ 100 \text{ kPa} \times (0.25799 - 0.00103)\text{m}^3/\text{kg} \times \text{kJ/(kPa} \cdot \text{m}^3)]$$
$$= 35.0 \text{ kJ}$$

$$X_2 = m\left[(u_2 - u_0) - T_0(S_2 - S_0) + p_0(U_2 - U_0)\right]$$
$$= 0.05 \text{ kg} \times [(2577.1 - 104.83)\text{kJ/kg}$$
$$- 298 \text{ K} \times (7.2810 - 0.3672)\text{kJ/(kg} \cdot \text{K)}$$
$$+ 100 \text{ kPa} \times (0.95986 - 0.00103)\text{m}^3/\text{kg} \times \text{kJ/(kPa} \cdot \text{m}^3)]$$
$$= 25.4 \text{ kJ}$$

That is to say, steam initially has an exergy content of 35 kJ, which drops to 25.4 kJ at the end of the process. In other words, if the steam were allowed to undergo a reversible process from the initial state to the state of the environment, it would produce 35 kJ of useful work.

(b) The exergy change for a process is simply the difference between the exergy at initial and final states of the process: $\Delta X = X_2 - X_1 = (25.4 - 35.0) \text{kJ} = -9.6 \text{ kJ}$.

That is to say, if the process between State 1 and State 2 were executed in a reversible manner, the system would deliver 9.6 kJ of useful work.

(c) The total exergy destroyed during this process can be determined from the exergy balance applied on the extended system (system + immediate surroundings) whose boundary is at the environment temperature of T_0 (so that there is no exergy transfer accompanying heat transfer to or from the environment):

$$\underbrace{X_{\text{in}} - X_{\text{out}}}_{\substack{\text{Net exergy transfer} \\ \text{by heat, work, and mass}}} - \underbrace{X_{\text{destroyed}}}_{\substack{\text{Exergy} \\ \text{destruction}}} = \underbrace{\Delta X_{\text{system}}}_{\substack{\text{Change} \\ \text{in exergy}}} - X_{\text{work,out}} - X_{\text{heat,out}}^0 - X_{\text{destroyed}}$$

$$= X_2 - X_1$$

$$X_{\text{destroyed}} = X_2 - X_1 - W_{\text{u,out}}$$

Where $X_{\text{u,out}}$ is the useful boundary work delivered as the system expands. By writing an energy balance on the system, the total boundary work done during the process is determined to be:

$$\underbrace{E_{\text{in}} - E_{\text{out}}}_{\substack{\text{Net energy transfer} \\ \text{by heat, work, and mass}}} = \underbrace{\Delta E_{\text{system}}}_{\substack{\text{Change in internal, kinetic,} \\ \text{potential, etc., energies}}}$$

$$-Q_{\text{out}} - W_{\text{b,out}} = \Delta U$$

$$W_{\text{b,out}} = -Q_{\text{out}} - \Delta U = -Q_{\text{out}} - m(u_2 - u_1)$$

$$= -2 \text{ kJ} - 0.05 \text{ kg} \times (2577.1 - 2793.7) \text{kJ/kg}$$

$$= 8.8 \text{ kJ}$$

This is the total boundary work done by the system, including the work done against the atmosphere to push the atmospheric air out of the way during the expansion process. The useful work is the difference between the two:

$$W_{\text{u}} = W - W_{\text{surr}} = W_{\text{b,out}} - p_0(V_2 - V_1) = W_{\text{b,out}} - p_0 m(U_2 - U_1)$$

$$= 8.8 \text{ kJ} - 100 \text{ kPa} \times 0.05 \text{ kg} \times (0.9599 - 0.25799) \text{m}^3/\text{kg} \times \left(\frac{1 \text{ kJ}}{1 \text{ kPa} \cdot \text{m}^3}\right) = 5.3 \text{ kJ}$$

Substituting, the exergy destroyed is determined to be:

$$X_{\text{destroyed}} = X_1 - X_2 - W_{\text{u,out}} = (35.0 - 25.4 - 5.3) \text{kJ} = 4.3 \text{ kJ}$$

That is to say, 4.3 kJ of work potential is wasted during this process. In other words, an additional 4.3 kJ of energy could have been converted to work during this process, but was not. The exergy destroyed could also be determined from:

$$X_{\text{destroyed}} = T_0 S_{\text{gen}} = T_0 \left[m(S_2 - S_1) + \frac{Q_{\text{surr}}}{T_0} \right]$$

$$= 298 \text{ K} \times \left[0.05 \text{ kg} \times (7.2810 - 7.1246) \text{kJ/(kg} \cdot \text{K)} + \frac{2 \text{ kJ}}{298 \text{ K}} \right] = 4.3 \text{ kJ}$$

Which is the same result obtained before.

(d) Noting that the decrease in the exergy of the steam is the exergy expended and the useful work output is the exergy recovered, the second-law efficiency for this process can be determined from:

$$\eta_{II} = \frac{\text{Exergy recovered}}{\text{Exergy expended}} = \frac{W_u}{X_1 - X_2} = \frac{5.3}{35.0 - 25.4} = 0.552$$

That is to say, 44.8% of the work potential of the steam is wasted during this process.

EXAMPLE 6-9

Exergy Destroyed During Stirring of a Gas.

An insulated rigid tank contains 2 lbm of air at 20 kPa and 60 ℃. A paddle wheel inside the tank is now rotated by an external power source until the temperature in the tank rises to 120 ℃ (Fig. 6-19). If the surrounding air is at T_0 = 60 ℃, determine (a) the exergy destroyed and (b) the reversible work for this process.

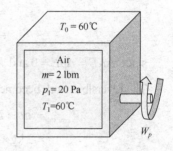

Fig.6-19 Schematic for Example 6-9

SOLUTION

The air in an adiabatic rigid tank is heated by stirring it with a paddle wheel. The exergy destroyed and the reversible work for this process are to be determined.

Assumptions: ① Air at about atmospheric conditions can be treated as an ideal gas with constant specific heats at room temperature. ② The kinetic and potential energies are negligible. ③ The volume of a rigid tank is constant, and thus there is no boundary work. ④ The tank is well insulated and thus there is no heat transfer.

We take the air contained within the tank as the system. This is a closed system since no mass crosses the system boundary during the process. We note that shaft work is done on the system.

(a) The exergy destroyed during a process can be determined from an exergy balance, or directly from $X_{destroyed} = T_0 S_{gen}$. We will use the second approach since it is usually easier. But first we determine the entropy generated from an entropy balance:

$$\underbrace{S_{in} - S_{out}}_{\substack{\text{Net entropy transfer} \\ \text{by heat and mass}}} + \underbrace{S_{gen}}_{\substack{\text{Entropy} \\ \text{generation}}} = \underbrace{\Delta S_{system}}_{\substack{\text{Change} \\ \text{in entropy}}}$$

$$0 + S_{gen} = \Delta S_{system} = m\left(c_V \ln\frac{T_2}{T_1} + R\ln\frac{V_2}{U_1}\right)(V_2 = 0)$$

$$S_{gen} = mc_V \ln\frac{T_2}{T_1}$$

Taking c_V = 0.172 kW/(kg · ℃) and substituting, the exergy destroyed becomes:

$$X_{\text{destroyed}} = T_0 S_{\text{gen}} = T_0 m c_V \ln\frac{T_2}{T_1}$$

$$= 60°C \times 2 \text{ kg} \times 0.172 \text{ kW/(kg}\cdot°C) \times \ln\frac{120}{60} = 14.31 \text{ kW}$$

(b) The reversible work, which represents the minimum work input $W_{\text{rev,in}}$ in this case, can be determined from the exergy balance by setting the exergy destruction equational to zero,

$$\underbrace{X_{\text{in}} - X_{\text{out}}}_{\substack{\text{Net exergy transfer} \\ \text{by heat, work, and mass}}} - \underbrace{X_{\text{destroyed}}}_{\substack{\text{Exergy} \\ \text{destruction}}}^{0(\text{reversible})} = \underbrace{\Delta X_{\text{system}}}_{\substack{\text{Change} \\ \text{in exergy}}}$$

$$W_{\text{rev,in}} = X_2 - X_1$$
$$= (E_2 - E_1) + p_0(V_2 - V_1) - T_0(S_2 - S_1)$$
$$= (U_2 - U_1) - T_0(S_2 - S_1)$$

since $\Delta KE = \Delta pE = 0$ and $V_2 = V_1$. Noting that $T_0(S_2 - S_2) = T_0 \Delta S_{\text{system}} = 14.31 \text{kW}$.

The reversible work becomes:

$$W_{\text{rev,in}} = mc_V(T_2 - T_1) - T_0(S_2 - S_1)$$
$$= 2 \text{ kg} \times 0.172 \text{ kW/(kg}\cdot°C) \times (120-60)°C - 14.31 \text{ kW}$$
$$= (20.64 - 14.31) \text{kW}$$
$$= 6.33 \text{ kW}$$

Therefore, a work input of just 6.33 kW would be sufficient to accomplish this process (raise the temperature of air in the tank from 60°C to 120°C) if all the irreversibilities were eliminated.

DISCUSSION

The solution is complete at this point. However, to gain some physical insight, we will set the stage for a discussion. First, let us determine the actual work (the paddle-wheel work W_{pw}) done during this process. Applying the energy balance on the system:

$$\underbrace{E_{\text{in}} - E_{\text{out}}}_{\substack{\text{Net energy transfer} \\ \text{by heat, work, and mass}}} = \underbrace{\Delta E_{\text{system}}}_{\substack{\text{Change in internal, kinetic,} \\ \text{potential, etc., energies}}}$$

$$W_{\text{pw,in}} = \Delta U = 20.64 \text{ kW} \left[\text{from part(b)}\right]$$

Since the system is adiabatic ($Q = 0$) and involves no moving boundaries ($W_b = 0$).

To put the information into perspective, 20.64 kW of work is consumed during the process, 14.31 kW of exergy is destroyed, and the reversible work input for the process is 6.33 kW. What does all this mean? It simply means that we could have created the same effect on the closed system (raising its temperature to 120°C at constant volume) by consuming 6.33 kW of work only instead of 20.64 kW, and thus saving 14.31 kW of work from going to waste. This would have been accomplished by a reversible heat pump.

To prove what we have just said, consider a Carnot heat pump that absorbs heat from the surroundings at $T_0 = 60\,°C$ and transfers it to the air in the rigid tank until the air temperature T rises from $60\,°C$ to $120\,°C$, as shown in Fig. 6-20. The system involves no direct work interactions in this case, and the heat supplied to the system can be shown in differential form as:

$\delta Q_H = dU = mc_U dT$, the coefficient of performance of a reversible heat pump is given by:

$$COP_{HP} = \frac{\delta Q_H}{\delta Q_{net,in}} = \frac{1}{1 - T_0/T}$$

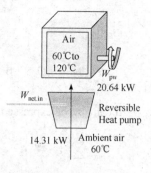

Thus,

$$\delta W_{net,in} = \frac{\delta Q_H}{COP_{HP}} = \left(1 - \frac{T_0}{T}\right) mc_V dT$$

Integrating, we get:

$$W_{net,in} = \int_1^2 \left(1 - \frac{T_0}{T}\right) mc_V dT$$

$$= mc_{V,avg}(T_2 - T_1) - T_0 mc_{V,avg} \ln \frac{T_2}{T_1}$$

$$= (20.64 - 14.31)\,kW = 6.33\,kW$$

Fig.6-20 The same effect on the system can be accomplished by a reversible heat pump that consumes only 1 kW of work

The first term on the right-hand side of the final expression is recognized as ΔU and the second term as the exergy destroyed, whose values were determined earlier. By substituting those values, the total work input to the heat pump is determined to be 6.33 kW, proving our claim. Notice that the system is still supplied with 20.64 kW of energy; all we did in the latter case was replace the 14.31 kW of valuable work with an equational amount of "useless" energy captured from the surroundings.

DISCUSSION

It is also worth mentioning that the exergy of the system as a result of 20.64 kW of paddle-wheel work done on it has increased by 6.33 kW only, that is, by the amount of the reversible work. In other words, if the system were returned to its initial state, it would produce, at most, 6.33 kW of work.

6.9 Exergy Balance: Control Volumes

The exergy balance relations for control volumes differ from those for closed systems in that they involve one more mechanism of exergy transfer: mass flow across the boundaries. As mentioned earlier, mass possesses exergy as well as energy and entropy, and the amounts of these three extensive properties are proportional to the amount of mass (Fig. 6-21). Taking the positive direction of heat transfer to be to the system and the positive direction of work transfer to be from the system, the general exergy balance relations can be expressed for a control volume more explicitly as:

$$X_{\text{heat}} - X_{\text{work}} + X_{\text{mass,in}} - X_{\text{mass,out}} - X_{\text{destroyed}} = (X_2 - X_1)_{\text{CV}} \quad (6\text{-}46)$$

Or

$$\sum\left(1-\frac{T_0}{T_k}\right)Q_k - \left[W - p_0(V_2 - V_1)\right] + \sum_{\text{in}} m\psi - \sum_{\text{out}} m\psi - X_{\text{destroyed}} = (X_2 - X_1)_{\text{CV}} \quad (6\text{-}47)$$

It can also be expressed in the rate form as:

$$\sum\left(1-\frac{T_0}{T_k}\right)Q_k - \left(W - p_0\frac{\mathrm{d}V_{\text{CV}}}{\mathrm{d}t}\right) + \sum_{\text{in}} m\psi - \sum_{\text{out}} m\psi - X_{\text{destroyed}} = \frac{\mathrm{d}X_{\text{CV}}}{\mathrm{d}t} \quad (6\text{-}48)$$

The exergy balance relation above can be stated as the rate of exergy change within the control volume during a process is equational to the rate of net exergy transfer through the control volume boundary by heat, work, and mass flow minus the rate of exergy destruction within the boundaries of the control volume.

6.9.1 Exergy Balance for Steady-Flow Systems

Most control volumes encountered in practice such as turbines, compressors, nozzles, diffusers, heat exchangers, pipes, and ducts operate steadily, and thus they experience no changes in their mass, energy, entropy, and exergy contents as well as their volumes. Therefore, $\frac{\mathrm{d}V_{\text{CV}}}{\mathrm{d}t}=0$ and $\frac{\mathrm{d}X_{\text{CV}}}{\mathrm{d}t}=0$ for such systems, and the amount of exergy entering a steady-flow system in all forms (heat, work, mass transfer) must be equational to the amount of exergy leaving plus the exergy destroyed.

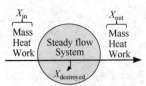

Fig.6-21. The exergy transfer to a steady-flow system is equational to the exergy transfer from it plus the exergy destruction within the system

Steady-flow:

$$\sum\left(1-\frac{T_0}{T_k}\right)Q_k - W + \sum_{\text{in}} m\psi - \sum_{\text{out}} m\psi - X_{\text{destroyed}} = 0 \quad (6\text{-}49)$$

For a single-stream (one-inlet, one-exit) steady-flow device, the relation above further reduces to:

$$\sum\left(1-\frac{T_0}{T_k}\right)Q_k - W + m(\psi_1 - \psi_2) - X_{\text{destroyed}} = 0 \quad (6\text{-}50)$$

Where the subscripts 1 and 2 represent inlet and exit states, m is the mass flowrate, and the change in the flow exergy is given by Eq. (6-21) as:

$$\psi_1 - \psi_2 = (H_1 - H_2) - T_0(S_1 - S_2) + \frac{V_1^2 - V_2^2}{2} + g(z_1 - z_2)$$

For the case of an adiabatic single-stream device with no work interactions, the exergy balance relation further simplifies to $X_{\text{destroyed}} = m(\psi_1 - \psi_2)$, which indicates that the specific exergy of the fluid must decrease as it flows through a work-free adiabatic device or remain the same ($\psi_1 = \psi_2$) in the limiting case of a reversible process regardless of the changes in other properties of the fluid.

EXAMPLE 6-10

There are four kinds of steam, respectively, saturated steam with 1.013 MPa, 6.868 MPa and 8.611 MPa pressure and superheated steam with 1.013 MPa and 573 K pressure. If these four kinds of steam are fully utilized, condensate water of 0.1013 MPa and 298 K will be discharged finally.

(1) Compare the effective energy per kilogram of steam with the heat released.

(2) Try to compare the effective energy and enthalpy of 1.013 MPa and 6.868MPa saturated water vapor.

(3) The rational utilization of steam is discussed according to the calculation results.

SOLUTION

(1) The effective energy of the four kinds of steam are shown in Table 6-1.

$$E_{x,ph} = -(H_0 - H) + T_0(S_0 - S)$$

Table 6-1 Enthalpy, entropy and effective energy of four kinds of water vapor

Name	Pressure p /MPa	Temperature T/K	Entropy S/[kJ/(kg·K)]	Enthalpy H /(kJ/kg)	$H-H_0$ /(kJ/kg)	$E_{x,ph}$ /(kJ/kg)	$\dfrac{E_{x,ph}}{H_0-H}$/%
H_2O	0.1013	298	0.367	104.8			
Saturated vapor	1.013	453	6.582	2671	2671	819	30.66
Superheated vapor	1.013	573	7.125	3052	2947	933	31.67
Saturated vapor	6.868	557.5	5.822	2670	2669	1044	39.10
Saturated vapor	8.611	573	5.708	2751	2646	1054	39.85

(2) As can be seen from the table, the heat released when 1.013 MPa and 6.868 MPa saturated water vapor in different states condenses into 298.15 K water (the enthalpy difference between them) is respectively 2671 kJ/kg and 2670 kJ/kg, The values are very close, but the effective energies are very different. The effective energy of 6.868 MPa saturated water vapor is 28% higher than that of 1.013 MPa, and the high-pressure steam has more utilization space. Therefore, it is a waste to use high temperature and high pressure steam to heat cold materials with low temperature.

(3) It can be known from the calculation results

① At the same pressure (1.013 MPa), the effective energy of superheated steam is greater than that of saturated steam;

② At the same temperature (573 K), the enthalpy of high-pressure steam is smaller than that of low-pressure steam;

③ At the same temperature (573 K), the effective energy value of high-pressure steam is larger than that of low-pressure steam, and the efficiency of heat conversion to work is also higher (40.78% for the former, 31.68% for the latter);

④ The heat released by saturated steam at 557.5 K and 453 K is basically the same, but the effective energy of high temperature steam is larger than that of low temperature steam.

CONCLUSION

The effective energy represents the power capacity of the system, and the larger the effective energy is, the larger the power capacity is. Therefore, the production should be reasonable selection of steam according to the situation, but not blindly choose high pressure steam as the heating source of the process, because its enthalpy value is smaller than low pressure steam, and the equipment cost is large, so does the general low pressure steam as the process heating. And high pressure steam can be used as power energy to improve the utilization rate of heat energy. In addition, reasonable recycling releasing heat should be taken in the process of production, such as the modernization of large synthetic ammonia plant, the use of waste heat recovery boiler high temperature gas heat, the superheat steam driving turbine to do work, turbine discharge waste gas heating heat source also can be used as a process, not only can achieve energy self-sufficient, but also can be sending steam or power, it has the vital significance to energy conservation and emissions reduction.

6.9.2 Second-Law Efficiency of Steady-Flow Devices

The second-law efficiency of various steady-flow devices can be determined from its general definition, η_{II} = (Exergy recovered)/ (Exergy expended). When the changes in kinetic and potential energies are negligible, the second-law efficiency of an adiabatic turbine can be determined from:

$$\eta_{II,turb} = \frac{W_{out}}{\psi_1 - \psi_2} = \frac{H_1 - H_2}{\psi_1 - \psi_2} = \frac{W_{out}}{W_{rev,out}} \text{ or } \eta_{II,turb} = 1 - \frac{T_0 S_{gen}}{\psi_1 - \psi_2} \tag{6-51}$$

Where $S_{gen} = S_2 - S_1$. For an adiabatic compressor with negligible kinetic and potential energies, the second-law efficiency becomes:

$$\eta_{II,comp} = \frac{\psi_2 - \psi_1}{W_{in}} = \frac{\psi_2 - \psi_1}{H_2 - H_1} = \frac{W_{rev,in}}{W_{in}} \text{ or } \eta_{II,comp} = 1 - \frac{T_0 S_{gen}}{H_2 - H_1} \tag{6-52}$$

Where again $S_{gen} = S_2 - S_1$. Note that in the case of turbine, the exergy resource utilized is steam, and the expended exergy is simply the decrease in the exergy of the steam. The recovered exergy is the turbine shaft work. In the case of compressor, the exergy resource is mechanical work, and the expended exergy is the work consumed by the compressor. The recovered exergy in this case is the increase in the exergy of the compressed fluid.

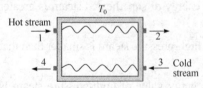

Fig.6-22 A heat exchanger with two unmixed fluid streams

For an adiabatic heat exchanger with two unmixed fluid streams, the exergy expended is the decrease in the exergy of the hot stream, and the exergy recovered is the increase in the exergy of the cold stream, provided that the cold stream is not at a lower temperature than the surroundings(Fig.6-22). Then the second-law efficiency of the heat exchanger becomes:

$$\eta_{II,HX} = \frac{m_{cold}(\psi_4 - \psi_3)}{m_{hot}(\psi_1 - \psi_2)} \text{ or } \eta_{II,HX} = 1 - \frac{T_0 S_{gen}}{m_{hot}(\psi_1 - \psi_2)} \tag{6-53}$$

It is losing some heat to its surroundings at T_0. If the temperature of the boundary (the outer surface of the heat exchanger) T_b is equational to T_0, the definition above still holds (except the entropy generation term needs to be modified if the second definition is used). However, if $T_b > T_0$, then the exergy of the lost heat at the boundary should be included in the recovered exergy:

$$\eta_{II,HX} = \frac{m_{cold}(\psi_4 - \psi_3) + Q_{loss}(1 - T_0/T_b)}{m_{hot}(\psi_1 - \psi_2)} = 1 - \frac{T_0 S_{gen}}{m_{hot}(\psi_1 - \psi_2)} \quad (6\text{-}54)$$

An interesting situation arises when the temperature of the cold stream remains below the temperature of the surroundings at all times. In that case, the exergy of the cold stream actually decreases instead of increasing. In such cases, it is better to define the second-law efficiency as the ratio of the sum of the exergy of the outgoing streams to the sum of the exergies of the incoming streams.

EXAMPLE 6-11

Second-Law Analysis of a Steam Turbine.

Steam enters a turbine steadily at 3 MPa and 450℃ at a rate of 8 kg/s and exits at 0.2 MPa and 150℃ (Fig. 6-23). The steam is losing heat to the surrounding air at 100 kPa and 25℃ at a rate of 300 kW, and the kinetic and potential energy changes are negligible. Determine (a) the actual power output, (b) the maximum possible power output, (c) the second-law efficiency, (d) the exergy destroyed, and (e) the exergy of the steam at the inlet conditions.

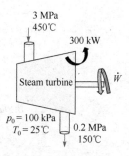

Fig.6-23 Schematic for Example 6-10

SOLUTION

A steam turbine operating steadily between specified inlet and exit states is considered. The actual and maximum power outputs, the second-law efficiency, the exergy destroyed, and the inlet exergy are to be determined.

Assumption: this is a steady-flow process since there is no change with time at any point and thus $\Delta m_{C_V} = 0$, $\Delta E_{C_V} = 0$, and $\Delta X_{C_V} = 0.2$. The loss of kinetic and potential energies is negligible.

We take the turbine as the system. This is a control volume since mass crosses the system boundary during the process. We note that there is only one inlet and one exit and thus $m_1 = m_2 = m$. Also, heat is lost to the surrounding air and work is done by the system.

The properties of the steam at the inlet and exit states and the state of the environment are:

$$\left. \begin{array}{l} p_1 = 3 \text{ MPa} \\ T_1 = 450℃ \end{array} \right\} \begin{array}{l} H_1 = 3344.9 \text{ kJ/kg} \\ S_1 = 7.0856 \text{ kJ/(kg·K)} \end{array}$$

$$\left. \begin{array}{l} p_2 = 0.2 \text{ MPa} \\ T_2 = 150℃ \end{array} \right\} \begin{array}{l} H_2 = 2769.1 \text{ kJ/kg} \\ S_2 = 7.2810 \text{ kJ/(kg·K)} \end{array}$$

$$\left. \begin{array}{l} p_0 = 100 \text{ kPa} \\ T_0 = 25℃ \end{array} \right\} \begin{array}{l} H_0 \cong H_{f=25℃} = 104.83 \text{ kJ/kg} \\ S_0 \cong S_{f=25℃} = 0.3672 \text{ kJ/(kg·K)} \end{array}$$

(a) The actual power output of the turbine is determined from the rate form of the energy balance:

$$\underbrace{E_{in} - E_{out}}_{\text{Rate of net energy transfer by heat, work, and mass}} = \underbrace{dE_{system}/dt}_{\text{Rate of change in internal, kinetic, potential, etc., energies}} = 0$$

$$E_{in} = E_{out}$$

$$mH_1 = W_{out} + Q_{out} + mH_2 \text{ (since } ke \cong pe \cong 0\text{)}$$

$$W_{out} = m(H_1 - H_2) - Q_{out} = 8 \text{ kg/s} \times (3344.9 - 2769.1) \text{ kJ/kg} - 300 \text{ kW} = 4306 \text{ kW}$$

(b) The maximum power output (reversible power) is determined from the rate form of the exergy balance applied on the extended system (system + immediate surroundings), the boundary between them is at the environment temperature of T_0, and by setting the exergy destruction term equational to zero:

$$\underbrace{X_{in} - X_{out}}_{\text{Rate of net exergy transfer by heat, work, and mass}} - \underbrace{X_{destroyed}}_{\substack{\text{Rate of exergy} \\ \text{destruction}}}^{0(\text{reversible})} = \underbrace{dX_{system}/dt}_{\substack{\text{Rate of change} \\ \text{in exergy}}}^{0(\text{steady})} = 0$$

$$\dot{X}_{in} = \dot{X}_{out}$$

$$mg\psi_1 = W_{rev,out} + X'_{heat} + m\psi_2 W_{rev,out} \& = m(\psi_1 - \psi_2)$$
$$= m\left[(h_1 - h_2) - T_0(S_1 - S_2) - \Delta ke^{\nearrow 0} - \Delta pe^{\nearrow 0}\right]$$

Note that exergy transfer with heat is zero when the temperature at the point of transfer is the environment temperature T_0. Substituting:

$$W_{rev,out} = 8 \text{ kg/s}\left[(3344.9 - 2769.1) \text{ kJ/kg} - 298 \text{ K} \times (7.0856 - 7.2810) \text{ kJ/(kg} \cdot \text{K)}\right]$$
$$= 5072 \text{ kW}$$

The second-law efficiency of a turbine is the ratio of the actual work delivered to the reversible work:

$$\eta_{II} = \frac{W_{out}}{W_{in}} = \frac{4306 \text{ kW}}{5072 \text{ kW}} = 0.849 \text{ or } 84.9\%$$

That is to say, 15.1% of the work potential is wasted during this process.

(c) The difference between the reversible work and the actual useful work is the exergy destroyed, which is determined to be:

$$X_{destroyed} = W_{rev,out} - W_{out} = (5072 - 4306) \text{ kW} = 776 \text{ kW}$$

That is to say, the potential to produce useful work is wasted at a rate of 776 kW during this process. The exergy destroyed could also be determined by first calculating the rate of entropy generation S.

(d) The exergy (maximum work potential) of the steam at the inlet conditions is simply the stream exergy and is determined from:

$$\psi_1 = (H_1 - H_0) - T_0(S_1 - S_0) + \frac{V_1^2}{2} + gz_1$$
$$= (H_1 - H_0) - T_0(S_1 - S_0)$$
$$= (3344.9 - 104.83)\,\text{kJ/kg} - 298\,\text{K} \times (7.0856 - 0.3672)\,\text{kJ/(kg}\cdot\text{K)}$$
$$= 1238\,\text{kJ/kg}$$

That is, not counting the kinetic and potential energies, every kilogram of the steam entering the turbine has a work potential of 1238 kJ. This corresponds to a power potential of 8 kg/s× 1238 kJ/kg = 9904 kW. Obviously, the turbine is converting 4306/9904 = 43.5% of the available work potential of the steam to work.

6.10 Chemical Process Energy Analysis and Aspen Plus

EXAMPLE 6-12

Air Compression
Pressurize 10 m³/h of air to 3 Mpa and 10 Mpa through a compressor, and calculate the exergy change of air and the exergy loss in this process
Aspen simulation is used for calculation. The operation steps are shown in the following QR code.

EXAMPLE 6-13

Heat transfer between ethylene glycol and ethanol. Glycol organic solvent is widely used in industry and has a high residual temperature. It can be used to exchange heat with ethanol aqueous solution by adding a heat exchanger.
Aspen simulation is used for calculation. The operation steps are shown in the following QR code.

EXAMPLE 6-14

Use distillation separate 50 kmol/hr methanol and 50 kmol/hr water based on Aspen Plus.
Calculate exergy losses and exergy conversion efficiency during distillation. Property methods choose the NRTL. Distillation columns model choose RadFrac.
The specific operation steps are shown in the following QR code.

EXERCISES

1. Heat 1kg air of 0.1 MPa and 127℃ reversibly to 427℃ at constant pressure, and try to

find: (1) effective energy and ineffective energy in heat (heat is supplied by the sensible heat of the heat source, so the temperature of the heat source is changing). (2) If the same amount of heat is added and released by a thermostatic heat source at 500℃, how much is the effective energy and ineffective energy in the heat? Assume that the ambient temperature is 25℃ and the average mass heat capacity of air is $C_p^{ig} = 1.004 \text{ kJ/(kg·K)}$.

2. A countercurrent heat exchanger, as shown, uses exhaust gas to heat air. The 0.1 MPa air is heated from 293 K to 398 K, and the flow rate of the air is 1.5 kg/s, The 0.13 MPa exhaust gas cools from 523 K to 368 K. The mass of air is constant pressure heat capacity 1.04 kJ/(kg·K), The mass constant pressure heat capacity of the exhaust gas is 0.84 kJ/(kg·K). It is assumed that the pressure and kinetic energy of air and waste gas passing through the heat exchanger can be ignored, and the heat dissipation loss of the heat exchanger is not included. The environmental state is 0.1 MPa and 20℃.

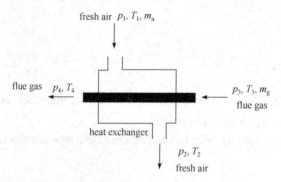

(1) Heat transfer process loss work;
(2) Thermodynamic efficiency of the process.

3. There is a counterflow heat exchanger that uses exhaust gas to heat the air. The 10^5 Pa air is heated from 20℃ to 125℃, the air flow rate is 1.5 kg/s, and the $1.3×10^5$ Pa exhaust gas is cooled from 250℃ to 95℃. The isobaric heat capacity of air is 1.04 kJ/(kg·K), the isobaric heat capacity of exhaust gas is 0.84 kJ/(kg·K), assuming that the pressure and kinetic energy changes of air and exhaust gas passing through the heat exchanger are negligible, and there is no heat exchange between the heat exchanger and the environment, and the environment is 10^5 Pa and 20℃.

(1) Exergy loss of irreversible heat transfer in the heat exchanger;
(2) The exergy efficiency of the heat exchanger.

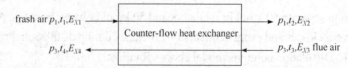

4. In production, fluid is transported through pipelines. When the temperature is not equal to the ambient temperature, the pipeline will dissipate heat (cold) to the environment. According to analysis, the heat dissipation loss of uninsulated steam pipelines is 9 times that of insulated pipelines. Therefore, the insulation of equipment and pipelines is a very important energy-saving measure.

A certain factory has pipes for delivering 90℃ hot water. Due to poor insulation, the water temperature has dropped to 70 ℃ by the time the unit is used. Try to find the heat loss and lost work in the process of reducing water temperature. The atmospheric temperature is 25 ℃, and the heat capacity of water at a constant pressure is 4.1868 kJ/(kg·K).

5. How much work is required to separate the air at 0.10133 MPa and 25 ℃ into pure nitrogen and pure oxygen at the same temperature and pressure?

6. Distillation is the most commonly used method for separating mutual liquid mixtures, and it is also the largest energy-consuming operation in chemical production. The volatility of various liquids is different, and rectification uses this to separate them. The figure is a schematic diagram of the operation of a rectification tower, in which f, d and b are the flow rates of the raw material, distillate and raffinate respectively. $H_f, S_f, H_d, S_d, H_b, S_b$ are the specific enthalpy and specific entropy of the raw material, distillate and residual liquid, respectively; Q_C is the heat released per unit time at the condensing temperature T_C; Q_R is at the reboiling temperature Q_R. The heat input per unit time. If the heat loss of the rectification tower is neglected and the temperature of the raw materials, distillate and residual liquid is basically the same, try to derive the calculation formula of the lost work during the operation of the rectification tower and discuss the factors affecting the lost work of the rectification tower.

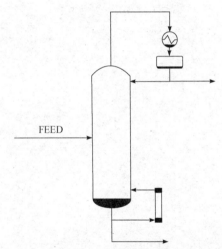

7. The steady-state flow of propane gas is throttled from 2 MPa and 400 K to 0.1 MPa. Try to estimate the final temperature of propane and the entropy change of the process. (Hint: the properties of propane can be calculated with a suitable generalized relationship)

8. In winter, the indoor and outdoor temperatures are 25 ℃ and −15 ℃ respectively. To change 298 K and 0.133 MPa water into 273 K and ice at the same pressure, (1) put the water in the refrigerator; (2) put it directly going outdoors, try to use thermodynamics to quantify which method is more reasonable? (The melting enthalpy of 0 ℃ ice is known to be 334.7 kJ/kg)

REFERENCES

[1] Chen G J. Chemical thermodynamics[M]. Beijing: Petroleum Industry Press, 2006.
[2] Chen X Z. Chemical thermodynamics[M]. 4th ed. Beijing: Chemical Industry Press, 2015.

[3] Cui K Q. Introduction to safety engineering and science[M]. Beijing: Chemical Industry Press, 2004.
[4] Feng X. Chemical thermodynamics[M]. Beijing: Chemical Industry Press, 2009.
[5] Gao G H. Advanced chemical thermodynamics[M]. Beijing: Tsinghua University Press, 2010.
[6] He L M. Engineering thermodynamics[M]. Beijing: Aviation Industry Press, 2004.
[7] Ma P S. Chemical thermodynamics (general type)[M]. 2nd ed. Beijing: Chemical Industry Press, 2009.
[8] Prausnitz. Molecular thermodynamics of fluid phase equilibrium[M]. Translated by Lu Xiaohua, Liu Honglai. 3rd ed. Beijing: Chemical Industry Press, 2006.
[9] Sandler S I. Chemical, biochemical, and engineering thermodynamics[M]. 4th ed. John Wiley & Sons, 2006.
[10] Shi J, Wang J D, Yu G Q, Chen M H. Handbook of chemical engineering[M]. 2nd ed. Beijing: Chemical Industry Press, 2002.
[11] Shi Y H. Chemical thermodynamics. [M]. 2nd ed .Shanghai: East China University of Science and Technology Press, 2013.
[12] SI Sandler. Using Aspen Plus in thermodynamics instruction: A step-by-step guide[M]. Wiley-AIChE, 2015.
[13] Cengel Y A, Boles M A. Thermoayann: an engineering approach[M]. McGraw-Hill, 2014.
[14] Zheng D X. Fluid and process thermodynamics[M]. 2nd ed. Beijing: Chemical Industry Press, 2010.
[15] Zhu Z Q, Wu Y T. Thermodynamics of chemical engineering[M]. 3rd ed. Beijing: Chemical Industry Press, 2012.

Chapter 7

Thermodynamic Processes and Cycles

Learning Objectives

- Solve energy and entropy equilibriums to fully characterize the processes.
- Apply entropy and energy equilibriums to larger chemical processes with multiple unit operations.
- Recognize the purposes of Rankine heat engines and refrigerators appreciate the role of each unit operation in these processes.
- Compare and evaluate process design based on different benchmarks, such as efficiency for heat engines.

This chapter introduces the combined operation of single unit, including the application of mass, energy and entropy balance in various systems, and the working principle and content of steam power cycle and refrigeration cycle are studied.

7.1 Chemical Process Design

Since entering the 21st century, China has been carrying out industrialization, which includes the field of chemical engineering and has made great achievements. Here is a simple example. If a continuous and stable chemical process is designed, 30 million pounds of ammonia can be produced every year. Ammonia can be produced by combining nitrogen and hydrogen through iron catalyst under high pressure. The first step might be to draw a basic block diagram like that shown in Fig. 7-1.

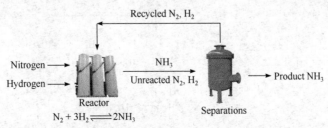

Fig.7-1 Preliminary schematic diagram of process for ammonia synthesis

Developing a complete design will require extensive engineering decisions. But now the boxes labeled "reactor" and "separation" is obtained, so we need to identify and design processes for each individual unit operation.

Where does nitrogen come from? One of the most common answers is "There's a lot of free nitrogen in the air, which makes up 80% of the atmosphere." This simple answer influences our view of the process in two ways.

(1) If air is used as raw material, oxygen enters the process together with other impurities.

(2) Air enters the process at normal temperature and pressure, but the reaction is carried out at "high" pressure.

Improvements to the block diagram: Fig. 7-2 shows another option. Nitrogen can be purified and compressed before entering the reactor, and a separation process can be designed in which compressed air is fed into the reactor after purification.

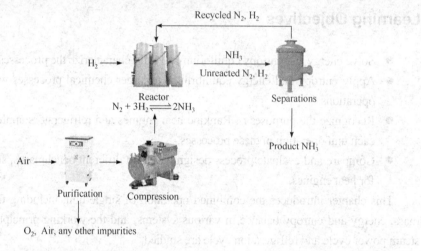

Fig.7-2 Schematic diagram of the ammonia synthesis process (nitrogen is purified before being fed to the reactor)

In general, the purpose of this discussion is not to draw any final conclusion on ammonia synthesis, but to consider what is the role of thermodynamics in the development of the whole process?

Fig. 7-3 shows another alternative scheme. Instead of designing a separation process to purify nitrogen, it is better to send compressed air directly to the reactor and allow oxygen and nitrogen to circulate in the process. The alternative scheme in Fig. 7-3 is simpler, requires less equipment, and therefore requires less investment.

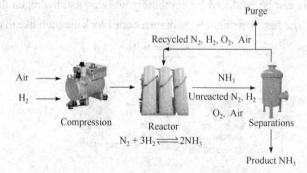

Fig.7-3 Schematic diagram of the ammonia synthesis process (compressed air is sent to the reactor without separation)

What is the role of thermodynamics in the development of complete processes?

The principles of mass, energy and entropy can be applied to simple systems, such as a single device. These same equilibrium equations can also be applied to any system, including large systems such as the whole chemical plant. The compilation and solution of the mass conservation of the system shown in Fig. 7-3 become complex due to the existence of scavenging gas. If there is no Purge unit, recording the oxygen balance of the system will show that oxygen enters the system but cannot leave. Oxygen and other inert materials can build up in the system, creating an unsafe pressure buildup. Through the design of each step in the research process, the thermodynamic principle is confirmed again.

7.2 Real Heat Engines

Carnot proposed the Carnot Cycle (Ramon Ferreiro Garcia et al., 2021). As an important theory of thermodynamics, Carnot Cycle solved the fundamental way to improve the thermal efficiency and laid the foundation for the second law of thermodynamics. An important prerequisite for the Carnot Cycle (M. Askinto et al., 2019) work is that all of these processes are reversible and lossless, which means they must be infinitely slow. This is of course an ideal hypothesis. The Carnot heat engine is a hypothetical, idealized case that can't be implemented in the real world.

However, the Carnot Cycle (Robert H Dickerson et al., 2019) is a useful theoretical construct because it represents the maximum conversion of heat into work that is possible for a heat engine operating between the hot and cold reservoirs at two specific temperatures.

The Rankine Cycle (Luan Baoqi et al., 2021) is a thermodynamic cycle that converts heat into work. Rankine Cycle absorbs heat from the outside, and its working medium, usually water, is used for heating work. The Rankine Cycle produces 90% of the world's electricity, including nearly all solar thermal, biomass, coal and nuclear power plants. The Rankine Cycle is the fundamental thermodynamic principle supporting the heat engine.

Rankine Cycle is an ideal cycle in which water vapor is used as the working medium. It mainly includes isentropic compression, isobaric heating, isentropic expansion and isobaric condensation process. Except that the process of water in the boiler is a normal pressure process, the other three processes are completely reversible. Fig. 7-4 shows the steps of the Carnot Cycle, and Fig. 7-5 shows the steps of the Rankine Cycle, each sketched on a T-$\hat{S}$ diagram for water.

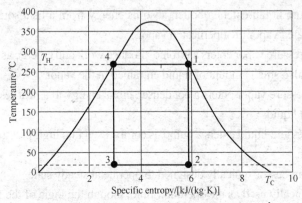

Fig.7-4 An example Carnot heat engine cycle, illustrated on the T-$\hat{S}$ plot for water

State 1 is the steam after isothermal heating, State 2 is the liquid/vapor mixture after adiabatic expansion, State 3 is the liquid/vapor mixture after isothermal cooling, and State 4 is the liquid after adiabatic compression. Consequently, the adiabatic steps are isentropic, and there is no temperature difference between the heat reservoir and the working fluid in the heating/cooling steps.

7.2.1 Comparing the Carnot Cycle with the Rankine Cycle

Two prominent applications for heat engines are powering vehicles and generating electricity. Circulation is a closed system, and the working fluid is continuously circulated through the four steps (Fig. 7-4).

Carnot Step 1: Isothermal Heating

A reversible heat engine requires that the working fluid is heated isothermally and at essentially the same temperature as in a high-temperature reservoir. If a gas expands at the same time, it can be isothermally heated. This simultaneous heating and expansion are feasible in a closed piston cylinder system. In fact, a real Carnot engine is not isothermal.

Carnot Step 2: Adiabatic, Reversible Expansion

Reversible and adiabatic operations are performed in the Carnot Cycle to convert heat into shaft work, so the heat engine close to reversible shall be selected as far as possible. If the mass entering the heat engine is saturated steam, the mass leaving is usually a liquid-steam mixture. However, if there is too much liquid in the turbine, the equipment is vulnerable to erosion damage. If the incoming steam is overheated to a higher temperature, the discharged steam may not contain liquid at all.

Carnot Step 3: Isothermal Cooling

The cooling step is carried out in the condenser. The operating temperature of the boiler needs significant temperature difference, that is, the operating temperature is lower than that of the high temperature accumulator, so that the required heat transfer occurs in a manageable volume heat exchanger.

Carnot Step 4: Adiabatic Compression

Compression is carried out in the pump. The increase of large pressure is accompanied by the increase of small temperature. The liquid leaving the pump is supercooled liquid.

The first steam power cycle with practical significance is Rankine Cycle. The Rankine Cycle works as follows:

> - High-pressure liquid enters a boiler, absorbs energy from a heat source, and emerges as either saturated vapor or superheated vapor.
> - The high-pressure vapor enters a turbine, produce work and emerge as low-pressure vapor, possibly with a small amount of liquid entrained in the vapor.
> - The low-pressure vapor enters a condenser, emits energy to a heat reservoir, and emerges as saturated liquid.
> - The low-pressure liquid enters a pump, is compressed into high-pressure liquid, and return to the boiler.
> - The entire process is typically designed to operate at steady state.
> - Water is typically used as the operating fluid, though the logic of the four steps is equally valid for other operating fluids.

State 1 is the steam leaving the boiler, State 2 is the liquid/vapor mixture leaving the turbine, State 3 is the liquid leaving the condenser, and State 4 is the liquid leaving the pump. State 1 and State 2 represent an alternative cycle in which the steam leaving the boiler is superheated rather than saturated(Fig.7-5).

Fig.7-5 An example Rankine heat engine cycle, illustrated on the T-$\hat{S}$ plot for water

7.2.2 Design Variations in the Rankine Heat Engine

This section will further explore the design considerations of Rankine heat engine (Luo Wenhua et al., 2021). Excessive amounts of liquid entrained in the vapor in a turbine can damage the equipment.

EXAMPLE 7-1

Saturated versus Superheated Steam Entering a Turbine.

A turbine (Fig. 7-6) has an efficiency of 80% and an outlet pressure of p=0.5 bar. The inlet stream is T=300 ℃, which can be saturated steam or superheated steam with p=3 bar, 5 bar, 10 bar or 30 bar.

A. Which of these inlet streams provides the most work without exceeding the maximum allowable liquid fraction of 10% in turbine?

B. What is the efficiency of a Rankine engine operating with this turbine when the inlet air flow is selected in part A?

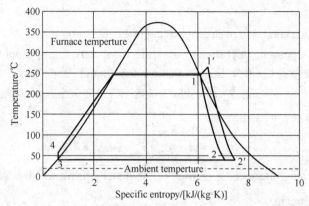

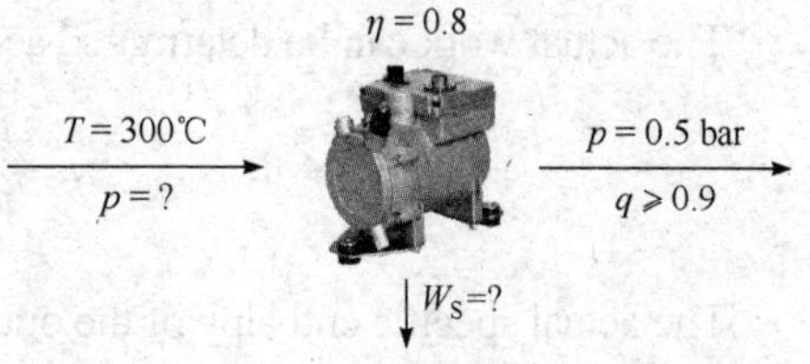

Fig.7-6 Turbine examined in Example 7-2

SOLUTION

Solution A

Step 1 Apply entropy and energy balances to reversible turbine

Define Stream 1 as entering the turbine and Stream 2 as exiting the turbine. As shown in Example 7-1, the entropy balance for a reversible turbine is

$$\hat{S}_{2,\text{rev}} - \hat{S}_1 = 0$$

$$\hat{S}_{2,\text{rev}} = \hat{S}_1$$

and the energy balance for a turbine is:

$$\frac{\dot{W}_{S,\text{turbine,rev}}}{\dot{m}} = \hat{H}_{2,\text{rev}} - \hat{H}_1$$

Step 2 Determine condition of stream leaving reversible turbine

Each of the five possible inlet streams has different specific entropy, but they are all known. At the outlet pressure p=0.5 bar, the saturated liquid and vapor specific entropies are $\hat{S}^L = 1.0912$ kJ/(kg·K) and $\hat{S}^V = 7.5930$ kJ/(kg·K). Consequently, if the existing substance is a liquid vapor mixture, it can be used:

$$\hat{S}_{2,\text{rev}} = (1 - q_{\text{rev}})\hat{S}^L + (q_{\text{rev}})\hat{S}^V$$

where $\hat{S}_{2,\text{rev}}$ representing the specific entropy of the outlet stream, which is equal to the specific entropy of the inlet stream.

Step 3 Solve for reversible work

Once q_{rev} is known, the outlet specific enthalpy can be determined using:

$$\hat{H}_{2,\text{rev}} = (1 - q_{\text{rev}})\hat{H}^L + (q_{\text{rev}})\hat{H}^V$$

And the results of Step 1 to Step 3 are summarized. As it turns out, the specific entropy $[7.7037 \text{ kJ/(kg·K)}]$ from the p=3 bar inlet stream is slightly larger than that of superheated steam at 0.5 bar and 100℃ $[7.6953 \text{ kJ/(kg·K)}]$, and the temperature of 101.7℃ is obtained by interpolation between the values of T=100℃ and 150℃. This temperature is used in the determination of $\hat{H}_{2,\text{rev}}$ and subsequent calculations. For the other four inlet streams, the outlet stream is actually a VLE mixture.

Step 4 Determine the conditions under which steam leaves the turbine

The actual work can be determined according to the application efficiency of 80%:

$$\eta = \frac{W_{S,\text{turbine,act}}}{W_{S,\text{turbine,rev}}} = 0.8$$

The actual specific enthalpy of the outlet stream can be found by solving the energy balance:

$$\hat{H}_{2,\text{act}} = \frac{W_{S,\text{turbine,act}}}{m} + \hat{H}_1$$

Note that in the case of inlet steam at p=5 bar, the actual outlet stream is actually a superheated steam, though it is a vapor-liquid equilibrium mixture in the reversible case. For the higher inlet pressures, the actual outlet stream is a VLE mixture, but with higher quality than the corresponding reversible turbine.

Solution B

Step 5 Apply energy balance to condenser

The VLE mixture entering the condenser, its $\hat{H}_{2,\text{act}} = 2416.8$ kJ/kg. Saturated liquid has $\hat{H} = 340.5$ kJ/kg at p=0.5 bar, so the heat removed in the condenser is:

$$\frac{Q_C}{m} = \hat{H}_{out} - \hat{H}_{in} = \hat{H}_3 - \hat{H}_2 = 340.5 \text{ kJ/kg} - 2416.8 \text{ kJ/kg} = -2076.3 \text{ kJ/kg}$$

Step 6 Determine pump work

The work added in the pump can be estimated:

$$\frac{W_{S,pump}}{m} = \int \hat{V} dp \approx \hat{V}(p_{out} - p_{in})$$

$$\frac{W_{S,pump}}{m} = 0.00103 \text{ m}^3/\text{kg} \times (30 \text{ bar} - 0.5 \text{ bar}) \times 10^5 \text{ Pa/bar} \times \left(\frac{1 \text{ N/m}^2}{1 \text{ Pa}}\right) \times 1 \text{ J/(N·m)} \times \frac{1 \text{ kJ}}{1000 \text{ J}}$$

$$\frac{\dot{W}_{S,pump}}{\dot{m}} = 3.04 \text{ kJ/kg}$$

Step 7 Apply energy balance to entire process

The overall energy balance for the entire process is solved to find $\frac{Q_H}{m}$ using:

$$0 = \frac{Q_C}{m} + \frac{Q_H}{m} + \frac{W_{S,turbine}}{m} + \frac{W_{S,pump}}{m}$$

$$0 = -2076.3 \text{ kJ/kg} + \frac{Q_H}{m} + (-577.5 \text{ kJ/kg}) + 3.04 \text{ kJ/kg}$$

$$\frac{Q_H}{m} = 2650.8 \text{ kJ/kg}$$

Step 8 Determine overall efficiency of complete process

The overall efficiency of the entire cycle can be determined by:

$$\eta_{H.E} = \frac{\frac{-W_{S,net}}{m}}{\frac{Q_H}{m}} = \frac{\frac{-(W_{S,turbine} + W_{S,pump})}{m}}{\frac{Q_H}{m}} = \frac{-(-577.5 \text{ kJ/kg} + 3.04 \text{ kJ/kg})}{2650.8 \text{ kJ/kg}}$$

$$\eta_{H.E} = 0.217$$

If the steam entering the turbine is to be superheated, a separate boiler and superheater can be employed.

The boiler and the superheater are two separate heat exchangers. The first produces the steam and the second heats the steam to the desired turbine inlet temperature. The advantage of this approach is that separate heat sources can be used. In the example shown in Fig. 7-7 (b), while the superheater heat source must be above 400℃. The scheme in Fig. 7-7 (a) is not desirable. In this scheme, all heat comes from a heat source above 400℃. Whether the "boiler" is a single equipment or two equipment, the energy balance is basically the same. The absolute heat added in Fig. 7-7 (a) and Fig. 7-7 (b) is the same. Table 7-1 is the estimates of the cost.

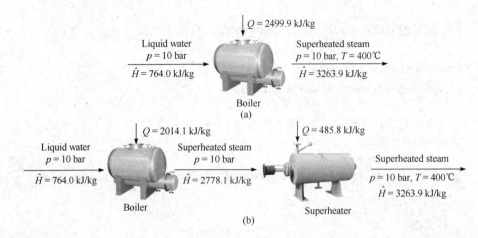

Fig7-7 Formation of steam at p=10 bar and T=400℃ by (a) a single boiler and (b) a boiler/superheater sequence

Table 7-1 Estimates of the cost of steam heating, as a function of temperature (Turton, 2009)

Heat Source	Steam Pressure(gage)/bar	Steam Temperature/℃	Utility Cost/($/GJ)
Low-pressure steam	5	160	14.05
Medium-pressure steam	10	184	14.83
High-pressure steam	41	254	17.70

Based on data from Turton, Bailie, Whiting and Shaewitz, Analysis, Synthesis and Design of Chemical Processes, 3rd ed. Prentice-Hall, 2009.

The single turbine can be replaced by multiple turbines with multistage heating. The purpose of superheating the steam entering the turbine is to reduce the fraction of steam that condenses in the turbine. It is possible to achieve the same effect by using two turbines with moderate pressure drops, rather than a single turbine with a large pressure drop, as shown in Fig. 7-8.

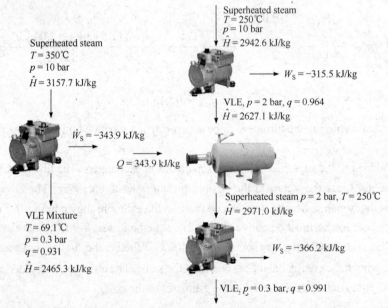

Fig.7-8 Comparison of (a) a single reversible turbine to (b) a sequence of two reversible turbines with inter-stage heating. The overall pressure drop in both cases is identical, and both designs maintain a liquid fraction (1−q) below 10% in the turbines

7.3 The Vapor-Compression Cycle

This section discusses design and analysis of refrigeration cycles (Zhao Jiayue et al., 2021). Refrigeration cycles are often used in industry and are also essential in daily life. The most common refrigerators in life are discussed below.

Household refrigerators are normally set to maintain a temperature at or below 4.44 ℃. There is a non-spontaneous cycle inside the refrigerator to transfer heat from low temperature to high temperature. This is the focus of our discussion in this chapter: Vapor-compression cycles. Fig. 7-9 shows a schematic diagram of vapor-compression cycle, in which the refrigerant circulates through a continuous steady-state process.

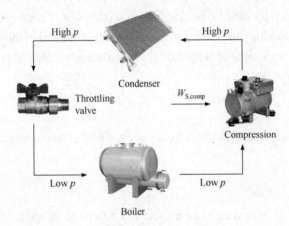

Fig7-9 Schematic diagram of vapor-compression refrigeration cycle

The effective cycle is quantified by the coefficient of performance (COP): $\text{COP} = \dfrac{Q_C}{W_S}$

Where $\dot{Q}_C$ is the rate at which substance absorbs heat in the refrigerator, $\dot{W}_S$ is the rate at which work is done outward. Why is it called "COP" rather than "efficiency"? Because according to practice, "efficiency" is a score. The value range is 0~1 (or 0~100%). The efficiency of a heat engine is always a fraction: it is a fraction of the heat converted into useful work.

Analyzing a Refrigeration Cycle with Simple Models

The Carnot heat engine, which is an idealized heat engine that is 100% reversible and follows four steps.
- Isothermal heating
- Adiabatic expansion
- Isothermal cooling
- Adiabatic compression

A reversible cycle following these steps is called a Carnot refrigerator.

When one is developing or working with new chemical products, one cannot expect that comprehensive data is going to be available.

7.4 Power Cycle and Refrigeration Cycle

7.4.1 Thermodynamic Cycles

Two important areas of application for thermodynamics are power generation and refrigeration. These two methods are usually completed by the system running on the thermodynamic cycle (Oh Jungmo et al., 2021). These measures include:

(1) The devices or systems used to produce a net power output are often called engines, and the thermodynamic cycles they operate on are called power cycles.

(2) The devices or systems used to produce a refrigeration effect are called refrigerators, air conditioners, or heat pumps, and the cycles they operate on are called refrigeration cycles.

Thermodynamic cycles can also be classified in another way: closed cycle and open cycle. In the closed cycle, the working fluid returns to the initial state at the end of the cycle and is recycled. In the open cycle, the working fluid is updated at the end of each cycle, rather than recycled.

Process

Any change that a system undergoes from one equilibrium state to another is called a process.

Cycle

A system is said to have undergone a cycle if it returns to its initial state at the end of the process. That is, for a cycle the initial and final states are identical.

Study of power cycles is an exciting and important part of thermodynamics. The actual cycle is very complex, but most power plants are carried out on the cycle. However, when the actual cycle is stripped of all the internal irreversibilities and complexities, we end up with a cycle that resembles the actual cycle closely but is made up totally of internally reversible processes. Such a cycle is called an ideal cycle (Fig. 7-10). The cycles discussed in this chapter are ideal cycles, and some of them apply to actual cycles. This chapter will make a detailed analysis of the cycle.

Heat engines are designed for the purpose of converting thermal energy to work.

In thermodynamics and engineering, a heat engine is a system that converts heat or thermal energy and chemical energy to mechanical energy, which can be used to do mechanical work. Heat engines operating in fully reversible cycles, such as the Carnot Cycle, have the highest thermal efficiency

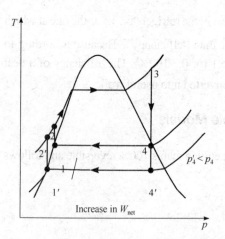

Fig.7-10 The analysis of many complex processes can be simplified to manageable levels in some idealized ways

among all heat engines operating at the same temperature. Carnot Cycle helps us better understand the mutual conversion between heat and work. However, most cycles encountered in practice are very different from the Carnot Cycle, so it is not suitable as a realistic model.

Thermal efficiency is the ratio of the net-work produced by the engine to the total heat input.

$$\eta = \frac{W_{net}}{Q_{in}} \tag{7-1}$$

Idealization and simplification commonly used in power cycle analysis can be summarized as follows:

(1) This cycle does not involve any friction. Therefore, the working fluid will not experience any pressure drop when flowing in pipelines or equipment (such as heat exchangers).

(2) All expansion and compression processes occur in a quasi-equilibrium manner.

(3) The pipes connecting the components of the system are adiabatic, and the heat transfer through them can be ignored.

Thermal efficiency

There are a lot of engines that have thermal efficiencies of 35% to 37%, which means that only 35% to 37% of the power produced by burning fuel is actually used, and the rest is wasted. Diesel engines have higher thermal efficiency than gasoline engines because they burn differently. Gasoline engines are ignited by spark plugs. Due to detonation, the compression ratio should not be too large, generally 9∶1~12∶1. Diesel engines are compression ignition, and the compression ratio can be set higher than gasoline engines. The higher the compression ratio is, the higher the thermal efficiency of the engine is.

The larger the compression ratio is, the longer the piston stroke in a single work is, and the greater the work is. The compression ratio of traditional gasoline engine is limited to "9∶1-12∶1", which will lead to engine detonation. Detonation occurs when the fuel is deflated before the piston is compressed to the ignition position, and the upward piston and the expanded gas are in the opposite position, resulting in severe engine vibration.

Because gasoline has a higher fuel point than diesel, it burns more quickly than diesel, making its pressure burn more difficult to control. Mazda does not cancel the spark plug setting, and the spark plug is able to control the ignition moment, big load case will also be able to use the traditional pattern of spark plug ignites, cold start has been guaranteed. This "extremely thin burning" problem is solved by "pressure burning". Once the petrol engine achieves extreme "lean burn" through "pressure combustion", it can be as efficient as Mazda claims, achieving a maximum fuel efficiency of over 50%. Mazda's pursuit of technology and innovative exploration of gasoline compression and combustion have brought new hope to traditional gasoline.

Nicolas Léonard Sadi Carnot (1796—1832) was a French military engineer and physicist, often described as the "father of thermodynamics".

He was the first person to connect heat and energy. Using the ideological method of "ideal experiment", he proposed the simplest but theoretically important engine cycle—Carnot Cycle, and created the ideal engine (Carnot engine) by assuming that the cycle is reversible under quasi-static conditions and has nothing to do with the working medium.

Each ideal cycle discussed in this chapter is related to a specific work-producing device and is an idealized version of the actual cycle.

The idealizations and simplifications commonly employed in the analysis of power cycles can be summarized as follows:
- The cycle does not involve any friction.
- All expansion and compression processes take place in a quasi-equilibrium manner.
- The pipes connecting the various components of a system are well insulated, and heat transfer through them is negligible.

7.4.2 Property Diagrams

On both p-V and T-S diagrams, the area enclosed by the process curve represents the network of the cycle.

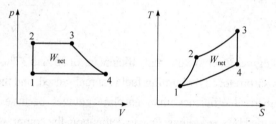

Fig.7-11 On p-V and T-S diagrams, the area enclosed by the process curve represents the net-work of the cycle

The p-V and T-S diagram is particularly useful as a visual aid in the analysis of ideal power cycles.

An ideal power cycle does not involve any internal irreversibilities, and so the only effect that can change the entropy of the working fluid during a process is heat transfer. On p-V and T-S graphs, the area enclosed by a loop process curve represents the net-work generated during this loop process (Fig. 7-11).

On a T-S diagram, a heat absorption process proceeds in the direction of increasing entropy, a heat release process proceeds in the direction of decreasing entropy, and an isentropic (internally reversible, adiabatic) process proceeds at constant entropy. The area under the process curve on a T-S diagram represents the heat transfer for that process. The difference between these two (the area enclosed by the cyclic curve) is the net heat transfer, which is also the net-work produced during the cycle. Therefore, on the T-S diagram, the ratio of the area enclosed by the cyclic curve to the area under the heat absorption process curve represents the thermal efficiency of the cycle.

Any modification that increases the ratio of these two areas will also increase the thermal efficiency of the cycle. The working fluid in the ideal dynamic cycle works in a closed loop, and the type of a single process that constitutes the cycle depends on a single device used to perform the cycle.

7.4.3 The Carnot Cycle and Its Value in Engineering

Carnot heat engine includes two isothermal reversible processes and two adiabatic reversible processes: isothermal heat addition, isentropic expansion, isothermal heat rejection, and isentropic compression.

Carnot Cycle can be performed in a closed system (piston cylinder device) or a stable flow system (using two turbines and compressors, as shown in Fig. 7-12). Gas or steam can be used as working fluids.

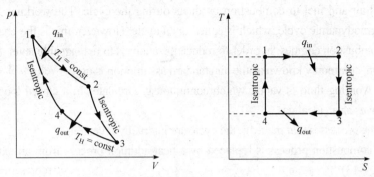

Fig.7-12 *p-V* and *T-S* Carnot Cycle

The Carnot Cycle is a theoretical thermodynamic cycle proposed by French physicist (Sadi Carnot) in 1824 and expanded upon by others in the 1830s and 1840s.

(1) Isothermal Expansion.

(2) Isentropic (reversible adiabatic) expansion of the gas.

(3) Isothermal Compression.

(4) Adiabatic reversible compression.

Carnot Cycle is the most effective thermodynamic cycle. It can convert the heat input by high-temperature heat source into work to the greatest extent. The Carnot Cycle is the most efficient cycle that can be executed between a heat sources of temperature T_H and temperature T_L, and its thermal efficiency is expressed as:

$$\eta_{th,Carnot} = 1 - \frac{T_L}{T_H} \tag{7-2}$$

However, the adiabatic reversible process is difficult to achieve in practice, so the Carnot heat engine is an ideal heat engine that cannot be realized. However, its efficiency can provide a relatively useful highest standard for the actual thermal efficiency. The real value of the Carnot Cycle comes from its being a standard against which the actual or the ideal cycles can be compared. Because reversible isothermal heat transfer is difficult to achieve in practice, it requires very large heat exchangers and takes a long time (a power cycle in a typical engine is completed within a fraction of a second). Therefore, it is not practical to build an engine that would operate on a cycle that closely approximates the Carnot Cycle.

The thermal efficiency increases with the increase of the average temperature of supplying heat to the system or the decrease of the average temperature of releasing heat from the system. However, the high temperature and low temperature heat sources that can be used in practice are not unlimited. The highest temperature in the cycle is limited by the maximum temperature that the components of the heat engine, such as the piston or the turbine blades, can withstand. The lowest temperature is limited by the temperature of the cooling medium utilized in the cycle such as a lake, a river, or the atmospheric air.

7.4.4 Air-standard Assumptions

In gas power cycles (ChengWeijun et al., 2009), the working fluid remains a gas throughout the entire cycle. During the engine combustion process, the composition of the working fluid changes from air and fuel to combustion products during the cycle. The working fluid is not a complete thermodynamic cycle, which is called open cycle. However, this is the characteristic of all internal combustion engines. In order to reduce the analysis to manageable level, the following approximation, commonly known as the air standard assumption is used:

> The working fluid is vapor, which continuously circulates in a closed loop and always behaves as an ideal gas.
> All the processes that make up the cycle are internally reversible.
> The combustion process is replaced by a heat-addition process from an external source (Fig. 7-13)
> The exhaust process is replaced by a heat-rejection process that restores the working fluid to its initial state.

Gas power cycle

Spark-ignition engines, diesel engines, and conventional gas turbines are familiar examples of devices that operate on gas cycles.

This simplified model enables us to study qualitatively the influence of major parameters on the performance of the actual engines.

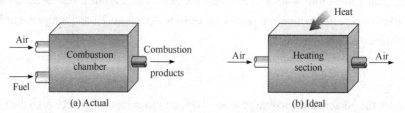

Fig.7-13 The combustion process is replaced by heating process in ideal cycle

Another assumption often used to simplify analysis is that air has a constant specific heat, which is determined at room temperature (25℃). When this assumption is used, the air-standard assumptions are called the cold-air-standard assumptions.

EXAMPLE 7-2

An air-standard cycle is executed in a closed system and is composed of the following four processes:

1-2 Isentropic compression from 100 kPa to 1 MPa.
2-3 p=constant heat addition in amount of 2800 kJ/kg.
3-4 V=constant heat rejection to 100 kPa.
4-1 p=constant heat rejection to initial state.
(a) Show the cycle on p-V and T-S diagrams.

(b) Calculate the maximum temperature in the cycle.
(c) Determine the thermal efficiency.
Assume constant specific heats at room temperature.

SOLUTION

The four processes of an air-standard cycle are described. The cycle is to be shown on p-V and T-S diagrams, and the maximum temperature in the cycle and the thermal efficiency are to be determined.

Assumptions:

The air-standard assumptions are applicable.

Kinetic and potential energy changes are negligible.

Air is an ideal gas with constant specific heats.

Properties:

The properties of air at room temperature are C_p=1.005 kJ/(kg · K), C_V=0.718 kJ/(kg · K), and k=1.4

Analysis:

(a) The cycle is shown on p-V and T-S diagrams in Fig.7-13.

(b) From the ideal gas isentropic relations and energy balance.

$$T_2 = T_1 \left(\frac{p_2}{p_2}\right)^{(k-1)/k} = 300 \text{ K} \times \left(\frac{1000 \text{ kPa}}{100 \text{ kPa}}\right)^{0.4/1.4} = 579.2 \text{ K}$$

$$q_{in} = H_3 - H_2 = C_p(T_3 - T_1)$$

$$2800 \text{ kJ/kg} = 1.005 \text{ kJ}/(\text{kg} \cdot \text{K}) \times (T_3 - 579.2 \text{ K})$$

$$T_{max} = T_3 = 3360 \text{ K}$$

(c) The temperature at State 4 is determined from the ideal gas relation for a fixed mass:

$$\frac{p_3 V_3}{T_3} = \frac{p_4 V_4}{T_4} \rightarrow T_4 = \frac{p_4}{p_3} T_3 = \frac{100 \text{ kPa}}{1000 \text{ kPa}} \times 3360 \text{ K} = 336 \text{ K}$$

The total amount of heat rejected from the cycle is

$$q_{out} = q_{34,out} + q_{41,out} = (U_3 - U_4) + (H_4 - H_1)$$
$$= C_V(T_3 - T_4) + C_p(T_4 - T_1)$$
$$= 0.718 \text{ kJ}/(\text{kg} \cdot \text{K}) \times (3360 - 336) \text{K} + 1.005 \text{ kJ}/(\text{kg} \cdot \text{K}) \times (336 - 300) \text{K}$$
$$= 2212 \text{ kJ/kg}$$

Then, the thermal efficiency is determined from its definition to be

$$\eta_{th} = 1 - \frac{q_{out}}{q_{in}} = 1 - \frac{2212 \text{ kJ/kg}}{2800 \text{ kJ/kg}} = 0.210 \text{ or } 21.0\%$$

7.4.5 Rankine Cycle: The Ideal Cycle for Vapor Power Cycles

The first steam power cycle with practical significance is Rankine Cycle.

Together with Rudolf Clausius and William Thomson (Lord Kelvin), he was a founding contributor to the science of thermodynamics. William John Macquorn Rankine (July 5, 1820—December 24, 1872) was a Scottish civil engineer, physicist, and mathematician. Along with Kelvin, he was one of the authors of the first law of thermodynamics. Rankin developed a complete theory that worked for steam engines, and indeed for all heat engines. After publishing these theories in the 1850s and 1860s, they were widely used in engineering and practice.

Many of the impracticalities associated with the Carnot Cycle can be eliminated by superheating the steam in the boiler and condensing it completely in the condenser, as shown schematically on a *T-S* diagram in Fig. 7-14.

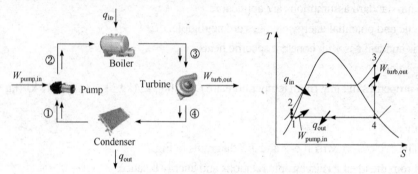

Fig.7-14 The simple ideal Rankine cycle

The ideal Rankine cycle consists of the following four processes:
1-2 Isentropic compression in a pump.
2-3 Constant pressure heat addition in a boiler.
3-4 Isentropic expansion in a turbine.

7.4.6 Energy Analysis of the Ideal Rankine Cycle

Deviation of actual vapor power cycle from idealized ones.

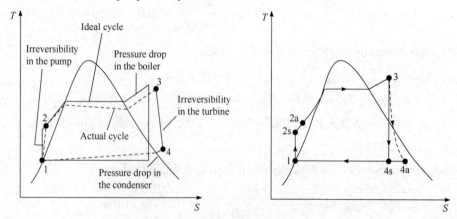

Fig.7-15 (a) The actual steam deviation comes from the dynamic cycle period of the ideal Rankine Cycle (b) Pump and ideal turbine irreversible Rankine Cycle

The actual vapor power cycle differs from the ideal Rankine Cycle, as illustrated in Fig. 7-15, as a result of irreversibilities in various components. Fluid friction and heat loss to the surroundings are the two common sources of irreversibilities.

How can we increase the efficiency of the Rankine Cycle?

Lowering the Condenser Pressure

The effect of lowering the condenser pressure on the Rankine Cycle efficiency is illustrated on a *T-S* diagram in Fig.7-16. For comparison purposes, the turbine inlet state is maintained the same. The colored area on this diagram represents the increase in net-work output as a result of lowering the condenser pressure from p_4 to p_4'. The heat input requirements also increase (represented by the area under curve $2' \rightarrow 2$), but this increase is very small. Thus, the overall effect of lowering the condenser pressure is an increase in the thermal efficiency of the cycle.

Superheating the Steam to High Temperatures

The average temperature at which heat is transferred to steam can be increased without increasing the boiler pressure by superheating the steam to high temperatures. The effect of superheating on the performance of vapor power cycles is illustrated on a *T-S* diagram in Fig. 7-17. The colored area on this diagram represents the increase in the net-work. The total area under the process curve $3 \rightarrow 3'$ represents the increase in the heat input. Thus, both the net-work and heat input increase as a result of superheating the steam to a higher temperature. The overall effect is an increase in thermal efficiency, since the average temperature at which heat is added increases.

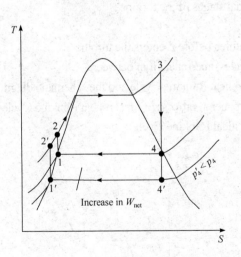
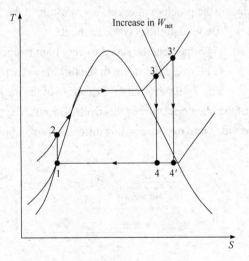

Fig.7-16 The effect of lowering the condenser pressure on the ideal Rankine Cycle

Fig.7-17 The effect of superheating the steam to higher temperatures on the ideal Rankine Cycle

Increasing the Boiler Pressure

Another way to increase the average temperature during heat addition is to increase the

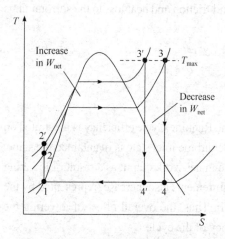

Fig.7-18 Effect of superheated steam to ideal state of higher temperature Rankine Cycle

working pressure of the boiler, which will automatically increase the temperature when boiling occurs. This, in turn, raises the average temperature at which heat is transferred to the steam and thus raises the thermal efficiency of the cycle.

The effect of increasing the boiler pressure on the performance of vapor power cycles is illustrated on a T-S diagram in Fig. 7-18. Notice that for a fixed turbine inlet temperature, the cycle shifts to the left and the moisture content of steam at the turbine exit increases.

Steam exists in the condenser in the form of saturated mixture, which corresponds to the pressure in the condenser. Therefore, reducing the working pressure of the condenser will automatically reduce the temperature of the steam and the heat will be discharged.

7.4.7 The Ideal Re-heat Rankine Cycle

We noted in the previous section that increasing the boiler pressure would improve the thermal efficiency of Rankine Cycle, but would also reduce the water content of steam. Then:

How can we take advantage of the increased efficiencies at higher boiler pressures without facing the problem of excessive moisture at the final stages of the turbine?

Two possibilities come to mind:

(1) Superheat the steam to very high temperatures before it enters the turbine.

(2) Expand the steam in the turbine in two stages, and reheat it in between.

Fig. 7-19 shows the T-S diagram of the ideal reheat Rankine Cycle and the schematic diagram of the plant operating in this cycle. Because the expansion process of ideal reheat Rankine Cycle is divided into two stages, it is different from simple ideal Rankine Cycle.

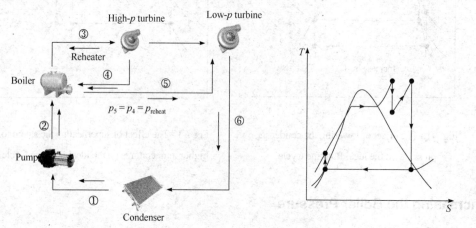

Fig.7-19 Ideal reheated Rankine Cycle

The average temperature of heat transfer during reheating increases with the increase of reheating stages is shown in Fig.7-20.

At the highest temperature, the expansion and reheat processes are close to isothermal processes, as shown in Fig. 7-20. However, two reheating stages are unrealistic. Theoretically, secondary reheating can improve efficiency, but the effect of primary reheating is only about half. If the steam turbine inlet pressure is not high enough, the reheating will cause the exhaust gas to overheat. This does not want to increase the average heat dissipation temperature, thereby reducing the cycle efficiency. The third stage reheat will increase its cycle efficiency by about half to achieve the second stage reheat. This benefit is too small to offset the increased cost and complexity.

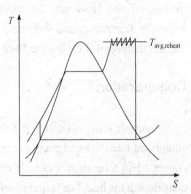

Fig.7-20 Average temperature of heat transfer during reheating increases with reheating times

The reheat cycle was introduced in the mid—1920s, but it was abandoned in the 1930s because of the operational difficulties. The steady increase in boiler pressures over the years made it necessary to reintroduce single reheat in the late 1940s and double reheat in the early 1950s.

7.4.8 The Ideal Regenerative Rankine Cycle

As shown in the T-S diagram of Fig. 7-21, heat is transferred to the working fluid in the process of 2-2′. In order to make up for this disadvantage, it is necessary to find a method to increase the liquid temperature before entering the boiler and leaving the pump (called water supply). The solution is that the heat in the expansion process is transferred to water steam, and the built-in turbine in the countercurrent heat exchanger is used for regeneration. However, this solution is impractical because it will increase the water content of the last stage of the steam turbine. Another solution is to complete the regeneration process by extracting or "pumping" air from various points of the turbine. This regeneration method not only improves the circulation efficiency, but also provides a convenient method for boiler deaeration (removing the air leaked from the condenser) to prevent boiler corrosion. Next, the regeneration of two feed water heaters is discussed.

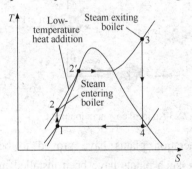

Fig.7-21 Heating of the first part of the boiler process occurs at relatively low temperatures

7.5 Second-Law Analysis of Vapor Power Cycles

The ideal Carnot Cycle is a totally reversible cycle, and thus it does not involve any irreversibilities. However, the ideal Rankine Cycles (simple, reheat, or regenerative) are only internally reversible, and they may involve irreversibilities external to the system, such as heat transfer through a finite temperature difference.

Cogeneration

A simple process-heating plant. In some industries that rely heavily on heat, many systems or equipment need to input energy in the form of heat, which is called process heat. Energy is usually converted in a heating device. If only the operation of the heating device is considered without considering the heat loss in the pipeline, all the heat transferred to the boiler steam will be used for the process heating device, as shown in in Fig. 7-22. However, the temperature in the furnace usually needs to be higher, which puts forward higher requirements for the quality of the furnace. Therefore, this method is not very good from the perspective of engineering economy. Plant power generation is carried out under the condition of meeting the process thermal requirements of some industrial processes. This kind of power plant is called cogeneration power plant.

Fig. 7-23 shows the ideal schematic diagram of steam turbine cogeneration device. Therefore, no heat is discharged from the plant as waste heat. In other words, all the energy transferred to the boiler steam is used as process heat or electrical energy. The utilization factor of thermal power plant is:

$$\varepsilon_\mu = \frac{W_{net} + Q_p}{Q_{in}} \tag{7-3}$$

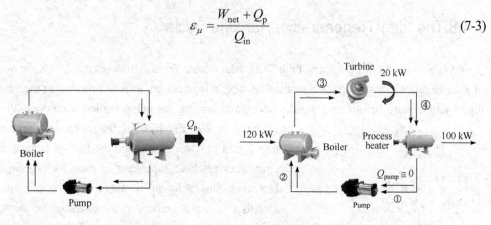

Fig.7-22 A simple process heating device   Fig.7-23 An ideal cogeneration plant

A cogeneration plant with adjustable loads. Cogeneration plants have proved to be economically very attractive. Consequently, more and more such plants have been installed in recent years, and more are being installed. Fig. 7-24 is a schematic diagram of a more practical (but more complex) cogeneration device. The heat discharged from the condenser represents the waste heat of the cycle. When the process heat requirement is high, all steam is transmitted to the process heating device, and no steam is transmitted to the condenser ($M=0$). In this mode, the residual heat is 0.

In the last century, power plants were mainly used to heat residents. With the rise of fuel prices, thermal power plants have become a common heating channel. Therefore, more and more such power plants have been installed in recent years.

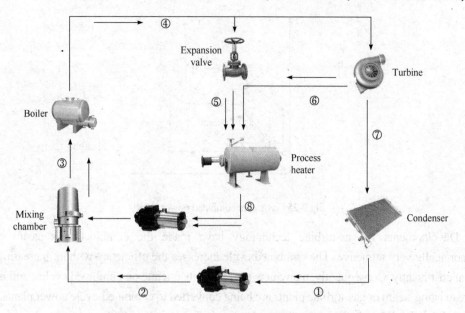

Fig.7-24 A cogeneration plant with adjustable loads

7.5.1 Combined Gas-Vapor Power Cycles

With the development of the world, people are no longer only satisfied with the demand for heating, and people begin to pursue efficient power generation. This has continuously promoted the improvement of traditional power plants. The most concerned combined cycle is the gas turbine (Brayton) cycle rather than the steam turbine (Rankine) cycle, and its thermal efficiency is higher than any cycle executed separately.

The continuous pursuit of higher thermal efficiency has led to the innovative transformation of traditional power plants. An advanced improvement: gas steam combined cycle, as shown in Fig. 7-25. During the cycle, energy is transferred from the exhaust gas to the steam in the heat exchanger used for boiler recovery.

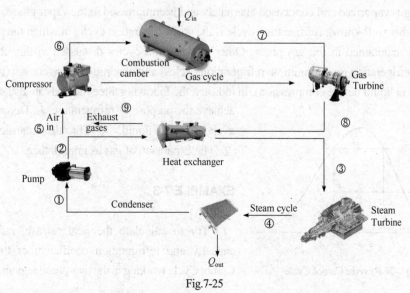

Fig.7-25

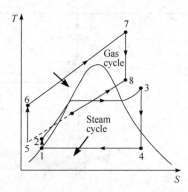

Fig.7-25 Gas-steam combined power plant

Developments in gas-turbine technology have made the combined gas steam cycle economically very attractive. The combined cycle increases the efficiency without increasing the initial cost greatly. Consequently, many new power plants operate on combined cycles, and many more existing steam or gas-turbine plants are being converted to combined-cycle power plants. It is reported that the thermal efficiency after conversion is far more than 40%.

7.5.2 Refrigeration Cycles

Refrigeration is indispensable in industry and life. The process of reducing the system temperature below ambient temperature is called "freezing or cooling". The refrigeration process usually uses the refrigeration device to transfer the heat of the low-temperature system to the high-temperature system at the cost of consuming mechanical energy or electromagnetic energy, thermal energy, solar energy, nuclear energy and other forms of energy, so as to obtain low temperature.

A major application area of thermodynamics is refrigeration, which is the transfer of heat from a lower temperature region to a higher temperature one. Devices that produce refrigeration are called refrigerators, and the cycles on which they operate are called refrigeration cycles. The most frequently used refrigeration cycle is the vapor-compression refrigeration cycle in which the refrigerant is vaporized and condensed alternately and is compressed in the vapor phase.

Another well-known refrigeration cycle is the gas refrigeration cycle, in which the refrigerant is always maintained in the gas phase. Other refrigeration cycles discussed in this chapter are cascade refrigeration (using multiple refrigeration cycles) and absorption refrigeration (refrigerant dissolved in liquid before compression). In industry, the following three methods are usually used to achieve the purpose of refrigeration:① Decompression evaporation of liquid. ② Throttle expansion of gas. ③ The expansion of gas as grandfather.

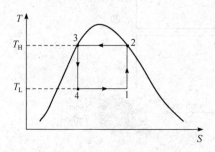

Fig.7-26 Reverse Carnot Cycle

EXAMPLE 7-3

Try to calculate the heat release, refrigeration capacity and refrigeration coefficient of the reverse Carnot Cycle working in the two-phase region(Fig.7-26).

SOLUTION

Schematic diagram of reverse Carnot Cycle in two-phase region. T_H is the temperature of high-temperature object and T_L is the temperature of low-temperature object. Since the reverse Carnot Cycle is a reversible external cycle, it can be obtained with the help of T-S diagram.

Circulating heat release

$$q_2 = T_H (S_1 - S_4)$$

$$q_0 = T_L (S_1 - S_4)$$

$$W_S = \Sigma q = q_2 + q_0 = (T_H - T_L) \times (S_1 - S_4)$$

Refrigeration coefficient of reverse Carnot cycle

$$\varepsilon_c = \frac{q_0}{W_S} = \frac{T_L}{T_H - T_L}$$

The results of Example 6-7 are analyzed.

> In the refrigeration cycle, the heat release of high-temperature objects is greater than the heat absorption of low-temperature objects. The difference is equal to the amount of heat converted by the energy consumed.
> For the refrigeration cycle operating in the same temperature range, the refrigeration coefficient of the reverse Carnot Cycle is the largest.

During refrigeration calculation, the working parameters of refrigerant and refrigeration cycle shall be determined first (evaporation temperature, condensation temperature, if there is supercooling, the supercooling temperature shall be reported). The evaporation temperature depends on the temperature of the cooled system, the condensation temperature depends on the temperature of the cooling medium (atmosphere or cooling water, etc.), and the necessary heat transfer temperature difference is considered. According to the given working parameters, the corresponding state points can be found out on the thermodynamic chart of the refrigerant, the enthalpy of each state point can be found or calculated, and then substituted into the corresponding calculation formula.

It is worth pointing out that it is most convenient to analyze and calculate the refrigeration cycle by using the pressure enthalpy diagram of refrigerant ($\ln p$-H diagram). Because the evaporation and condensation processes are isobaric processes, it is represented by a horizontal straight line on the $\ln p$-H diagram, and the throttling expansion is an isoenthalpy process, which is represented by a vertical line. Moreover, the refrigeration capacity q_0, heat release q_2 and power consumption W_S can be represented by corresponding horizontal distance, which is very intuitive.

The performance of vapor compression refrigeration cycle is closely related to the thermodynamic properties of refrigerant. The selection of refrigerant shall meet the following requirements,

> Low boiling point at atmospheric pressure. Low boiling point can not only obtain low refrigeration temperature, but also make the evaporation pressure higher than the atmospheric pressure at a certain refrigeration temperature to prevent air from entering the refrigeration device.

> The condensation pressure at normal temperature shall be as low as possible to reduce the pressure resistance and sealing requirements of the condenser.
> The latent heat of vaporization is large, which reduces the circulation of refrigerant and reduces the size of compressor.
> With higher critical temperature and lower solidification temperature, most of the exothermic process takes place in the two-phase region.
> It is chemically stable, non flammable, non decomposing and non corrosive.

The commonly used refrigerants are ammonia, chlorofluorocarbons, carbon dioxide, ethane, ethylene, etc. It should be noted that R11, R12, R113, R114 and R115 refrigerants in HCFCs have been found to seriously damage the ozone layer in the atmosphere. In order to protect the environment, the Montreal conference in 1987 drafted an agreement to protect the ozone layer and put forward the process of limiting the production of these five hydrochlorofluorocarbons. Therefore, the development of pollution-free alternatives has attracted the attention of countries all over the world.

7.5.3 Refrigerators and Heat Pumps

The transfer of heat from a low-temperature region to a high-temperature one requires special devices called refrigerators. Refrigerators are cyclic devices, and the working fluids used in the refrigeration cycles are called refrigerants.

Another device that transfers heat from a low-temperature medium to a high-temperature one is the heat pump.

Refrigerators and heat pumps are essentially the same devices. However, the goal of the heat pump is to keep the heating space high temperature. This is accomplished by absorbing heat from a low-temperature source, such as well water or cold outside air in winter, and supplying the heat to a warmer medium such as a house (Fig. 7-27).

The performance of refrigerators and heat pumps is expressed in terms of the COP, defined as:

$$\text{COP}_R = \frac{\text{Desired Output}}{\text{Required Input}} = \frac{\text{Cooling Effect}}{\text{Work Input}} = \frac{Q_L}{W_{net,in}}$$

(7-4)

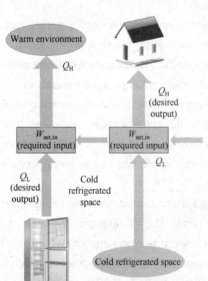

Fig.7-27 The objective of a refrigerator is to remove heat (Q_L) from the cold medium; the objective of a heat pump is to supply heat (Q_H) to a warm medium

$$\text{COP}_{HP} = \frac{\text{Desired Output}}{\text{Required Input}}$$
$$= \frac{\text{Heating Effect}}{\text{Work Input}} = \frac{Q_H}{W_{net,in}}$$

(7-5)

$$\text{COP}_{HP} = \text{COP}_R + 1$$

(7-6)

7.5.4 The Reversed Carnot Cycle

A refrigerator or heat pump that operates on the reversed Carnot Cycle is called a Carnot refrigerator or a Carnot heat pump.

$$\text{COP}_{R,\text{Carnot}} = \frac{1}{\dfrac{T_H}{T_L} - 1} \qquad (7\text{-}7)$$

$$\text{COP}_{HP,\text{Carnot}} = \frac{1}{1 - \dfrac{T_L}{T_H}} \qquad (7\text{-}8)$$

Schematic of a Carnot refrigerator and *T-S* diagram of the reversed Carnot Cycle, as shown in Fig. 7-28.

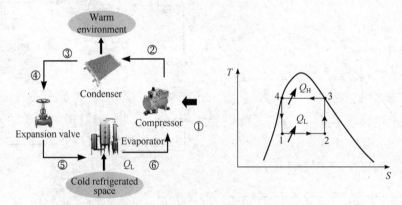

Fig.7-28 Schematic of a Carnot refrigerator and *T-S* diagram of the reversed Carnot Cycle

The reversed Carnot Cycle is the most efficient refrigeration cycle operating between two specified temperature levels.

Many of the impracticalities associated with the reversed Carnot Cycle can be eliminated by vaporizing the refrigerant completely before it is compressed and by replacing the turbine with a throttling device, such as an expansion valve or capillary tube. The ideal vapor compression refrigeration cycle, and it is shown schematically and a *T-S* diagram in Fig. 7-29.

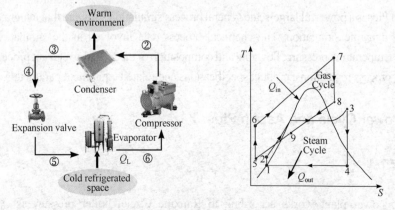

Fig.7-29 Schematic and *T-S* diagram for the ideal vapor-compression refrigeration cycle

A diagram frequently used in the analysis of vapor-compression refrigeration cycles is the *p-H* diagram, as shown in Fig. 7-30.

Actual Vapor-Compression Refrigeration Cycle

The actual vapor compression refrigeration cycle is different from the ideal vapor compression refrigeration cycle in several aspects, mainly due to the irreversibility of various components (Fig. 7-31). Two common sources of irreversibility are fluid friction (resulting in pressure drop) and heat transfer with the surrounding environment.

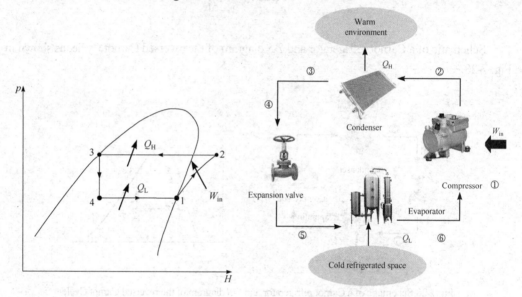

Fig.7-30 The *p-H* diagram of an ideal vapor compression refrigeration cycle

Fig.7-31 Schematic and *T-S* diagram for the actual vapor-compression refrigeration cycle

7.6 Application of Thermodynamic Processes and Cycles in Aspen Plus

Aspen Plus is a powerful large-scale general process simulation software that integrates chemical design and dynamic simulation. The chemical process data involved in the simulation generally include the temperature, pressure, flow rate and composition of the feed, the relevant process operating conditions, process regulations, product specifications and related equipment parameters.

Steam Power Cycle and Aspen Plus

EXAMPLE 7-4

Steam power plant works according to Rankine Cycle, boiler pressure is 40×10^5 Pa, producing 440°C superheated steam, steam turbine outlet pressure is 0.04×10^5 Pa, steam flow is 60 t/h.

(1) The amount of heat absorbed per hour by superheated steam from the boiler.

(2) The humidity of low pressure wet steam and the heat released by low pressure wet steam in the condenser per hour.

(3) The theoretical power produced by the turbine and the theoretical power consumed by the pump.

(4) Thermal efficiency of cycle.

SOLUTION

The cycle is presented on a *T-S* diagram as shown below.

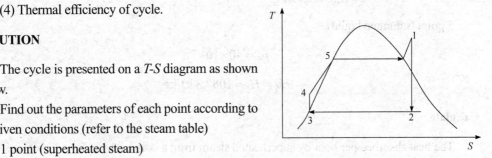

Find out the parameters of each point according to the given conditions (refer to the steam table)

1 point (superheated steam)

$$p_1 = 40 \times 10^5 \text{ Pa}$$

$$H_1 = 3307.1 \text{ kJ/kg}$$

$$T_1 = 440°C$$

$$S_1 = 6.9041 \text{ kJ/(kg} \cdot \text{K)}$$

2 point (wet steam)

$$p_2 = 0.04 \times 10^5 \text{ Pa}$$

$$S_2 = S_1 = 6.9041 \text{ kJ/(kg} \cdot \text{K)}$$

$$H_g = 2554.4 \text{ kJ/kg}$$

$$H_L = 121.46 \text{ kJ/kg}$$

$$S_g = 8.4746 \text{ kJ/(kg} \cdot \text{K)}$$

$$S_L = 0.4226 \text{ kJ/(kg} \cdot \text{K)}$$

$$V_L = 1.0040 \text{ cm}^3/\text{kg}$$

Set the dryness at 2 as x

$$8.4746x + (1-x) \times 0.4226 = 6.9041$$

$$x = 0.805$$

$$H_2 = [2554.4 \times 0.805 + (1 - 0.805) \times 121.46] \text{ kJ/kg} = 2079.98 \text{ kJ/kg}$$

3 point (saturated liquid)

$$p_3 = 0.04 \times 10^5 \text{ Pa}$$

$$H_3 = H_L = 121.46 \text{ kJ/kg}$$

4 point (unsaturated water)

$$\begin{aligned} H_4 &= H_3 + W_P = H_3 + V(p_4 - p_3) \\ &= [121.46 + 0.001004 \times (40 - 0.04) \times 10^5 \times 10^{-3}] \text{ kJ/kg} \\ &= 125.472 \text{ kJ/kg} \end{aligned}$$

5 point (saturated liquid)

$$p_5 = 40 \times 10^5 \text{ Pa}$$

$$H_5 = H_1 = 1087.3 \text{ kJ/kg}$$

Calculate

The heat absorbed per hour by superheated steam from a boiler

$$Q_1 = m(H_1 - H_4) = [60 \times 10^3 \times (3307.1 - 125.472)] \text{ kJ/h} = 190.9 \times 10^6 \text{ kJ/h}$$

The heat released by low pressure wet steam in the condenser

$$Q_2 = m(H_1 - H_4) = [60 \times 10^3 \times (2079.98 - 121.46)] \text{ kJ/h} = 117.51 \times 10^6 \text{ kJ/h}$$

The low pressure wet steam humidity is

$$1 - x = 1 - 0.805 = 0.195$$

The turbine makes the theoretical power:

$$p_T = mW_S = -m(H_1 - H_2) = \left[\frac{-60 \times 10^3}{3600} \times (3307.1 - 2079.98)\right] \text{ kW} = -20452 \text{ kW}$$

The theoretical power consumed by the pump

$$N_P = mW_P = m(H_4 - H_3) = \left[\frac{60 \times 10^3}{3600} \times (125.472 - 121.46)\right] \text{ kW} = 66.87 \text{ kW}$$

or

$$\begin{aligned} N_P &= m\left[-\int V dp\right] = mV(p_4 - p_3) \\ &= \left[\frac{60 \times 10^3}{3600} \times 0.01004 \times (40 - 0.04) \times 10^5 \times 10^{-3}\right] \text{ kW} = 66.87 \text{ kW} \end{aligned}$$

Cyclic theoretical thermal efficiency:

$$\eta = \frac{-3600 \times (p_T + N_P)}{Q_1} = \frac{-3600 \times (-20452 + 66.87)}{190.9 \times 10^6} = 0.384$$

Similarly, this example can be simulated in Aspen Plus.
The operation steps are shown in the following QR code.

Summary

It can be seen from the calculation results and simulation results that the theoretical thermal efficiency of the calculation results is greater than that of the simulation results, because the simulation results consider the actual situation and are the actual thermal efficiency of the cycle.

EXAMPLE 7-5

The air state at the compressor outlet is $p_1 = 9.12$ MPa (90 atm), $T_1 = 300$ K, If $p_2 = 0.203$ MPa (2 atm) is expanded by the following two expansion methods.

(1) Throttling expansion;

(2) For adiabatic expansion of external work, it is known that the isentropic efficiency of the expander is $\eta = 0.8$.

Take the ambient temperature as 25°C, and try to find the temperature of the two expanded gases, the work done by the expander and the work lost in the expansion process.

SOLUTION

(1) Throttling expansion.

For the throttling expansion process, the lost work is calculated according to the following formula. Firstly, the ΔS_{sur} is calculated. the throttling expansion is regarded as an adiabatic process, $Q_{sur} \to 0$, so

$$W_L = T_0 \Delta S_{sys} + T_0 \Delta S_{sur} = T_0 \left(\Delta S_{sys} + \Delta S_{sur} \right) = T_0 \Delta S_t$$

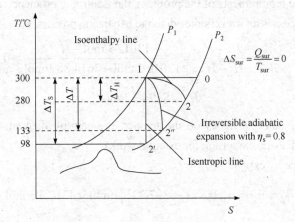

T-S diagram of air

Then calculate ΔS_{sys} of the system. According to the temperature-entropy diagram of air.

For the throttling expansion process, the lost work is calculated according to the following formula. Firstly, the ΔS_{sur} is calculated. the throttling expansion is regarded as an adiabatic process, $Q_{sur} \to 0$, so

$$W_L = T_0 \Delta S_{sys} + T_0 \Delta S_{sur} = T_0 \left(\Delta S_{sys} + \Delta S_{sur} \right) = T_0 \Delta S_t$$

$$\Delta S_{\text{sur}} = \frac{Q_{\text{sur}}}{T_{\text{sur}}} = 0$$

Then calculate ΔS_{sys} of the system. According to the temperature-entropy diagram of air

$p_1 = 9.12$ MPa, $T_1 = 300$ K, $H_1 = 13012$ J/mol, $S_1 = 87.03$ J/(mol · K)

It is obtained from the intersection of the isoenthalpy line of H_1 and the isobaric line of p_2.

$T_2 = 280$ K (Temperature after throttling expansion)

$S_2 = 118.41$ J/(mol/K)

$\Delta S_{\text{sys}} = S_2 - S_1 = 31.38$ J/mol

Work loss during throttling expansion.

$$W_L = T_0 \Delta S_t = T_0 \left(\Delta S_{\text{sys}} + \Delta S_{\text{sur}} \right) = \{(273.15 + 25) \times [(118.41 - 87.03) + 0]\} \text{ J/mol} = 9351.2 \text{ J/mol}$$

(2) For adiabatic expansion of external work.

If the expansion process is reversible, the intersection of isentropic line at compressor outlet State 1 and p_2 isobaric line is the corresponding enthalpy value.

$H'_{2S} = 7614.88$ J/mol, $T'_{2S} = 98$ K (Temperature after reversible adiabatic expansion)

Reversible adiabatic expansion work

$$W_R = \Delta H = H'_{2S} - H_1 = (7614.88 - 13012) \text{ J/mol} = -5397.12 \text{ J/mol}$$

According to the requirements of the problem, the isentropic efficiency of the process is 0.8. It shows that this process is an irreversible adiabatic expansion process, so

$$\eta_S = \frac{W_S}{W_R} = \frac{H_1 - H_2}{H_1 - H'_{2S}} = \frac{13012 \text{ J/mol} - H_2}{13012 \text{ J/mol} - 7614.88 \text{ J/mol}} = 0.8$$

$H_2 = 8694.3$ J/mol

According to H_2 and p_2 values on the temperature-entropy diagram of air, $T_2 = 133$ K (Temperature after adiabatic expansion for external work)

Actual work done by expander

$$W_S = \eta_S W_R = [0.8 \times (-5397.12)] \text{ J/mol} = 4317.7 \text{ J/mol}$$

Loss work of adiabatic expansion

$$W_L = W_R - W_S = (5397.12 - 4317.7) \text{ J/mol} = 1079.42 \text{ J/mol}$$

Process	T_2/K	ΔT	Work duty/(J/mol)	Lost work/(J/mol)
Throttling expansion	280	−20	0	9355.947
Adiabatic expansion for external work	133	−167	4317.7	1079.42

Similarly, this example can be simulated in Aspen Plus.

The operation steps are shown in the following QR code.

SIMULATION CASE

CASE 1

A simple process for the liquefaction of propane. This process starts with vapor propane at ambient conditions (298 K and 1 bar), which is compressed to 15 bar, cooled back down to 298 K, expanded through an adiabatic (i.e., Joule-Thomson) valve to 1 bar, and then the resulting gaseous and liquid streams are separated.

SOLUTION

Aspen simulation is used for calculation. The specific operation steps are shown in the following QR code.

CASE 2

For a steam power cycle device, the boiler pressure is 4 MPa, the working pressure of the condenser is 0.004 MPa, and the temperature of superheated steam entering the turbine is 500 ℃. Assuming an ideal Rankine cycle, the thermal efficiency and steam consumption of the cycle are calculated.

SOLUTION

In this example, the boiler and condenser are simulated by Heater module in Aspen Plus, the turbine is simulated by Compr module, and the working fluid pump is simulated by Pump module. The calculation process using Aspen-Plus is shown in the following QR code.

EXERCISES

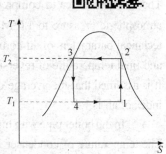

1. The power of a heat pump working according to the reverse Carnot Cycle is 10 kW, the ambient temperature is $-13\,℃$, and the heating temperature required by the user is 95 ℃. Ask for heat supply.

2. The air state at the outlet of a compressor in a factory is $p_1 = 1.01\,\text{MPa}(10\,\text{atm})$, $T_1 = 325\,\text{K}$. Now it needs to expand to $p_2 = 0.203\,\text{MPa}(2\,\text{atm})$. Calculate the lost work of throttling expansion. The ambient temperature is 25 ℃, when $p_1 = 1.01\,\text{MPa}(10\,\text{atm})$, $T_1 = 325\,\text{K}$, $H_1 = 535.55\,\text{kJ/kg}$, $S_1 = 3.226\,\text{kJ/(kg·K)}$.

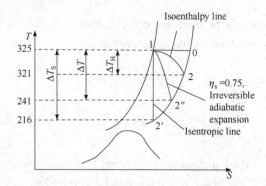

3. The cycle of a nuclear submarine with steam power cycle is shown in the figure. The boiler sucks heat Q from the nuclear reactor with a temperature of 400 ℃ to generate superheated steam with a pressure of 7 MPa and a temperature of 360 ℃ (Point 1). The superheated steam is discharged under the pressure of 0.008 MPa after expansion and work of the steam turbine (Point 2). The exhaust gas releases heat at constant pressure to the ambient temperature $T_2 = 20℃$ in the condenser and becomes saturated water (Point 3), It is then pumped back to the boiler (Point 4) to complete the cycle, and the rated power of the steam turbine is known to be 15×10^4 kW, the steam turbine makes irreversible adiabatic expansion, and its equivalent efficiency is 0.75, while the water pump can be considered as reversible absolute compression, and the solution is:

(1) Mass flow of steam in this power cycle.
(2) Humidity of exhaust gas at turbine outlet.

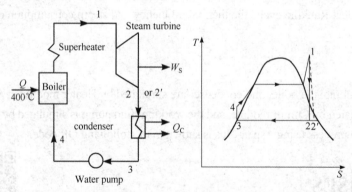

4. A factory needs at least 100 m³ of 1.0 MPa compressed air for production every day. Therefore, in order to compress the air at room temperature (taking 20 ℃) from 0.1 MPa of atmospheric pressure to 1.0 MPa, technicians are required to select and determine the required technical parameters of air compressor through calculation, and calculate the power consumption and final temperature of reversible isothermal compression and reversible adiabatic compression. It is assumed that the average compression factor at the inlet and outlet is 1.05 and the adiabatic index of air is 1.4.

5. In the boiler pipe with high pressure up to 6180 kPa and temperature does not exceed 400 ℃. The temperature of condenser is operated under 65.6 ℃. If the pressure of the boiler is raised to 6180 kPa and reheat is added to make it do single-phase expansion, ask:

(a) How many reheat cycles are required?
(b) What is the maximum temperature in the final superheater?

(c) What is the overall cycle efficiency (Note: pumps and turbines can be assumed to be adiabatic and reversible)?

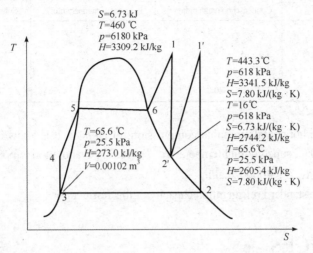

6. For the refrigeration device with R-22 as refrigerant, the working conditions of the cycle are as follows: the condensation temperature is 20 ℃, the undercooling degree $\Delta T = 5$ ℃, the evaporation temperature is -20 ℃, and the dry saturated steam enters the compressor. Try to find:

(1) The unit refrigerating capacity of this cycle.

(2) The power consumption and refrigeration efficiency coefficient per kilogram of refrigerant, and compare them with those without subcooling (other working conditions are the same). In p-H figure: $H_1 = 397.7$ kJ/kg, $H_2 = 433.1$ kJ/kg, $H_4 = 223.8$ kJ/kg, $H_4' = 218.7$ kJ/kg.

7. A steam compression refrigeration cycle is shown in figure. Its evaporation temperature is -20 ℃, condensation temperature is 20 ℃, the original working medium is R12. Now, protect the ozone layer, the replacement R-134a as the working medium. Try to calculate the refrigeration efficiency coefficient of two kinds of working medium. ($H_1 = 134.75$ kJ/kg, $H_2 = 140.3$ kJ/kg, $H_5 = H_4 = 104.6$ kJ/kg, $H_1' = 387$ kJ/kg, $H_2' = 417$ kJ/kg, $H_5' = H_4' = 230$ kJ/kg)

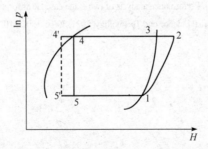

8. A chemical plant has a steam compression refrigeration unit, which uses ammonia as the refrigerant, with refrigeration capacity of 10^5 kJ/h, evaporation temperature of -15 ℃, condensation temperature of 30 ℃. A compressor is set to do reversible adiabatic compression. T-S diagram and $\ln p$-H diagram are as follows, (when the evaporation temperature is -15 ℃, $H_1 = 1664$ kJ/kg, $S_1 = 9.021$ kJ/(kg·K), $V_1 = 0.508$ m³/kg, $H_2 = 1800$ kJ/kg, $H_5 = H_4 = 566.93$ kJ/kg, $H_5^{sv} = 1664$ kJ/kg, $H_5^{sl} = 349.89$ kJ/kg)

9. Set the required refrigeration capacity as $Q_0 = 40000$ kcal/h, try to find G, ε, N_T. If

known:

Refrigerant	Condensation temperature/℃	Evaporation temperature/℃	Subcooling temperature/℃
Ammonia	30	−15	25
Ammonia	30	−35	25
Ammonia	30	−35	non-over cold
Freon -12	30	−15	25

10. There is a horizontal ammonia compression refrigerator. It is known that: condensation temperature: $T=30℃$; supercooling temperature: $T_3=+25℃$; evaporation temperature: $T_0=-30℃$; refrigeration capacity: $Q_0=40000$ kcal/h;

Try to find the standard refrigeration capacity of this refrigerator.

REFERENCES

[1] Askin M, Salti M, Aydogdu O. Polytropic Carnot heat engine[J]. Modern Physics Letters A, 2019, 34(24):1009.

[2] Chen W J. Research on the principle of power MEMS quasi gas power cycle engine[D]. Chongqing: Chongqing University, 2008.

[3] Dickerson R H, Mottmann J. The Stirling cycle and Carnot's theorem[J]. European Journal of Physics, 2019, 40(6):065103.

[4] Garcia R F. Approaching an efficient and feasible Carnot engine by Carnot cycle analysis[J]. Journal of Energy Research and Reviews, 2021: 27-45.

[5] Luan B Q, Wang W, Dong Z Z, et al. Feasibility analysis of waste heat recovery technology of diesel locomotive diesel engine based on organic Rankine cycle[J]. Railway Locomotives and Motor Cars, 2021(11): 16-20.

[6] Luo W H, Chen W, Jiang A G, et al. Comparative analysis of thermodynamic performance of organic Rankine cycle system for recovering waste heat of marine diesel engine[J]. China Mechanical Engineering. 2022: 1-7.

[7] Oh J, Noh K, Lee C A. Theoretical study on the thermodynamic cycle of concept engine with Miller cycle[J]. Processes, 2021:9.

[8] Zhao J Y, Zhang J J, Sun Y K, et al. Performance analysis of two-stage cascade absorption refrigeration cycle driven by waste heat of fishing boat engine[J]. Thermal Science and Technology, 2021, 20(05): 502-510.

Chapter 8

Chemical Reaction Equilibrium

Learning Objectives

- Mole balance is applied to chemical reaction system.
- The equilibrium constant of the reaction is defined and quantified as a function of temperature.
- The reaction degree is determined by combining the mole balance and the equilibrium principle.
- Model by means of chemical reaction balance fugacity model is suitable for physical situation. It explains the pressure and the phase behavior of ideal and actual solutions.
- Equilibrium model is applied to the simultaneous multiple reaction system.

Chemical Reaction Equilibrium

Through chemical reactions, converting cheap and readily available raw materials to more valuable products, is an important job in chemical manufacturing. The chemical engineer must have a clear understanding of whether chemical reactions can proceed in the direction of product, the limits to which they may proceed, and the conditions that may affect the limits of the reaction, under conditions of given temperature, pressure, and composition. Because these problems are the process of design, the conditions of production and economic accounting.

The basic chemical reaction is connected to many chemicals, so we can't model them in a way that doesn't work without the "thermodynamics of mixtures" foundation.

Using the reaction as an example:

$$A + B \longleftrightarrow P$$

What happens when pure A and pure B mix? If the system is allowed to achieve equilibrium, we can use the balanced theory to determine the number of the existence of A, B and P. The answer highly depends on temperature and pressure.

The rate of the reaction is also an important factor in design and analyzing the course of chemical processes. Some rapid chemical reactions reach to balance, and they can be able to model like instant reactions. There are very slow chemical reactions, and they can be simulated as a reaction that doesn't happen. Although the product is easy to obtain in the balance process. Rate, like equilibrium, can be highly dependent upon temperature and pressure. The goal of maximizing rate is sometimes competing with and contradictory to the goal of maximizing the extent of the reaction that is reached at equilibrium. However, the rate of reaction is not something we can determine from thermodynamics. Therefore, our focus in this chapter is to determine the composition of reaction mixtures at equilibrium, but we recognize that there is more to the story in modeling and analyzing chemical reactions. Typical chemical engineering courses include a separate course on reaction kinetics and chemical reactor design.

8.1 Motivational Example: Propylene from Propane

Worldwide, the annual output of compound propylene (CH$_3$CH=CH$_2$) is more than 100 million tons, which is the highest yield of organic chemical products in the world and one of the top five chemical products in the world. Propylene is not widely used as an "end product". It is originally an intermediate. It is produced and then consumed in the synthesis of other chemicals (Afeefy et al., 2012). Most directly, it is polymerized into different types of polypropylenes. This section examines the process of producing propylene by vapor phase chemical reaction:

$$C_3H_8 \longleftrightarrow C_3H_6 + H_2$$

Another question: if 1 mol propane, 0.5 mol propylene and 0.5 mol hydrogen are placed in an airtight container, does the reaction proceed forward or backward? At that time, we simply said that the answer depends on T and p. Quantitative solutions are not yet on the cards. Since then, we have learned two basic concepts that can be applied.

- In Chapter 3, we know that what the spontaneous process does is to lower the Gibbs free energy (G) of the system, and when the Gibbs free energy of the system is minimized, the system reaches equilibrium.
- In Chapter 3, We define the chemical potential μ of a pure compound as the partial molar Gibbs free energy of the compound. Then, we learned to quantify the chemical potentials of the components of a mixture, both ideal and real.

EXAMPLE 8-1

A container that initially contains 1 mol propane; 0.5 mol propylene and 0.5 mol hydrogen. The container is kept at a constant T=258 K and p=1×10^5 Pa, and the following reversible reactions occur:

$$C_3H_8(g) \longleftrightarrow C_3H_6(g) + H_2(g)$$

If the reaction continues to equilibrium, what is the mole fraction of these three gases at the end of the reaction?

SOLUTION

Step 1 Write and simplify mole balances for each compound

Chapter 8 Chemical Reaction Equilibrium

We need to know the amount of material remaining in the reactor at the end of the reaction—the final values of $y_{C_3H_8}$, $y_{C_3H_6}$ and y_{H_2}. It's necessary to write and simplify the mole balance for each compound.

In this case, there's no information on the speed of progress, so we're going to use an approach that doesn't take time into account. The equation that shows mass balance independent of time is as follows(The intial and final state are shown in Table 8-1):

$$M_{\text{final}} - M_{\text{initial}} = \sum_{j=1}^{j=J} m_{j,\text{in}} - \sum_{k=1}^{k=K} m_{k,\text{out}} \tag{8-1}$$

Table 8-1 The system is closed and the process is isothermal and isobaric

	Initial state	Final state
T/K	298.15	298.15
p/Pa	1×10^5	1×10^5
$N_{C_3H_8}$/mol	1	
$N_{C_3H_6}$/mol	0.5	
N_{H_2}/mol	0.5	

This expression is the total mass of the system. We assume that system mass is conserved. However, the number of moles of a compound is not a conserved quantity in a chemical reaction. Therefore, a mole equilibrium expression similar to the above must also account for compounds produced or consumed by chemical reactions:

$$N_{\text{final}} - N_{\text{initial}} = \sum_{j=1}^{j=J} n_{j,\text{in}} - \sum_{k=1}^{k=K} n_{k,\text{out}} + N_{\text{gen}} \tag{8-2}$$

In Eq. (8-2), the N_{gen} can either positive or negative. Negative suggests that the compounds are consumed by chemical responses.

In the current example, the reaction occurred in a closed-circuit system——no physical entry or exit. So, the mole balance of propylene is shown below,

$$N_{C_3H_6,\text{final}} - N_{C_3H_6,\text{initial}} = N_{C_3H_6,\text{gen}} \tag{8-3}$$

Because our goal is to find out the final state of the system, so it'll be more convenient.

$$N_{C_3H_6,\text{final}} = N_{C_3H_6,\text{initial}} + N_{C_3H_6,\text{gen}} \tag{8-4}$$

An analogous expression can be written for hydrogen and for propane:

$$N_{H_2,\text{final}} = N_{H_2,\text{initial}} + N_{H_2,\text{gen}} \tag{8-5}$$

$$N_{C_3H_8,\text{final}} = N_{C_3H_8,\text{initial}} + N_{C_3H_8,\text{gen}} \tag{8-6}$$

At first glance it looks like we don't really have we've written three equations ($N_{C_3H_8,\text{gen}}$, $N_{C_3H_6,\text{gen}}$ and $N_{H_2,\text{gen}}$), but we've introduced three new unknown problems, so we're obviously not close to solving $N_{C_3H_8,\text{final}}$, $N_{C_3H_6,\text{final}}$ and $N_{H_2,\text{final}}$. However, we can generate a propylene molecule through a chemical reaction between stoichiometry $N_{C_3H_8,gen}$, $N_{C_3H_6,gen}$ and $N_{H_2,gen}$, where hydrogen is also generated and propane is consumed.

$$N_{C_3H_6,gen} = N_{H_2,gen} \tag{8-7}$$

$$N_{C_3H_6,gen} = -N_{C_3H_8,gen} \tag{8-8}$$

Thus, Eq. (8-4) to Eq. (8-8) represent five equations in six unknowns. In Step 3, a sixth equation is obtained through the calculation of an equilibrium constant. First, to simply identify the nomenclature of mass balance, we introduce the extent of reaction. The reaction level is formatted by Section 8.2. It refers to the number of moles of the compound produced by the reaction divided by the stoichiometric coefficient of the compound in the reaction:

$$\xi = \frac{N_{C_3H_6,gen}}{+1} = \frac{N_{H_2,gen}}{+1} = \frac{N_{C_3H_8,gen}}{-1} \tag{8-9}$$

The -1 for propane represents the fact that in the reaction as written, propane is consumed, not generated.

What we can use to represent the N_{gen} in the Eq. (8-4) to Eq. (8-6) and insert in the initial value known to the mole number of each compound.

$$N_{C_3H_6,final} = 0.5 + \xi \tag{8-10}$$

$$N_{H_2,final} = 0.5 + \xi \tag{8-11}$$

$$N_{C_3H_8,final} = 1 - \xi \tag{8-12}$$

Therefore, if the extent of reaction j can be determined, the exact contents of the vessel at equilibrium ("final") can be found using Eq. (8-10) to Eq. (8-12).

Step 2 Write expressions for mole fractions of each compound

We're asked to find the mole fraction of each compound. Eq. (8-9) to Eq. (8-11) give the final number of moles for each compound and summing these gives the total number of moles present. Thus,

$$N_{tot,final} = N_{C_3H_6,final} + N_{C_3H_8,final} + N_{H_2,final}$$

$$N_{tot,final} = 2 + \xi \tag{8-13}$$

It means that the mole fraction of each of these things in the final equilibrium state can be expressed as

$$y_{C_3H_6} = \frac{N_{C_3H_6,final}}{N_{tot,final}} = \frac{0.5 + \xi}{2 + \xi} \tag{8-14}$$

$$y_{H_2} = \frac{N_{H_2,final}}{N_{tot,final}} = \frac{0.5 + \xi}{2 + \xi} \tag{8-15}$$

$$y_{C_3H_8} = \frac{N_{C_3H_8,final}}{N_{tot,final}} = \frac{1 - \xi}{2 + \xi} \tag{8-16}$$

Step 3 Evaluate the equilibrium constant for the reaction

Perhaps you have learned in the chemical courses that the balance of the reaction can be defined by the Gibbs free energy, using

$$K_p = \exp\left(-\frac{\Delta \underline{G}_T^\ominus}{RT}\right) \tag{8-17}$$

in which $\Delta \underline{G}_T^\ominus$ is the standard change in the Gibbs free energy of the reaction. For this example,

$$\Delta \underline{G}_T^\ominus = \Delta \underline{G}_{f,\text{products}}^\ominus - \Delta \underline{G}_{f,\text{reactants}}^\ominus \tag{8-18}$$

$$\Delta \underline{G}_T^\ominus = \Delta \underline{G}_{f,\text{propylene}}^\ominus + \Delta \underline{G}_{f,\text{hydrogen}}^\ominus - \Delta \underline{G}_{f,\text{propane}}^\ominus$$

We find the following data is available at $T=298$ K and $p=1\times 10^5$ Pa:

For propane, $\Delta \underline{G}_f^\ominus = -23.4$ kJ/mol

For propylene, $\Delta \underline{G}_f^\ominus = 62.76$ kJ/mol

For hydrogen, $\Delta \underline{G}_f^\ominus = 0$ kJ/mol

Step 4 will make some qualitative observations of the important data so that you can use them to find the value of the equilibrium constant.

Aside——the reactant is more stable than the products

The reaction is written as if propane is being converted into propylene and hydrogen.

However, this is a reversible reaction, where the container initially contains 1 mol "reactants" (propane) and 1 mol "products" (propylene and hydrogen), so the reaction can go in either direction. In Chapter 3 we learned to associate low values of G with stability. The data show that the reactants are more stable than the products, so we now expect the reaction to go in the opposite direction. From the point of view of getting a numerical answer, it doesn't matter, the fact that the opposite reaction is favorable will be reflected in negative values of $N_{C_3H_6,\text{gen}}$ and $N_{H_2,\text{gen}}$.

At equilibrium the container contains pure propane. We learned in Chapter 3 that any spontaneous process goes in the direction of minimum G. Since $\Delta \underline{G}_f^\ominus$ is negative for propane and positive for propylene, when propylene reacts with hydrogen to make propane, the Gibbs free energy of the system goes down. G is minimal when all the propylene is converted to propane.

If the above reasoning is followed, all chemical reactions will react completely. What it does wrong is it ignores the entropy of mixing. The above $\Delta \underline{G}_f^\ominus$ values are the $\Delta \underline{G}_f^\ominus$ values of pure propane, propylene and hydrogen at $T=258$ K and $p=1\times 10^5$ Pa. While the $\Delta \underline{G}_f^\ominus$ of pure propane is lower than that of pure propylene or pure hydrogen, the G of the mixture of propane, propylene and hydrogen is lower than that of pure propane of the same mass due to the presence of mixing entropy. The equilibrium constant represents the entropy of mixing.

Step 4 Compute numerical value of equilibrium constant

Using the data listed in Step 3, we can calculate the standard the Gibbs free energy for the reaction

$$\Delta \underline{G}_R^\ominus = \Delta \underline{G}_{f,\text{propylene}}^\ominus + \Delta \underline{G}_{f,\text{hydrogen}}^\ominus - \Delta \underline{G}_{f,\text{propane}}^\ominus$$

$$\Delta \underline{G}_R^\ominus = 62.76 \text{ kJ/mol} + 0 - (-23.4 \text{ kJ/mol}) = 86.16 \text{ kJ/mol} \tag{8-19}$$

$$\Delta \underline{G}_R^\ominus = 86160 \text{ kJ/mol}$$

By inserting this value of $\Delta \underline{G}_f^\ominus$ into Eq. (8-16), we can change the equilibrium constant quantity to

$$K_p = \exp\left(-\frac{\Delta \underline{G}_T^\ominus}{RT}\right) = \exp\left[-\frac{86160 \text{ J/mol}}{8.314 \text{ J/(mol·K)} \times 298.15 \text{ K}}\right] = 8.03 \times 10^{-16} \tag{8-20}$$

Step 5 Relate equilibrium constant to mole fraction of each compound

For a gas phase reaction, the equilibrium constant can be related to the partial pressures of the reactants and products as

$$K_p = \frac{p_{C_3H_6} p_{H_2}}{p_{C_3H_8}} \tag{8-21}$$

This is only valid if you have an ideal gas. In Section 8.3.1, we will explain the theoretical basis of Eq. (8-21) and derive a general chemical reaction equilibrium model. If propane, hydrogen and propylene are considered ideal gas mixtures during the modeling, the model is strictly simplified to Eq. (8-21). In this case, the pressure is 1×10^5 Pa, so it's reasonable to assume that the ideal gas behavior.

Eq. (8-21) has three unknown partial pressures. We know that the partial pressure of a gas is equal to the mole fraction times the total pressure, so

$$K_p = \frac{(y_{C_3H_6} p)(y_{H_2} p)}{(y_{C_3H_8} p)} \tag{8-22}$$

Pay attention to units when using equilibrium constants. In Eq. (8-20), K_p seems to be dimensionless, but in Eq. (8-22), the right-hand side seems to be dimensional——it has units of pressure. This obvious difference is explained in Example 8-4, in which Eq. (8-21) and Eq. (8-22) are valid only if the pressure is expressed in terms of Pa. Here $p=1\times 10^5$ Pa, so Eq. (8-22):

$$K_p = \frac{(y_{C_3H_6} p)(y_{H_2} p)}{(y_{C_3H_8} p)} = \frac{y_{C_3H_6} y_{H_2}}{y_{C_3H_8}} \tag{8-23}$$

Step 6 Solve for Eq. (8-23)

We can introduce the expressions for mol fraction into Eq. (8-23):

$$K_p = \frac{y_{C_3H_6} y_{H_2}}{y_{C_3H_8}}$$

$$K_p = \frac{\dfrac{0.5+\xi}{2+\xi} \times \dfrac{0.5+\xi}{2+\xi}}{\dfrac{1-\xi}{2+\xi}} \tag{8-24}$$

which simplifies to

$$K_p = \frac{0.5+\xi}{2+\xi} \times \frac{0.5+\xi}{1-\xi} \qquad (8\text{-}25)$$

In Step 3, the value of K_p is known, so Eq. (8-25) is one variable and can be solved numerically or quadratic. Thus,

$$\xi = -0.5 \text{ mol}$$

Notice that $(0.5+\xi)$ appears on the right-hand side of Eq. (8-25), so if $K_p = 0$, then $\xi = -0.5$ mol. Although K_p is not equal to zero, it is very small, so we calculate j to be different from –0.5 only if we carry 10 places. So $y_{C_3H_8} = 1$ and $y_{C_3H_6} = y_{H_2} = 0$.

In some applications, it is important to accurately estimate the value, even if the mole fraction of the compound is very small, especially when highly toxic or other dangerous impurities are present. In this case, the simplest way to accurately estimate the equilibrium mole fractions of hydrogen and propylene is to return to the Eq. (8-23). Recognize $y_{C_3H_8} = 1$, Eq. (8-23).

$$K_p = \frac{y_{C_3H_6} y_{H_2}}{y_{C_3H_8}}$$

$$8.03 \times 10^{-16} = \frac{y_{C_3H_6} y_{H_2}}{1} \qquad (8\text{-}26)$$

Based on the Eq. (8-14) and Eq. (8-15), we learn that the mole fraction of propylene and hydrogen must be equal not only at equilibrium, but also at any point in the process. Thus,

$$8.03 \times 10^{-16} = \frac{y_{C_3H_6} y_{C_3H_6}}{1} \qquad (8\text{-}27)$$

and $y_{C_3H_6} = y_{H_2} = 2.83 \times 10^{-8}$.

However, this last step is not necessary for many applications. If the goal is to produce propylene, the reaction will not be successful at this temperature and pressure.

Problems such as Example 8-1 are basic in an introductory chemistry course, but chemical engineering requires more in-depth knowledge of chemical reaction equilibria. Eq. (8-21) only work if we model the system as an ideal mixture of real gases. Eq. (8-21) is more accurate at 1×10^5 Pa, but there is a large error at higher pressure, because higher pressure will lead to wrong results. Other issues and points raised in Example 8-1 include: Does the equilibrium constant have units? Note the apparent size of the pressure to the right of Eq. (8-21) and Eq. (8-22), whereas Eq. (8-20) indicates that the equilibrium constant is dimensionless.

- Computing equilibrium constant need to understand the Gibbs free energy of formation, which is normally T=258 K and p=1×10^5 Pa. How can we explain the reactions that occur at other temperatures and pressures?
- Under the condition of the ideal gas approximation is not reality, how do we establish chemical reaction equilibrium model?

This chapter will discuss several of examples, some of which involve multiple reactions occurring simultaneously. From Section 8.3 we will study the equilibrium of chemical reactions. First, we should realize that it would be impossible for us to solve Example 8-1 because we do not have an exact mole balance for each reactant and product. We study reaction stoichiometry in Section 8.2.

8.2 Chemical Reaction Stoichiometry

Although the focus of this chapter is on the balance of chemical reactions, we should first write down the mole balance of each substance and explain its reaction stoichiometry (Chase et al., 1998). Because the propane example shows that there is no need to apply the reaction equilibrium model in a useful way. In the case of propane, we examined the chemical reaction:

$$C_3H_8(g) \longleftrightarrow C_3H_6(g) + H_2(g)$$

This example shows how to correlate the "generation" terms in the mole balance of each species (see Step 2 in Example 8-1). But because the stoichiometric ratio of the reaction is 1 : 1 : 1, the steps are simplified.

When there is no nuclear fission, the elements in the chemical reaction are conserved, and mass is conserved throughout the system.

The chemical equation can be expressed as follows:

$$|v_1|A_1 + |v_2|A_2 + \cdots \rightleftharpoons |v_3|A_3 + |v_4|A_4 + \cdots$$

where $|v_i|$ = Stoichiometric coefficient

A_i = Chemical formula

Regardless of the true direction of the reaction, the components on the left side of the reaction formula are usually called reactants, and the components on the right side are called products. The sign of $|v_i|$ is defined as: the $|v_i|$ of the reactant is taken as a negative value, and the $|v_i|$ of the product is taken as a positive value. React to

$$2SO_2(g) + O_2(g) \longleftrightarrow 2SO_3(g)$$

in which measurement coefficient is

$$v_{SO_2} = -2 \quad v_{O_2} = -1 \quad v_{SO_3} = 2$$

When the reaction is carried out, the amount of substance involved in the reaction varies strictly in proportion to the measured coefficients.

$$\frac{N_{SO_3,gen}}{2} = \frac{N_{O_2,gen}}{-1} \tag{8-28}$$

That is to say, the ratio of the amount of the reacted substance to the measurement coefficient of the various substances participating in the reaction is equal. Let this ratio be extent of reaction (ξ) and its definition is expressed by the following Eq. (8-29).

$$\xi = \frac{N_{SO_2,gen}}{-2} = \frac{N_{O_2,gen}}{-1} = \frac{N_{SO_3,gen}}{2} \tag{8-29}$$

It is stipulated that stoichiometric coefficients are positive for products and negative for reactants.

Step 1 of Example 8-1 illustrates how ξ is a useful parameter. We can express all unknown mole fractions as a function of ξ, so that when we apply an equilibrium constant, ξ is the only unknown in the equation. Section 8.3 details the application of equilibrium criteria in chemical reactions. In this section, we demonstrate the application of the parameter ξ in the time-independent (Section 8.2.1) and time-dependent (Section 8.2.2) mole balance.

8.2.1 Extent of Reaction and Time-Independent Mole Balances

In the previous chapters, we have proved that the equation of equilibrium for matter, energy or entropy can be written in a form related to time or a form independent of time. The form of time-independent mole equilibrium is shown in Eq. (8-2):

$$N_{\text{final}} - N_{\text{initial}} = \sum_{j=1}^{j=J} n_{j,\text{in}} - \sum_{k=1}^{k=K} n_{k,\text{out}} + N_{\text{gen}} \qquad (8\text{-}30)$$

In effect, reaction stoichiometry is used to account for inter-dependencies in the N_{gen} terms of the mole balances for reactants and products. The definition of extent of reaction generalizes as given here.

For a reaction of the form

$$v_A A + v_B B \longleftrightarrow v_C C + v_D D$$

the extent of reaction ξ is

$$\xi = \frac{N_{A,\text{gen}}}{v_A} = \frac{N_{B,\text{gen}}}{v_B} = \frac{N_{C,\text{gen}}}{v_C} = \frac{N_{D,\text{gen}}}{v_D} \qquad (8\text{-}31)$$

where $v_C > 0$, $v_D > 0$, $v_A < 0$, $v_B < 0$.

The above formula can also be applied to multiple reactions. In a system with multiple Reactions, ξ_1 would quantify the moles of each compound generated by Reaction 1 specifically, ξ_2 would refer to the moles generated by Reaction 2 and so on. The usage is explained in detail in Example 8-2.

EXAMPL 8-2

In the catalytic combustion of carbon, the following sequence of two gas reactions can occur:

R1 $\qquad\qquad 2C + O_2 \longleftrightarrow 2CO$

R2 $\qquad\qquad 2CO + O_2 \longleftrightarrow 2CO_2$

In a closed container, it initially contains 4 mol carbon and 8 mol oxygen (Table 8-2). Assume that only these two chemical reactions occur in this container. According to the extents of the two reactions, write the expression for the mole fraction of each substance, ξ_1 and ξ_2.

Table 8-2 Ammonia combustion process

	Initial state	Final state
N_C/mol	4	?
N_{O_2}/mol	8	?
N_{CO}/mol	0	?
N_{CO_2}/mol	0	?

SOLUTION

Step 1 Molar balance of each species

Eq. (8-2) gives the most general form of mole balance as:

$$N_{final} - N_{initial} = \sum_{j=1}^{j=J} n_{j,in} - \sum_{k=1}^{k=K} n_{k,out} + N_{gen} \tag{8-32}$$

As in Example 8-1, this is a closed system. In this case, the only way to change the number of moles of the compound is to produce (or consumption) through a chemical reaction.

Then the mole balance of each individual species simplifies to

$$N_{final} = N_{initial} + N_{gen} \tag{8-33}$$

The five compounds in the reaction can all be written in this form of mole balance (special, $N_{initial} = 0$). The generation term is the main factor that distinguishes the five mole balances, and it reflects how they participate in the reaction.

Step 2 Apply definition of extent of reaction

Eq. (8-31) gives the definition of the extent of reaction, ξ, as the number of moles of a compound generated by a reaction divided by the stoichiometric coefficient of that compound in the reaction. This definition is applicable, regardless of how many chemical reactions are occurring in the system. For example, in R1, the stoichiometric coefficient for carbon is –2 (the negative value signifying it is a reactant) so applying Eq. (8-31), the extent of reaction R1 is

$$\xi_1 = \frac{N_{C,gen}}{-2} \tag{8-34}$$

which can be rearranged as

$$N_{C,gen} = -2\xi_1 \tag{8-35}$$

Similarly, CO is a product in R1, and the stoichiometric coefficient is +2, so

$$\xi_1 = \frac{N_{CO,gen,1}}{2}$$

$$N_{CO,gen,1} = 2\xi_1 \tag{8-36}$$

But this represents only the generation of CO by R1. Unlike ammonia, CO also participates in reaction R2 as a reactant, and its stoichiometric coefficient of –2:

$$N_{CO,gen,2} = -2\xi_2 \tag{8-37}$$

Thus, the total number of moles of CO generated in the process is given by:

$$N_{CO,gen} = 2\xi_1 - 2\xi_2 \tag{8-38}$$

We can conduct a similar accounting for each of the five compounds. It is convenient to summarize the information in table form, as shown in Table 8-3.

Table 8-3 Mole balances and reaction stoichiometry

	C	O_2	CO	CO_2	Total Moles
$N_{initial}$ /mol	4	8	0	0	12
N_{gen} (R1)	$-2\xi_1$	$-\xi_1$	$2\xi_1$		ξ_1
N_{gen} (R2)		$-\xi_2$	$-2\xi_2$	$2\xi_2$	$-\xi_2$
N_{final}	$4-2\xi_1$	$8-\xi_1-\xi_2$	$2\xi_1-2\xi_2$	$2\xi_2$	$12+\xi_1-\xi_2$
y	$\dfrac{4-2\xi_1}{12+\xi_1-\xi_2}$	$\dfrac{8-\xi_1-\xi_2}{12+\xi_1-\xi_2}$	$\dfrac{2\xi_1-2\xi_2}{12+\xi_1-\xi_2}$	$\dfrac{2\xi_2}{12+\xi_1-\xi_2}$	

Step 3 Write expressions for mole fractions of each compound

We can find the total number of moles of the system by summing the mole balances for the five individual compounds. Recall that for a compound (C) the mole fraction y_C is by definition N_C/N_{total}. The mole fractions of each compound are also shown in Table 8-3.

Constructing a stoichiometric table analogous to that shown in Table 8-3 is a key step in the solution of typical problems involving chemical reactions. Example 8-2 investigated a closed system—a simple batch reactor. While the batch reactor is a very practical system to study, large-scale chemical processes are most often designed to operate continuously. Consequently, the next section applies the principles of stoichiometry to an open-system reactor operating under steady state conditions.

8.2.2 Extent of Reaction and Time-Dependent Material Balances

Time-dependent material, energy, and entropy balances were introduced in previous chapters, and applied to problems in which either known information or the desired results are expressed as rates. A time-dependent mole balance is of the following form:

$$\frac{dN}{dt} = \sum_{k=1}^{k=K} n_{k,in} - \sum_{k=1}^{k=K} n_{k,out} + N_{gen} \tag{8-39}$$

with N_{gen} representing the rate at which the compound is generated or consumed by a reaction. For a reaction of the form

$$\nu_A A + \nu_B B \longleftrightarrow \nu_C C + \nu_D D$$

We can modify our definition of the reaction stage to relate the rates of the individual compounds produced by the reaction. As a result,

$$\dot{\xi} = \frac{N_{A,gen}}{\nu_A} = \frac{N_{B,gen}}{\nu_B} = \frac{N_{C,gen}}{\nu_C} = \frac{N_{D,gen}}{\nu_D} \qquad (8\text{-}40)$$

8.3 The Equilibrium Criterion Applied to a Chemical Reaction

In this section, we will explore the theoretical basis behind the equilibrium constants of chemical reactions. Most readers are familiar with equilibrium constants from introductory chemistry courses (Domalski et al., 2012). For example, the form of chemical reaction mentioned in the freshman chemistry textbook chemistry principles is as follows:

$$\nu_A A + \nu_B B \longleftrightarrow \nu_C C + \nu_D D$$

The extent of reaction ξ is given by

$$\xi = \frac{n_A^{gen}}{\nu_A} = \frac{n_B^{gen}}{\nu_B} = \frac{n_C^{gen}}{\nu_C} = \frac{n_D^{gen}}{\nu_D}$$

The equilibrium state is described by an equation of the form:

$$K = \frac{[C]^{\nu_C}[D]^{\nu_D}}{[A]^{\nu_A}[B]^{\nu_B}} \qquad (8\text{-}41)$$

[C], [D], [A], and [B] represent appropriate measures of concentration, such as partial pressures for a gas phase process or molarities for a reaction occurring in solution. Eq. (8-41) is useful for solving problems like the one posed in Example 8-1, but is only valid for ideal gases and ideal liquid solutions. While researcher acknowledges this limitation, quantitative accounting for departures from ideal solution behavior is beyond the scope of a typical freshman chemistry course. However, in previously chapters, we have learned to model deviations from the ideal gas and ideal solution models, largely in the context of phase equilibrium. In this section, we will see how to apply the same principles to reaction equilibrium, devise a model for chemical reaction equilibrium, and demonstrate its equivalence to Eq. (8-41) for the ideal solution case.

8.3.1 The Equilibrium Constant

Our discussions of chemical equilibrium began in Chapter 3, when we established that spontaneous processes progress in a direction that decreases the Gibbs free energy and that a system is at equilibrium when the Gibbs free energy reaches a minimum. The change in Gibbs free energy for a system can be written as

$$dG = -SdT + Vdp + \sum_i U_i dN_i \qquad (8\text{-}42)$$

This equation is valid whether changes in the number of moles of a compound (dn_i) changes due to substances entering or leaving the system, or due to chemical reactions occurring within the system.

Our goal is to develop a method for modeling chemical reactions under specific conditions, thus if a chemical reaction takes place in a closed system. As in Example 8-1, the system is at constant temperature and pressure. In this case, Eq. (8-42) is reduced to

$$dG = \sum_i U_i dN_i \qquad (8\text{-}43)$$

Since we are modeling a closed system, changes in the number of moles of a species can only occur by chemical reaction, which means they are related to each other by stoichiometry. While values of dN_i are different in different terms of the summation, we can use Eq. (8-31) to relate the change in moles of each compound to the extent of the reaction:

$$dG = \sum_i U_i \nu_i d\xi \qquad (8\text{-}44)$$

and $d\xi$ can be divided out of the summation as

$$\frac{dG}{d\xi} = \sum_i U_i \nu_i \qquad (8\text{-}45)$$

We know that G is minimized at equilibrium, which means that at equilibrium the derivative of G is zero:

$$\frac{dG}{d\xi} = \sum_i U_i \nu_i = 0 \qquad (8\text{-}46)$$

We will find the equation more convenient to use if we apply the definition of the mixture fugacity of component i:

$$\sum_i U_i \nu_i = \sum_i \nu_i \left[G_i^\ominus + RT \ln\left(\frac{\hat{f}_i}{f_i^\ominus}\right) \right] = 0 \qquad (8\text{-}47)$$

The Gibbs free energy, like enthalpy and internal energy, is measured with respect to the reference state. Here $G_i^\ominus$ is the molar Gibbs free energy of pure compound i at the chosen reference state, and $f_i^\ominus$ is the fugacity of pure compound i at that same reference state. It is instructional to separate the two terms of the summation in Eq. (8-47). Thus, at equilibrium:

$$\sum_i \nu_i G_i^\ominus + \sum_i \nu_i \left[RT \ln\left(\frac{\hat{f}_i}{f_i^\ominus}\right) \right] = 0 \qquad (8\text{-}48)$$

Here, we recall a point that was raised in the motivational example. The first term in Eq. (8-48) represents the Gibbs free energy of pure components——reactants and products. We know that at equilibrium the Gibbs free energy is minimal, so when the Gibbs free energy of the products is lower than the Gibbs free energy of the reactants, does the reaction continue until at least one of the reactants is completely consumed? The answer is "No", because comparing the Gibbs free energy of pure products to the Gibbs free energy of pure reactants omits the phenomenon of entropy of mixing. The fugacity term in Eq. (8-48) incorporates the entropy of mixing, because mixture fugacity is derived from the partial molar Gibbs free energy. Eq. (8-48) can be rearranged as

$$-\sum_i \nu_i \underline{G}_i^\ominus = \sum_i \nu_i \left[RT \ln\left(\frac{\hat{f}_i}{f_i^\ominus}\right) \right] \tag{8-49}$$

First, consider the term on the left-hand side. We will define this summation as the standard Gibbs free energy of reaction $\Delta \underline{G}_T^\ominus$:

$$\Delta \underline{G}_T^\ominus = \sum_i \nu_i \underline{G}_i^\ominus \tag{8-50}$$

The $\underline{G}_i^\ominus$ represents the Gibbs free energy of compound i at a reference state. Conventionally, the reference state used is $\underline{G} = 0$ for elements at $T=298.15$ K and $p=1\times10^5$ Pa. The change in the Gibbs free energy when a compound is formed from elements at $T=278.15$ K and $p=1\times10^5$ Pa is called $\Delta \underline{G}_f^\ominus$.

By introducing the standard Gibbs free energy of the reaction, we can simplify this Eq. (8-49) as

$$-\frac{\Delta \underline{G}_T^\ominus}{RT} = \sum_i \nu_i \ln\left(\frac{\hat{f}_i}{f_i^\ominus}\right) \tag{8-51}$$

Taking the exponential of both sides yields

$$\exp\left(\frac{-\Delta \underline{G}_T^\ominus}{RT}\right) = \prod_i \left(\frac{\hat{f}_i}{f_i^\ominus}\right)^{\nu_i} \tag{8-52}$$

Here we will use Eq. (8-53) to define the equilibrium constant K_T for a chemical reaction:

$$K_T = \exp\left(\frac{-\Delta \underline{G}_T^\ominus}{RT}\right) = \prod_i \left(\frac{\hat{f}_i}{f_i^\ominus}\right)^{\nu_i} \tag{8-53}$$

where T is the temperature.

$\Delta \underline{G}_T^\ominus$ is the change in standard molar Gibbs free energy for the reaction at the temperature T.

K_T is the equilibrium constant of the reaction at the temperature T.

$\hat{f}_i$ is the mixture fugacity of component i.

$f_i^\ominus$ is the fugacity of pure component i in its reference state at temperature T.

ν_i is the stoichiometric coefficient of compound i in the reaction, with the convention that it is negative for reactants and positive for products.

Conceptually, we can apply Eq. (8-53) using any reference state we wish. The crucial point is that, for each compound i, $f_i^\ominus$ on the right-hand side must be computed for the same reference state that is used in determining its standard Gibbs free energy $\underline{G}_i^\ominus$ (Perry et al., 1997).

In practice, it is conventional to define the reference state for gaseous compounds as the ideal gas state at $p=1\times10^5$ Pa. This is convenient for two reasons:
- When the reference state is chosen as an ideal gas state, $f_i^\ominus = p^\ominus = 1\times10^5$ Pa.
- Values of $\underline{G}_i^\ominus$ are often not available at the temperature. Section 8.3.3 illustrates the calculations used in such a case. Selecting the ideal gas state as the reference state allows

us to use published values of the ideal gas heat capacity C_p^* in these calculations.

At the same time, the pure liquid at $p=1\times10^5$ Pa is usually the reference state of the liquid compound.

The simplest case to analyze is one in which the actual reaction conditions are identical to the reference state. The motivational example, Example 8-1 demonstrates the connection between the generalized reaction equilibrium expression derived in this section:

$$K_T = \exp\left(\frac{-\Delta G_T^\ominus}{RT}\right) = \prod_i \left(\frac{\hat{f}_i}{f_i^\ominus}\right)^{\nu_i}$$

And the K_p expression that is likely familiar from introductory chemistry courses:

$$K_p = \frac{p_{C_2H_4} p_{H_2}}{p_{C_2H_6}}$$

8.3.2 Accounting for the Effects of Pressure

The previous section derived Eq. (8-53), the equilibrium criterion for chemical reactions:

$$K_T = \exp\left(\frac{-\Delta G_T^\ominus}{RT}\right) = \prod_i \left(\frac{\hat{f}_i}{f_i^\ominus}\right)^{\nu_i}$$

When the compound is at $T=298.15$ K and $p=1\times10^5$ Pa, the standard the Gibbs free energy of the reaction ($\Delta G_T^\ominus$) of the reaction is calculated from the Gibbs free energy of the individual reactants and products. The ideal gas state at $p=1\times10^5$ Pa is commonly used as a reference state. We can apply Eq. (8-53) at pressures other than $p=1\times10^5$ Pa. We must simply be mindful of the distinction between the actual pressure p and the reference pressure $p^\ominus$, and the mixture fugacity coefficient $\hat{f}_i$ using an approach that is appropriate for the system and state.

When the actual pressure p is not equal to the reference pressure $p^\ominus$, the pressures do not cancel out. The rearrangement combined with the pressure term is of great significance, and the equation is shown below:

$$K_T \left(\frac{p^\ominus}{p}\right)^2 = \frac{(y_{CO})(y_{H_2})^3}{(y_{CH_4})(y_{H_2O})} \tag{8-54}$$

The effect of pressure on equilibrium conversion in this gas phase reaction is apparent in Eq. (8-54), in which the equilibrium constant is multiplied by $(p^\ominus/p)^2$. The equilibrium criterion in Eq. (8-53) is

$$K_T = \prod_i \left(\frac{\hat{f}_i p}{f_i^\ominus}\right)^{\nu_i}$$

and can be re-expressed in terms of the fugacity coefficient as:

285

$$K_T = \prod_i \left(\frac{y_i \hat{\phi}_i p}{f_i^{\ominus}}\right)^{\nu_i} \tag{8-55}$$

If the ideal gas state is used as the reference state, the reference state fugacity is calculated by reference pressure:

$$K_T = \prod_i \left(\frac{y_i \hat{\phi}_i p}{p^{\ominus}}\right)^{\nu_i} \tag{8-56}$$

The pressure terms can be factored out and moved to the left-hand side.

$$K_T \left(\frac{p^{\ominus}}{p}\right)^{\Sigma \nu_i} = \prod_i \left(y_i \hat{\phi}_i\right)^{\nu_i} \tag{8-57}$$

In the reaction, there are more moles of gas on the product side than the reactant side. The sum of the stoichiometric coefficients $\Sigma \nu_i$ is positive. The equilibrium constant K_T is divided by $p^{\Sigma \nu_i}$ meaning that increasing pressure will shift the equilibrium toward the reactants. If there are more moles of reactant than product, then $\Sigma \nu_i$ is negative, meaning increasing pressure will increase equilibrium conversion. As you increase the pressure, the equilibrium moves to the side of the reaction with fewer moles of gas. Throughout the chapter, we have used the ideal gas state at $p^{\ominus} = 1 \times 10^5$ Pa as a reference state in modeling gas phase reaction equilibrium. It illustrates the use of pure liquid at $p=1\times 10^5$ Pa as a reference state when modeling liquid phase reactions. The reaction equilibrium criterion, as it applies to liquids:

$$K_T = \prod_i \left(\frac{\gamma_i x_i f_i}{f_i^{\ominus}}\right)^{\nu_i} \tag{8-58}$$

We learned that the "Poynting correction factor" for a pure compound is often negligible. It further demonstrates that even if the correction factors of the reactants and the products are significant, they can largely cancel each other out. Consequently, the effect of pressure on reaction equilibrium in the liquid phase is often assumed negligible, in which case:

$$K_T = \prod_i \left(\gamma_i x_i\right)^{\nu_i} \tag{8-59}$$

8.3.3 Accounting for Changes in Temperature

The equilibrium constant, as defined in Eq. (8-53), is independent of pressure but is dependent upon temperature. For modeling chemical reaction equilibrium, the simplest case occurs when a measurement of the equilibrium constant at the temperature of interest is available. If $\underline{G}^{\ominus}$ for each compound is known at the temperature of interest, then $\Delta \underline{G}_T^{\ominus}$ and the equilibrium constant K_T can be computed directly using Eq. (8-53). Often, however, values of $\underline{G}^{\ominus}$ are only available at a specific reference temperature (conventionally, 298.15 K) and one must estimate $\Delta \underline{G}_T$ at a different temperature (Poling et al., 2001).

EXAMPL 8-3

The compound chloromethane can be synthesized by the chemical reaction:

$$CH_4 + Cl_2 \longleftrightarrow CH_3Cl + HCl$$

The reaction takes place as a free radical mechanism and usually requires high temperature. The reaction is carried out in a closed vessel at $p = 1 \times 10^5$ Pa and constant temperature. The reactor initially contained two moles of methane and one mole of chlorine. Assuming that this is the only chemical reaction that occurs, determine the degree of reaction at equilibrium temperature:

A. 300 K
B. 1000 K

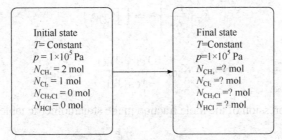

Fig.8-1 Synthesis of chloromethane modeled

SOLUTION

Step 1 Construct stoichiometric table

The reactants are present in stoichiometric ratios. Although it is common for raw materials to enter the true chemical process in stoichiometric ratios, in some cases there are compelling reasons not to do so. The theory we have developed in this chapter is by no means dependent on stoichiometric ratios. Here, methane is initially twice as much as chlorine, but they're being consumed at the same rate. The complete stoichiometry is shown in Table 8-4.

Table 8-4 Stoichiometric table

	CH_4	Cl_2	CH_3Cl	HCl	Total Moles
$N_{initial}$/mol	2	1	0	0	3
N_{gen}	$-\xi$	$-\xi$	$+\xi$	$+\xi$	0
N_{final}	$2-\xi$	$1-\xi$	ξ	ξ	3
y	$\dfrac{2-\xi}{3}$	$\dfrac{1-\xi}{3}$	$\dfrac{\xi}{3}$	$\dfrac{\xi}{3}$	

Step 2 Find equilibrium constant at 300 K

As usual with gas reactions, we use the ideal gas state of $p = 1 \times 10^5$ Pa as the reference state. It is given by applying Eq. (8-50):

$$\Delta G^\ominus_{300} = \sum_i v_i G^\ominus_i = \Delta G^\ominus_{f,CH_3Cl} + \Delta G^\ominus_{f,HCl} - \Delta G^\ominus_{f,CH_4} \tag{8-60}$$

$$\Delta G^\ominus_{300} = (-62.8855 \text{kJ/mol}) + (-95.2864 \text{kJ/mol}) - (-50.45 \text{kJ/mol}) - (0)$$

$$\Delta \underline{G}_{300}^{\ominus} = -107.72 \text{kJ/mol}$$

The definition of equilibrium constant in Eq. (8-52) is simplified as follows:

$$K_{300} = \exp\left(\frac{-\Delta \underline{G}_{300}^{\ominus}}{RT}\right) = \exp\left[\frac{107.720 \text{J/mol}}{8.38 \text{J/(mol·K)} \times 300 \text{K}}\right] \quad (8\text{-}61)$$

$$= 4.06 \times 10^{18}$$

Step 3 Find equilibrium extent of reaction

The actual pressure is $p=1\times 10^5$ Pa, so we assume ideal gas behavior ($\hat{\phi}_i = 1$) and Eq. (8-57) simplifies:

$$K_T \left(\frac{p^{\ominus}}{p}\right)^{\Sigma v_i} = \prod_i (y_i \hat{\phi}_i)^{v_i}$$

$$K_{300} = \frac{y_{\text{CH}_3\text{Cl}} y_{\text{HCl}}}{y_{\text{CH}_4} y_{\text{Cl}_2}} \quad (8\text{-}62)$$

Substitute the expression of the mole fraction in the stoichiometric table, and obtain:

$$4.06 \times 10^{18} = \frac{\dfrac{\xi}{3} \times \dfrac{\xi}{3}}{\dfrac{2-\xi}{3} \times \dfrac{1-\xi}{3}} \quad (8\text{-}63)$$

Because the equilibrium constant is so large, the extent of reaction is $\xi \sim 1$; the limiting reagent chlorine is essentially entirely consumed. However, this result is specific to $T=300$ K.

Step 4 Express mathematical effect of T on equilibrium constant

The equilibrium constant K_T is found from ($\Delta \underline{G}_T^{\ominus}/RT$), and we can only calculate $\Delta \underline{G}_T^{\ominus}$ directly at 300 K. Consequently, we first determine how the quantity ($\Delta \underline{G}_T^{\ominus}/RT$) is affected by temperature changes.

$$d(\Delta \underline{G}_T^{\ominus}/RT) = \left[\frac{\partial(\Delta \underline{G}_T^{\ominus}/RT)}{\partial p}\right]_T dp + \left[\frac{\partial(\Delta \underline{G}_T^{\ominus}/RT)}{\partial T}\right]_p dT \quad (8\text{-}64)$$

$\Delta \underline{G}_T^{\ominus}$ is defined relative to a reference pressure; here $p^{\ominus} = 1 \times 10^5$ Pa is the reference pressure regardless of the temperature. Therefore, we can treat pressure as a constant and Eq. (8-64) can be simplified as

$$d(\Delta \underline{G}_T^{\ominus}/RT) = \left[\frac{\partial(\Delta \underline{G}_T^{\ominus}/RT)}{\partial T}\right]_p dT \quad (8\text{-}65)$$

$$d(\Delta \underline{G}_T^{\ominus}/RT) = \frac{1}{RT}\left(\frac{\partial \Delta \underline{G}_T^{\ominus}}{\partial T}\right)_p + \Delta \underline{G}_T^{\ominus}\left(\frac{-1}{RT^2}\right) dT \quad (8\text{-}66)$$

From the fundamental relation $\underline{G}$. we get $(\partial \underline{G}/\partial T)_p = S$. Based on this result, we can calculate the sum of the $\underline{G}$ values of the single pure compound $\Delta \underline{G}_T^{\ominus}$.

$$\mathrm{d}\left(\Delta \underline{G}_T^\ominus / RT\right) = \frac{1}{RT}\left(-\Delta \underline{S}_T^\ominus\right) + \Delta \underline{G}_T^\ominus \left(\frac{-1}{RT^2}\right)\mathrm{d}T \qquad (8\text{-}67)$$

When we introduce the definition $\underline{G} = \underline{H} - T\underline{S}$, we can get the following simplification:

$$\mathrm{d}\left(\Delta \underline{G}_T^\ominus / RT\right) = \frac{1}{RT}\left(-\Delta \underline{S}_T^\ominus\right) + \left(\Delta \underline{H}_T^\ominus - T\Delta \underline{S}_T^\ominus\right)\left(\frac{-1}{RT^2}\right)\mathrm{d}T$$

$$\frac{\mathrm{d}\left(\Delta \underline{G}_T^\ominus / RT\right)}{\mathrm{d}T} = -\left(\frac{\Delta \underline{H}_T^\ominus}{RT^2}\right) \qquad (8\text{-}68)$$

Step 5 Determine standard enthalpy of reaction at reference T

The quantity $\Delta \underline{H}_T^\ominus$ is the standard enthalpy of a reaction. $\Delta \underline{H}_T^\ominus$ is defined as:

$$\Delta \underline{H}_T^\ominus = \sum_i \nu_i \underline{H}_i^\ominus \qquad (8\text{-}69)$$

We will refer to the values of the standard Gibbs free energy and standard enthalpy of the reaction at 300 K as $\Delta \underline{G}_T^\ominus$ and $\Delta \underline{H}_T^\ominus$. R stands for "reference", which is the parameter value at the reference temperature of 300 K.

Since the reference temperature is 300 K, $\Delta \underline{G}_R^\ominus$ is the $\Delta \underline{G}_{300}^\ominus$ from Step 2. $\Delta \underline{H}_R^\ominus$ can be obtained from Eq. (8-70):

$$\Delta \underline{H}_R^\ominus = \sum_i \nu_i \underline{H}_i^\ominus = \Delta \underline{H}_{f,CH_3C}^\ominus + \Delta \underline{H}_{f,HCl}^\ominus - \Delta \underline{H}_{f,CH_4}^\ominus - \Delta \underline{H}_{f,Cl_2}^\ominus \qquad (8\text{-}70)$$

$$\Delta \underline{H}_R^\ominus = \sum_i \nu_i \underline{H}_i^\ominus = (-80.7512 \text{ kJ/mol}) + (-92.3074 \text{ kJ/mol})$$

$$- (-74.8936 \text{ kJ/mol}) - (0)$$

$$\Delta \underline{H}_R^\ominus = -98.165 \text{ kJ/mol}$$

Step 6 Determine relationship between standard enthalpy of reaction and T

The value of $\underline{H}_i^\ominus$ for each individual compound is known at 300 K. The value of each compound at any temperature can be quantified through the relationship $\mathrm{d}H = C_p \mathrm{d}T$. It is possible to obtain $\Delta \underline{H}_T^\ominus$ by summation of the contribution of each individual compound:

$$\Delta \underline{H}_T^\ominus = \Delta \underline{H}_R^\ominus + \int_{T_n = 300 \text{ K}}^{T = 1000 \text{ K}} \sum_i \nu_i C_{p,i} \mathrm{d}T \qquad (8\text{-}71)$$

We will define the summation as ΔC_p since in effect it represents the difference in heat capacity between the products and reactants:

$$\Delta C_p = \sum_i \nu_i C_{p,i} \qquad (8\text{-}72)$$

For this example, it becomes

$$\Delta C_p = C_{p,CH_4} + C_{p,Cl_2} - C_{p,CH_3Cl} - C_{p,HCl} \qquad (8\text{-}73)$$

In this case, C_p^* of each compound is known, and is shown in Table 8-5 along with ΔC_p.

Table 8-5 The calculation of ΔC_p for this reaction

	N	A	$B \times 10^3$	$C \times 10^5$	$D \times 10^8$	$E \times 10^{11}$
CH_4	−1	4.568	−8.975	3.631	−3.407	1.091
Cl_2	−1	3.0560	5.3708	−0.8098	0.5693	−0.15256
CH_3Cl	+1	3.578	−1.750	3.071	−3.714	1.408
HCl	+1	3.827	−2.936	0.879	−1.031	0.439
$\Sigma C_{p,\text{reactants}}$		7.624	−3.6042	2.8212	−2.8377	0.93844
$\Sigma C_{p,\text{products}}$		7.405	−4.686	3.95	−4.745	1.847
ΔC_p		−0.219	−1.0818	1.1288	−1.9073	0.90856

Table 8-5 shows the calculation of ΔC_p for this reaction. The A term is very small for ΔC_p compared to the values of A for the individual components. The key to the calculation is to note how the B, C, D, and E alternates in signs, resulting in a lot of cancellations when the ΔC_p is computed.

Step 7 Apply simplifying assumption to find $\Delta \underline{G}_T^\ominus$

When $\Delta \underline{H}_R^\ominus$ and ΔC_p are known, Eq. (8-69) can be used to calculate $\Delta \underline{H}_T^\ominus$ at any temperature. However, the approximation that $\Delta C_p = 0$ implies that $\Delta \underline{H}_T^\ominus$ is constant; $\Delta \underline{H}_T^\ominus = \Delta \underline{H}_R^\ominus$. This approximation greatly simplifies Eq. (8-68) as

$$\frac{d\left(\dfrac{\Delta G}{RT}\right)}{dT} = -\left(\frac{\Delta H_T^\ominus}{RT^2}\right)$$

$$d\left(\frac{\Delta \underline{G}_T^\ominus}{RT}\right) = -\left(\frac{\Delta H_T^\ominus}{RT^2}\right)dT \tag{8-74}$$

This can be integrated from the reference temperature (T_R) to the target temperature (T):

$$\int_{T_R=300\,\text{K}}^{T=1000\,\text{K}} d\left(\frac{\Delta \underline{G}_T^\ominus}{RT}\right) = \int_{T_R=300\,\text{K}}^{T=1000\,\text{K}} -\left(\frac{\Delta H_T^\ominus}{RT^2}\right)dT \tag{8-75}$$

$$\frac{\Delta G_T^\ominus}{RT} - \frac{\Delta G_R^\ominus}{RT_R} = \left(\frac{\Delta \underline{H}_T^\ominus}{R}\right)\left(\frac{1}{T} - \frac{1}{T_R}\right)$$

$$\frac{\Delta \underline{G}_T^\ominus}{RT} = \frac{\Delta G_R^\ominus}{RT_R} + \left(\frac{\Delta \underline{H}_T^\ominus}{R}\right)\left(\frac{1}{T} - \frac{1}{T_R}\right) \tag{8-76}$$

Substitute the known values into the formulaic simplification:

$$\frac{\Delta G_{1000}^\ominus}{1000\,\text{K}} = \frac{-107.72\,\text{kJ/mol}}{300\,\text{K}} + (-98.165\,\text{kJ/mol})\left(\frac{1}{1000\,\text{K}} - \frac{1}{300\,\text{K}}\right) \tag{8-77}$$

$$\Delta G_{1000}^\ominus = -130.0\,\text{kJ/mol}$$

The standard Gibbs free energy for this reaction, which is negative, is greater at 1000 K than

Chapter 8 Chemical Reaction Equilibrium

at 300 K. We can now use this value to determine an equilibrium constant at 1000 K.

Step 8 Determine K_{1000}

Calculation of the equilibrium constant is analogous to Eq. (8-61). Thus,

$$K_{1000} = \exp\left(\frac{-\Delta \underline{G}^{\ominus}_{1000}}{RT}\right) = \exp\left[\frac{130.000 \text{ J/mol}}{8.38 \text{ J/(mol} \cdot \text{K)} \times 1000 \text{ K}}\right] = 6.18 \times 10^6 \qquad (8\text{-}78)$$

$$K_{1000} = 6.18 \times 10^6$$

Calculation of the equilibrium extent of reaction is identical to Eq. (8-65), but using the numerical value of K_{1000}:

$$6.18 \times 10^6 = \frac{\dfrac{\xi}{3} \times \dfrac{\xi}{3}}{\dfrac{2-\xi}{3} \times \dfrac{1-\xi}{3}} \qquad (8\text{-}79)$$

K_{1000} has a larger value, so the ξ value is essentially 1. The mole number of chlorines at equilibrium $(1-\xi)$ is 10^{-6}.

Eq. (8-68) shows a method to simulate the effect of temperature on reaction equilibrium:

$$\frac{\text{d}\left(\Delta \underline{G}^{\ominus}_T / RT\right)}{\text{d}T} = -\left(\frac{\Delta \underline{H}^{\ominus}_T}{RT^2}\right)$$

That's the Van't Hoff equation. There are no assumptions or approximations involved in the derivation of the Van't Hoff equation, but two different strategies are usually employed to progress from this point to a solution. The "shortcut" approach is shown in Example 8-3, where the standard enthalpy of the reaction is assumed to be a constant with respect to temperature; $\Delta \underline{H}^{\ominus}_R$ is calculated at a reference temperature but assumed valid at all temperatures. That's the assumption that the reactants have the same heat capacity as the products:

$$\Delta C_p = \sum_i v_i C_{p,i} = 0 \qquad (8\text{-}80)$$

This assumption produces the result in Eq. (8-76), which we will call the short cut Van't Hoff equation:

$$\frac{\Delta \underline{G}^{\ominus}_T}{RT} = \frac{\Delta \underline{G}^{\ominus}_R}{RT_R} + \left(\frac{\Delta \underline{H}^{\ominus}_R}{R}\right)\left(\frac{1}{T} - \frac{1}{T_R}\right)$$

Using this shortcut, K_T can be quickly estimated at any temperature if $\Delta \underline{G}^{\ominus}_R$ and $\Delta \underline{H}^{\ominus}_R$ are known at the reference temperature T_R. Conceptually, T_R can be any temperature for which there is data. In practice, the temperature is usually 298.15 K.

The general way to solve the Van't Hoff equation is to consider $\Delta \underline{H}^{\ominus}_R$ as a function of temperature and use Eq. (8-81) to illustrate this fact through the heat capacity of each compound.

$$\Delta \underline{H}^{\ominus}_T = \Delta \underline{H}^{\ominus}_R + \int_{T_R}^{T} \Delta C_p \text{d}T \qquad (8\text{-}81)$$

When the heat capacity is expressed as $C_p = A + BT + CT^2 + DT^3 + ET^4$, this becomes

$$\Delta \underline{H}_R^\ominus + \Delta A(T-T_R) + \frac{\Delta B}{2}(T^2-T_R^2) + \frac{\Delta C}{3}(T^3-T_R^3)$$
$$+ \frac{\Delta D}{4}(T^4-T_R^4) + \frac{\Delta E}{5}(T^5-T_R^5) \tag{8-82}$$

Definitions of ΔA, ΔB, ΔC, ΔD, and ΔE are similar to ΔC_p. The constant $\Delta \underline{H}_R^\ominus$ and the various T_R terms can be combined into a single constant J for

$$\Delta \underline{H}_T^\ominus = J + \Delta A T + \frac{\Delta B}{2}T^2 + \frac{\Delta C}{3}T^3 + \frac{\Delta D}{4}T^4 + \frac{\Delta E}{5}T^5 \tag{8-83}$$

with

$$J = \Delta \underline{H}_R^\ominus - \Delta A T_R - \frac{\Delta B}{2}T_R^2 - \frac{\Delta C}{3}T_R^3 - \frac{\Delta D}{4}T_R^4 - \frac{\Delta E}{5}T_R^5 \tag{8-84}$$

When the expression for $\Delta \underline{H}_T^\ominus$ from Eq. (8-83) is introduced into Eq. (8-68), the resulting integration is

$$\frac{\Delta G_T^\ominus}{RT} = \frac{\Delta G_R^\ominus}{RT_R} + \frac{J}{R}\left(\frac{1}{T}-\frac{1}{T_R}\right) - \frac{\Delta A}{R}\ln\left(\frac{T}{T_R}\right) - \frac{\Delta B}{2R}(T-T_R)$$
$$- \frac{\Delta C}{6R}(T^2-T_R^2) - \frac{\Delta D}{12R}(T^3-T_R^3) - \frac{\Delta E}{20R}(T^4-T_R^4) \tag{8-85}$$

The shortcut in Example 8-3 assumes that the heat capacity of the reactants is equal to the heat capacity of the products, that is, $\Delta C_p = 0$.

Example 8-3 also demonstrates that for an endothermic reaction, the equilibrium constant increases as temperature increases. In many cases, the designer needs to strike a balance between maximizing the equilibrium conversion of reactants to products and maximizing the reaction rate. Example 8-3 illustrates a case in which increasing temperature will increase the equilibrium conversion in addition to increasing the reaction rate. Therefore, if the goal is to make propylene, there is no trade-off in this case. The calculations shown in Example 8-3 may tell the designer that the reaction should take place at the highest temperature within a safe and practical range.

8.3.4 Reference States and Nomenclature

In this chapter, we use the symbol $\underline{G}^\ominus$ to represent the molar Gibbs free energy of a compound at a standard state, but in carrying out the calculations, we use $\Delta \underline{G}_f^\ominus$. This section aims to clarify the difference between $\Delta \underline{G}_f^\ominus$ and $\underline{G}^\ominus$. $\underline{G}^\ominus$ represents the standard molar Gibbs free energy at reference state. Conceptually it could be any reference state (Song et al., 2010).

$\Delta \underline{G}_f^\ominus$ is the standard molar Gibbs free energy of formation; the change in Gibbs free energy when the compound is formed from elements at $T = 298$ K and $p = 1\times 10^5$ Pa. It is the same property, but we use this symbol to indicate that we have chosen a specific reference state in which $\underline{H} = \underline{G} = 0$ for all elements at $T = 298$ K and $p = 1\times 10^5$ Pa.

Similarly, in an energy balance, we use the symbol $\underline{H}$ to represent molar enthalpy of a stream as it enters or leaves a system. In many instances, $\underline{H}$ can either be set equal to $\Delta \underline{H}_f^\ominus$ (if the stream is at the standard temperature and pressure) or $\Delta \underline{H}_f^\ominus$ can be used as a starting point in computing $\underline{H}$. But if we use $\Delta \underline{H}_f^\ominus$, then we must use "$\underline{H}$ =0 for elements at T=298 K and p=1×10^5 Pa" as the reference state throughout the energy balance.

8.4 Multiple Reaction Equilibrium

Unlike the examples in Section 8.3, it is common for multiple chemical reactions to occur simultaneously in a real chemical process reactor. When multiple reactions are present, each is modeled individually as illustrated in Section 8.3. An equilibrium constant is calculated for each reaction, and the mixture fugacity of each participating compound are quantified using a method appropriate for the relevant phase, pressure, and temperature (Zumdahl et al., 2009). The equilibrium condition of the system is one in which the equilibrium criterion:

$$K_T = \exp\left(\frac{-\Delta G_T^\ominus}{RT}\right) = \prod_i \left(\frac{\hat{f}_i}{f_i^\ominus}\right)^{\nu_i}$$

We illustrate multiple-reaction equilibrium through an expansion of Example 8-3.

EXAMPLE 8-4

Example 8-3 illustrated the chlorination of methane to form chloromethane, Example 8-3 also revealed that at equilibrium, the products were strongly favored:

$$(R1): CH_4 + Cl_2 \longleftrightarrow CH_3Cl + HCl$$

In subsequent reactions, however, chlorine replaces hydrogen to form propylene chloride, chloroform, and carbon tetrachloride:

$$(R2): CH_3Cl + Cl_2 \longleftrightarrow CH_2Cl_2 + HCl$$

$$(R3): CH_2Cl_2 + Cl_2 \longleftrightarrow CHCl_3 + HCl$$

$$(R4): CHCl_3 + Cl_2 \longleftrightarrow CCl_4 + HCl$$

A steady-state reactor is maintained at $p = 1\times 10^5$ Pa and $T = 1000\,°C$. Assuming the product stream leaving the reactor is at equilibrium, find out what the reactor feed contains if any of the following: 8 mol/s of chlorine and 2 mol/s of methane.

Use the shortcut method to account for the effects of temperature.

SOLUTION

Step 1 Construct stoichiometric table

As shown in Example 8-2, when we analyze a system containing multiple reactions, we can define a separate "reaction range" for each reaction. In this case, we are using time-dependent mole balances, so the extents of reaction will be expressed as rates: $\dot{\xi}_1, \dot{\xi}_2, \dot{\xi}_3$, and the definition in Eq. (8-39) is applied to each reactant and product in each reaction, and the individual "generation" terms are summed to determine the total generation for a compound. For example, the stoichiometric coefficient of chloroform is +1 in R3 and −1 in R4, so the generation of chloroform can be quantified as $+\dot{\xi}_3 - \dot{\xi}_4$ (Table 8-6).

Table 8-6 Stoichiometric table for sequential chlorination of methane

	CH_4	Cl_2	CH_3Cl	HCl	CH_2Cl_2	$CHCl_3$	CCl_4	Total
In/(mol/s)	8	2						10
Gen(R1, mol/s)	$-\xi_1$	$-\xi_1$	$+\xi_1$	$+\xi_1$				0
Gen(R2, mol/s)		$-\xi_2$	$-\xi_2$	$+\xi_2$	$+\xi_2$			0
Gen(R3, mol/s)		$-\xi_3$		$+\xi_3$	$-\xi_3$	$+\xi_3$		0
Gen(R4, mol/s)		$-\xi_4$		$+\xi_4$		$-\xi_4$	$+\xi_4$	0
Out (mol/s)	$8-\xi_1$	$2-\xi_1-\xi_2-\xi_3-\xi_4$	$\xi_1-\xi_2$	$\xi_1+\xi_2+\xi_3+\xi_4$	$\xi_2-\xi_3$	$\xi_3-\xi_4$	ξ_4	10
y	$\dfrac{8-\xi_1}{10}$	$\dfrac{2-\xi_1-\xi_2-\xi_3-\xi_4}{10}$	$\dfrac{\xi_1-\xi_2}{10}$	$\dfrac{\xi_1+\xi_2+\xi_3+\xi_4}{10}$	$\dfrac{\xi_2-\xi_3}{10}$	$\dfrac{\xi_3-\xi_4}{10}$	$\dfrac{\xi_4}{10}$	

Note that since all of the reactions have $\Sigma\nu=0$, the total numbers of moles of gas entering and leaving the reactor are same, regardless of the magnitude of the reactions.

Step 2 Calculate standard Gibbs free energy and standard enthalpy of each reaction

The calculation of $\Delta\underline{G}_T^\ominus$ and $\Delta\underline{H}_T^\ominus$ for each reaction is shown in Table 8-7.

Step 3 Estimate the rate constant of each reaction at $T=1000\,°C$

Table 8-7 The bottom row shows the Gibbs free energy and enthalpy changes of the reactions at the reference temperature of 298.15 K; thus these are $\Delta\underline{H}_R^\ominus$ and $\Delta\underline{G}_R^\ominus$. To estimate the rate constant K_{1273}, we begin by applying Eq. (8-76), the shortcut Van't Hoff equation:

$$\frac{\Delta\underline{G}_T^\ominus}{RT} = \frac{\Delta\underline{G}_R^\ominus}{RT_R} + \left(\frac{\Delta\underline{H}_R^\ominus}{R}\right)\left(\frac{1}{T} - \frac{1}{T_R}\right) \qquad (8-86)$$

Table 8-7 Summary of contributions of each compound to the standard Gibbs free energy and enthalpy for R1 to R4

	R1		R2		R3		R4	
	$\Delta H_R^\ominus$ / (kJ/mol)	$\Delta G_R^\ominus$ / (kJ/mol)	$\Delta H_R^\ominus$ / (kJ/mol)	$\Delta G_R^\ominus$ / (kJ/mol)	$\Delta H_R^\ominus$ / (kJ/mol)	$\Delta G_R^\ominus$ / (kJ/mol)	$\Delta H_R^\ominus$ / (kJ/mol)	$\Delta G_R^\ominus$ / (kJ/mol)
CH_4	−(−74.894)	−(−50.45)						
Cl_2	−(0)	−(0)	−(0)	−(0)	−(0)	−(0)	−(0)	−(0)
CH_3Cl	−92.307	−95.296	−92.307	−95.296	−92.307	−92.296	−92.307	−95.296

Continued

	R1	R2	R3	R4
HCl	−80.751	−62.886	−(−80.751)	−(−62.886)			
CH$_2$Cl$_2$		−95.395	−68.869	−(−95.395)	−(−68.869)		
CHCl$_3$			−103.345	−68.534	−(−103.345)	−(−68.534)
CCl$_4$						−95.8136	−60.6261
Σ	−98.165	−107.722	−106.951	−101.27	−100.257	−94.9516	−84.776	−87.3876

For R1, for example,

$$\frac{\Delta G_T^\ominus}{RT} = \frac{(-107.222 \text{ J/mol})}{8.38 \text{ J/(mol·K)} \times 298.15 \text{ K}} + \frac{-98.165 \text{ J/mol}}{8.38 \text{ J/(mol·K)}} \times \left(\frac{1}{1273.15 \text{ K}} - \frac{1}{298.15 \text{ K}}\right)$$

$$\frac{\Delta G_T^\ominus}{RT} = -13.13$$

It is given by applying Eq. (8-53):

$$K_{1273.15} = \exp(-13.13) = 5.035 \times 10^5 \tag{8-87}$$

The equilibrium constants of other reactions are estimated by similar methods and summarized in Table 8-8.

Step 4 Apply the equilibrium criterion to each reaction

The equilibrium criterion in its most general form is shown in Eq. (8-53) as

$$K_T = \prod_i \left(\frac{\hat{f}_i}{f_i^\ominus}\right)^{\nu_i}$$

Table 8-8 Equilibrium constants for sequential chlorination of methane at 1000℃

	$\Delta G_R^\ominus$/(kJ/mol)	$\Delta H_R^\ominus$/(kJ/mol)	$\dfrac{\Delta G_T^\ominus}{RT}$	$K_{1273.15}$
R1	−107.722	−98.165	−13.129	5.035×10^5
R2	−101.27	−106.951	−7.812	2470
R3	−94.9516	−100.257	−7.331	1527
R4	−87.3786	−84.776	−9.059	8596

Step 5 Apply the equilibrium criterion to each reaction

The equilibrium criterion in its most general form is shown in Eq. (8-53) as

$$K_T = \prod_i \left(\frac{\hat{f}_i}{f_i^\ominus}\right)^{\nu_i}$$

This is happening at $p = 1 \times 10^5$ Pa, so we've simplified the behavior and the equilibrium criterion for an ideal gas.

$$K_T = \prod_i \left(\frac{y_i p}{p^\ominus}\right)^{\nu_i} \tag{8-88}$$

Which, since both the system pressure and the reference pressure are 1×10^5 Pa, it is further simplified as

$$K_T = \prod_i (y_i)^{\nu_i} \tag{8-89}$$

An equation in the form of Eq. (8-89) is written for each of the four reactions:

$$K_{R1,1273} = \frac{y_{HCl} y_{CH_3Cl}}{y_{CH_4} y_{Cl_2}} \tag{8-90}$$

$$K_{R2,1273} = \frac{y_{HCl} y_{CH_2Cl_2}}{y_{CH_3Cl} y_{C_2}} \tag{8-91}$$

$$K_{R3,1273} = \frac{y_{HCl} y_{CHCl_3}}{y_{CH_2Cl_2} y_{Cl_2}} \tag{8-92}$$

$$K_{R4,1273} = \frac{y_{HCl} y_{CCl_4}}{y_{CHCl_3} y_{Cl_2}} \tag{8-93}$$

Step 6 Solve for equilibrium composition

When the mole fraction expressions from Table 8-6 and the equilibrium constants from Table 8-8 are inserted into Eq. (8-90) through Eq. (8-93), the result is

$$5.035 \times 10^5 = \frac{\left(\dfrac{\dot\xi_1 + \dot\xi_2 + \dot\xi_3 + \dot\xi_4}{10}\right)\left(\dfrac{\dot\xi_1 - \dot\xi_2}{10}\right)}{\left(\dfrac{2 - \dot\xi_1}{10}\right)\left(\dfrac{8 - \dot\xi_1 - \dot\xi_2 - \dot\xi_3 - \dot\xi_4}{10}\right)} \tag{8-94}$$

$$2470 = \frac{\left(\dfrac{\dot\xi_1 + \dot\xi_2 + \dot\xi_3 + \dot\xi_4}{10}\right)\left(\dfrac{\dot\xi_2 - \dot\xi_3}{10}\right)}{\left(\dfrac{\dot\xi_1 - \dot\xi_2}{10}\right)\left(\dfrac{8 - \dot\xi_1 - \dot\xi_2 - \dot\xi_3 - \dot\xi_4}{10}\right)} \tag{8-95}$$

$$1527 = \frac{\left(\dfrac{\dot\xi_1 + \dot\xi_2 + \dot\xi_3 + \dot\xi_4}{10}\right)\left(\dfrac{\dot\xi_3 - \dot\xi_4}{10}\right)}{\left(\dfrac{\dot\xi_2 - \dot\xi_3}{10}\right)\left(\dfrac{8 - \dot\xi_1 - \dot\xi_2 - \dot\xi_3 - \dot\xi_4}{10}\right)} \tag{8-96}$$

$$8596 = \frac{\left(\dfrac{\dot\xi_1 + \dot\xi_2 + \dot\xi_3 + \dot\xi_4}{10}\right)\left(\dfrac{\dot\xi_4}{10}\right)}{\left(\dfrac{\dot\xi_3 - \dot\xi_4}{10}\right)\left(\dfrac{8 - \dot\xi_1 - \dot\xi_2 - \dot\xi_3 - \dot\xi_4}{10}\right)} \tag{8-97}$$

At this point, we've completed the "thermodynamic" part of the problem——the rest is pure math. Eq. (8-94) to Eq. (8-97) represent a quaternion system of first order equations that can be solved by four equations. Once the extent of each reaction is known, the mole fractions are computed(Table 8-9). In practice, a typical non-linear equation solver will probably not converge to the solutions of Eq. (8-94) to Eq. (8-97) without good initial guesses.

Table 8-9 Equilibrium compositions for sequential chlorination of methane at $T = 1000\,°C$

Compound	Mole Fraction	Compound	Mole Fraction
CH_4	1.4×10^{-10}	CH_2Cl_2	4.12×10^{-4}
Cl_2	1.15×10^{-2}	$CHCl_3$	9.59×10^{-3}
CH_3Cl	2.8×10^{-7}	CCl_4	0.1857
HCl	0.7932		

Example 8-4 shows a reaction system in which four chlorination reactions occur sequentially. Because the equilibrium constants for all four reactions are large, the product formation in all of them is favorable, and intuitively expect that the chlorination progresses almost to completion. When the system started with four moles of chlorine for every mole of methane: about 98% of the methane at equilibrium was converted to carbon tetrachloride. This can be understood by recognizing that there isn't enough chlorine to convert all the methane into carbon tetrachloride, and while all four reactions have large equilibrium constants, R1 has the largest.

8.5 Summary

- The extent of a reaction, ξ, is defined as the moles of a compound generated (or consumed) by a reaction divided by the stoichiometric coefficient of the compound, and can be expressed on either an absolute $\xi = \dfrac{N_{i,\text{gen}}}{v_i}$ or rate $\dot\xi = \dfrac{\dot N_{i,\text{gen}}}{v_i}$ basis.

- The equilibrium constant for a reaction is a function of temperature, and its value is given by

$$K_T = \exp\left(\frac{-\Delta G_T^\ominus}{RT}\right)$$

- The equilibrium composition of a reacting system is described by

$$K_T = \prod_i \left(\frac{\hat f_i}{f_i^\ominus}\right)^{v_i}$$

One can model chemical reaction equilibrium by combining this equilibrium expression with reaction stoichiometry.

- The equilibrium constants are quantified using models applicable to system conditions, which also apply to chemical reactions occurring in liquid or vapor phases at high or low

pressures. The Van't Hoff equation can be used to calculate the equilibrium constant at temperatures other than the reference temperature.
- To model equilibrium in a system where multiple reactions are taking place simultaneously, the system involves applying an equilibrium constant to constrain each reaction.

8.6 Chemical Reaction Equilibrium Simulation

This chapter deals with the use of Aspen Plus for the calculation of chemical reaction equilibrium. The reactor of interest here is RGibbs since it can efficiently calculate chemical equilibrium in multiphase, multi-reaction systems; REquil can also be used, but only for single-phase systems. The underlying fundamental principle is that at equilibrium at constant temperature and pressure, the Gibbs energy should be a minimum. The calculation then is to vary the number of moles of each species in each phase subject to the stoichiometric constraints and find a solution that minimizes the total Gibbs energy of the system. In this way, a general minimization algorithm can be used to solve all chemical reaction equilibrium problems.

Example 8-1 is simulated by Aspen.

The operation steps are shown in the following QR code.

The result for the mole fraction of propane is 1 (compared to 1 in the textbook), the mole fraction of propylene is 2.39×10^{-8} (compared to 2.83×10^{-8}) and 2.39×10^{-8} for hydrogen (compared to 2.83×10^{-8}). Therefore, the agreement is quite good, but not exact.

Example 8-4 is simulated by Aspen.

The operation steps are shown in the following QR code.

EXAMPLE 8-5

At 373 K and 1 atm, acetic acid was esterified as follows to produce ethyl acetate.

$$CH_3COOH(l) + C_2H_5OH(l) \longrightarrow CH_3COOC_2H_5(l) + H_2O(l)$$

Assuming that the initial amounts of acetic acid and ethanol are both 1mol, and the relevant data at 298.15 K are in Table 8-10:

Table 8-10 Data for each component at 298 K

Material	CH_3COOH	C_2H_5OH	$CH_3COOC_2H_5$	H_2O
$\Delta H_f^\ominus$ /(J/mol)	−484500	−277690	−463250	−285830
$\Delta G_f^\ominus$ /(J/mol)	−389900	−174780	−318280	−237129

Calculate the molar fraction of ethyl acetate in the reaction mixture at equilibrium.
The response of $\Delta H_{298}^{\ominus}$ and $\Delta G_{298}^{\ominus}$ is

$$\Delta H_{298}^{\ominus} = (-463250 - 285830 + 484500 + 277690)\,\text{J} = 13110\,\text{J}$$

$$\Delta G_{298}^{\ominus} = (-318280 - 237129 + 389900 + 174780)\,\text{J} = 9271\,\text{J}$$

$$\ln K_{298} = \frac{-\Delta G_{298}^{\ominus}}{RT} = \frac{-9271}{8.314 \times 298.15} = -3.740$$

$$K_{298} = 0.0238$$

Assuming that the change in standard enthalpy of the reaction is not constant with temperature from 298.15 K to 373.15 K, then

$$\ln \frac{K_{373}}{K_{298}} = \frac{-\Delta H_{298}^{\ominus}}{R}\left(\frac{1}{373.15\,\text{K}} - \frac{1}{298.15\,\text{K}}\right) = \frac{-13110}{8.314}\left(\frac{1}{373.15} - \frac{1}{298.15}\right) = 1.0630$$

$$K_{373} = 0.0238 \times e^{1.0630} = 0.0238 \times 2.895 = 0.0689$$

Assuming that the reaction mixture is an ideal solution

$$K = \prod \chi_i^{\nu_i} = \frac{\chi_{CH_3COOC_2H_5}\,\chi_{H_2O}}{\chi_{CH_3COOH}\,\chi_{C_2H_5OH}}$$

It can be seen from the reaction formula

$$\nu_{CH_3COOC_2H_5} = -1$$

$$\nu_{C_2H_5OH} = -1$$

$$\nu_{CH_3COOH} = 1$$

$$\nu_{H_2O} = 1$$

then

$$\frac{dn_{CH_3COOC_2H_5}}{-1} = \frac{dn_{C_2H_5OH}}{-1} = \frac{dn_{CH_3COOH}}{1} = \frac{dn_{H_2O}}{1} = d\varepsilon$$

Integrate the above from the initial state to the final state ($\varepsilon = 0$, $n_{C_2H_5OH} = 1$, $n_{CH_3COOC_2H_5} = 1$, $n_{CH_3COOH} = dn_{H_2O} = 0$)

$$n_{CH_3COOH} = 1 - \varepsilon \qquad \chi_{CH_3COOH} = \frac{1-\varepsilon}{2}$$

$$n_{C_2H_5OH} = 1 - \varepsilon \qquad \chi_{C_2H_5OH} = \frac{1-\varepsilon}{2}$$

$$n_{CH_3COOC_2H_5} = \varepsilon \qquad \chi_{CH_3COOC_2H_5} = \frac{\varepsilon}{2}$$

$$n_{H_2O} = \varepsilon \qquad \chi_{H_2O} = \frac{\varepsilon}{2}$$

$$\sum n_i = 2$$

Substitute the above values into equation $K = \prod \chi_i^{v_i}$

$$K = \left(\frac{1-\varepsilon}{2}\right)^2 = 0.0689$$

$$\varepsilon = 0.208$$

$$\chi_{CH_3COOC_2H_5} = \frac{\varepsilon}{2} = 0.104$$

Compared with the molar fraction of ethyl acetate 0.33 measured in the experiment, the deviation is larger, indicating that the assumption of ideal solution in the calculation is unreasonable.

Aspen simulation is used for calculation. The operation steps are shown in the following QR code.

EXAMPLE 8-6

Calculate the equilibrium composition of synthetic ammonia reaction at 427℃ and 30.39 MPa. The known reactants are 75 % H_2 and 25 % N_2 (mole fraction), K_f=0.0091, and the standard state of each component is 1 atm (0.1013 MPa).

$$\frac{1}{2}N_2 + \frac{3}{2}H_2 \longleftrightarrow NH_2$$

Assuming that the reaction mixture is an ideal gas solution, the fugacity coefficients of each component are calculated in Table 8-11:

Table 8-11 The fugacity coefficients of each component

Vapor	H_2	N_2	NH_3
ϕ_i	1.10	1.15	0.90

Thus:

$$K_\phi = \frac{0.90}{1.15^{1/2} \times 1.10^{3/2}} 0.72$$

The moles and total moles of each component before and after the reaction were calculated.

$$\frac{1}{2}N_2 + \frac{3}{2}H_2 \rightleftharpoons NH_3$$

Before reaction: $\frac{1}{2}$, $\frac{3}{2}$, 0.

After reaction: $\left(\frac{1}{2} - \frac{1}{2}\alpha\right)$, $\left(\frac{3}{2} - \frac{3}{2}\alpha\right)$, α.

The total number of moles after reaction is:

$$\left(\frac{1}{2}-\frac{1}{2}\alpha\right)+\left(\frac{3}{2}-\frac{3}{2}\alpha\right)+\alpha=2-\alpha$$

The mole fraction of each component is:

$$y_{NH_3} = \frac{\alpha}{2-\alpha} \qquad y_{N_2} = \frac{\frac{1}{2}\times(1-\alpha)}{2-\alpha} \qquad y_{H_2} = \frac{\frac{3}{2}\times(1-\alpha)}{2-\alpha}$$

In this reaction:

$$v_{N_2} = -\frac{1}{2} \qquad v_{H_2} = -\frac{3}{2} \qquad v_{N_2} = 1$$

The balance consists:

$$y_{NH_3} = \frac{0.589}{2-0.589} = 41.6\%$$

$$y_{N_2} = \frac{\frac{1}{2}-\frac{0.589}{2}}{2-0.589} = 14.6\%$$

$$y_{H_2} = \frac{\frac{3}{2}-\frac{3}{2}\times 0.589}{2-0.589} = 43.8\%$$

Aspen simulation is used for calculation. The operation steps are shown in the following QR code.

EXERCISES

1. A series of the following reactions occur

$$CH_4 + H_2O \rightleftharpoons CO + 3H_2$$

It is assumed that the initial content of each substance is 1 mol CH_4, 2 mol H_2O, 1 mol CO and 5 mol H_2. Try to find the functional expression of the amount of substance n_i and mole fraction y_i.

2. Set up a system in which the following two reactions occur simultaneously:

$$CH_4 + H_2O \rightleftharpoons CO + 3H_2 \qquad (1)$$

$$CH_4 + 2H_2O \rightleftharpoons CO_2 + 4H_2 \qquad (2)$$

In the formula, numbers (1) and (2) represent j in formula

$$dn_i = \sum_j v_{ij} d\varepsilon_j \quad (i=1,2,\ldots,N)$$

If the initial amount of each substance is 3 mol CH_4 and 5 mol H_2O, and the initial amount of CO, CO_2 and H_2 is zero, try to determine the functional expressions of n_i and y_i for ε_1 and ε_2.

3. Try to calculate the equilibrium constants of the following reactions at 298 K.

$$C_2H_4(g) + H_2O(l) \rightleftharpoons C_2H_5OH(l)$$

The standard status of each component is specified as shown in the table below.

Component	Specified standard status
$C_2H_4(g)$	Pure gas 0.1013 MPa, 298 K
$H_2O(l)$	Pure liquid 0.1013 MPa, 298 K
$C_2H_5OH(l)$	Pure liquid 0.1013 MPa, 298 K

4. Try to calculate the equilibrium composition of synthetic ammonia reaction at 427℃ and 30.39 MPa. The known reactants are 75% H_2 and 25% N_2 (mole fraction). The K_f literature value of the reaction is 0.0091, in which the standard state of each component is 1 atm (0.1013 MPa).

5. The reaction formula for gas phase hydration of ethylene to ethanol is

$$C_2H_4(g) + H_2O(g) \rightleftharpoons C_2H_5OH(g)$$

The reaction temperature and pressure are 523 K and 3.45 MPa respectively, and $K_f = 8.15 \times 10^{-3}$ is known. The equilibrium conversion of ethanol is calculated when the initial ratio of water vapor to ethylene is 1 and 7 respectively.

6. Under 373 K and atmospheric pressure, acetic acid is esterified according to the following reaction to obtain ethyl acetate.

$$CH_3COOH(l) + C_2H_5OH(l) \longrightarrow CH_3COOC_2H_5(l) + H_2O(l)$$

Assuming that the initial amounts of acetic acid and ethanol are both 1 mol, the relevant data at 298K are as follows.

Substance	Acetic acid	Ethanol	Ethyl acetate	Water
$\Delta H_f^\ominus$/(J/mol)	−484500	−377690	−463250	−285830
$\Delta G_f^\ominus$/(J/mol)	−389900	−174780	−318280	−237129

Try to calculate the mole fraction of ethyl acetate in the reaction mixture at equilibrium.

7. Water gas shift reaction

$$CO(g) + H_2O(g) \longrightarrow CO_2(g) + H_2(g)$$

Try to find the equilibrium reaction degree of the reaction under the following different conditions. It is assumed that the reaction mixture is an ideal gas.

(1) The initial quantities of H_2O and CO are both 1mol, the temperature is 1100 K, and the pressure is 0.1 MPa;

(2) The initial quantities of H_2O, CO and CO_2 are all 1 mol, and other conditions are the

same as (1);

(3) The temperature is 1650 K, and other conditions are the same as (1);

(4) The pressure is 1 MPa, and the other conditions are the same as (1);

(5) The reactant also contains 2 mol N_2, and the other conditions are the same as (1).

8. Try to determine the degree of freedom F of the following systems.

(1) Two mutually soluble and non-reactive substances are in vapor-liquid equilibrium and form an azeotropic mixture;

(2) Partial decomposition system of $CaCO_3$;

(3) System of partial decomposition of NH_4Cl;

(4) A gas phase system composed of CO, CO_2, H_2, H_2O and CH_4 and in a chemical equilibrium.

9. In the reaction system of catalytic oxidation of ammonia to NO, the following group of reactions occur

$$4NH_3 + 5O_2 \rightleftharpoons 4NO + 6H_2O \quad (A)$$

$$4NH_3 + 3O_2 \rightleftharpoons 2N_2 + 6H_2O \quad (B)$$

$$4NH_3 + 6NO \rightleftharpoons 5N_2 + 6H_2O \quad (C)$$

$$2NO + O_2 \rightleftharpoons 2NO_2 \quad (D)$$

$$2NO \rightleftharpoons N_2 + O_2 \quad (E)$$

$$N_2 + 2O_2 \rightleftharpoons 2NO_2 \quad (F)$$

Try to determine the number of independent reactions in this system and write down the main reactions that express this system.

10. Try to calculate the equilibrium composition of CH_4, H_2O, CO, CO_2 and H_2 systems at 1000 K and 0.1013 Mpa. It is known that before the reaction contains 2 mol CH_4 and 3mol H_2O, the $\Delta G_f^\ominus$ value of each substance at 1000 K is: $\Delta G_{fCH_4}^\ominus = -19297$ J/mol, $\Delta G_{fH_2O}^\ominus = -192682$ J/mol, $\Delta G_{fCO}^\ominus = -200677$ J/mol, $\Delta G_{fCO_2}^\ominus = -396037$ J/mol.

REFERENCES

[1] Afeefy H Y, Liebman J F, Stein S E. Neutral thermochemical data[M]. NIST Standard Reference Database Number 69, 20899.

[2] Chase M W, Jr. NIST-JANAF thermochemical tables[M]. 4th ed. J. Phys. Chem. Ref. Data, Monograph 9, 1998: 1-1951.

[3] Domalski E S, Hearing E D. Condensed phase heat capacity data[M]. NIST Standard Reference Database Number 69, 20899.

[4] Perry R H, Green D W. Perry's chemical engineers' handbook[M]. 7th ed. McGraw-Hill, New York, 1999.

[5] Poling B E, Prausnitz J M, O'Connell J P. The properties of gases and liquids. [M]. 5th ed. McGraw-Hill, New York, 2001.

[6] Song C, Lu K, Subramani V. Hydrogen and syngas production and purification technologies[M]. John Wiley & Sons, 2009.

[7] Zumdahl S S. Chemical principles [M]. 6th ed. Houghton Mifflin Company, 2009.